3D Printing and Additive Manufacturing Technologies

L. Jyothish Kumar · Pulak M. Pandey
David Ian Wimpenny
Editors

3D Printing and Additive Manufacturing Technologies

Springer

Editors
L. Jyothish Kumar
Additive Manufacturing Society of India
Bengaluru, Karnataka
India

Pulak M. Pandey
Department of Mechanical Engineering
Indian Institute of Technology Delhi
New Delhi
India

David Ian Wimpenny
Component Technology
The Manufacturing Technology Centre
Coventry, Warwickshire
UK

ISBN 978-981-13-0304-3 ISBN 978-981-13-0305-0 (eBook)
https://doi.org/10.1007/978-981-13-0305-0

Library of Congress Control Number: 2018939143

Printed on acid-free paper

This Springer imprint is published by the registered company Springer Nature Singapore Pte Ltd.
The registered company address is: 152 Beach Road, #21-01/04 Gateway East, Singapore 189721, Singapore

Preface

We would like to thank the readers for taking the time to read this book, Advances in *3D Printing and Additive Manufacturing Technologies—II.*

3D printing also known as additive manufacturing is growing and growing at a rapid phase. The technology has potential applications in aerospace, automotive, biomedical, and engineering sectors. The technology helps in reducing the lead time for manufacturing of complex components as well as reduction of material waste compared to conventional manufacturing process.

The purpose of this book is to provide the details of the latest advancements in research and developments of 3D printing and additive manufacturing processes. This book will be useful for industrial experts, entrepreneurs, university professors, and research scholars. The chapters are written by experts from and across industry and academia.

Bengaluru, India	L. Jyothish Kumar
New Delhi, India	Pulak M. Pandey
Coventry, UK	David Ian Wimpenny

Acknowledgements

We express our sincere thanks to all the researchers, academicians, manufacturing organizations, and R&D centers for their key inputs and informational support.

Contents

Finite Element Analysis of Melt Pool Characteristics in Selective Laser Spot Melting on a Powder Layer 1
Akash Aggarwal and Arvind Kumar

Thermal Transport Phenomena in Multi-layer Deposition Using Arc Welding Process 15
Anshul Yadav, Aniruddha Ghosh and Arvind Kumar

Comparison of Bonding Strength of Ti–6Al–4V Alloy Deposit and Substrate Processed by Laser Metal Deposition 29
L. Jyothish Kumar and C. G. Krishnadas Nair

Study on Rayleigh–Bénard Convection in Laser Melting Process 39
Kurian Antony and T. R. Rakeshnath

Enhancing Surface Finish of Fused Deposition Modelling Parts 45
M. S. Khan and J. P. Dash

Development and Analysis of Accurate and Adaptive FDM Post-finishing Approach 59
Mohammad Taufik and Prashant K. Jain

Toolpath Generation for Additive Manufacturing Using CNC Milling Machine 73
Narendra Kumar, Prashant K. Jain, Puneet Tandon and Pulak M. Pandey

Modelling of Heat Transfer in Powder Bed Based Additive Manufacturing Process Using Lattice Boltzmann Method 83
Priya Gupta, Anshul Yadav, Arvind Kumar and Niraj Sinha

Effect of Process Parameters on Mechanical Properties of Solidified PLA Parts Fabricated by 3D Printing Process 95
Jagdish Khatwani and Vineet Srivastava

Metal Powder Based Additive Manufacturing Technologies—Business Forecast 105
Vadlamannati Sriram, Vipin Shukla and Soumitra Biswas

Design and Development of Drug Delivery System for Chronic Wound Using Additive Manufacturing 119
Mohan Pushparaj, Rajesh Ranganathan and Sivakumar Ganesan

Design and Development of Orthosis for Clubfoot Deformity 127
Chandrasekeran Vivek and Rajesh Ranganathan

Optimization of Selective Laser Sintering Process Parameters on Surface Quality 141
M. Akilesh, P. R. Elango, A. Achith Devanand, R. Soundararajan and P. Ashoka Varthanan

Reconstruction of Damaged Parts by Integration Reverse Engineering (RE) and Rapid Prototyping (RP) 159
Anuj V. Dongaonkar and Rajesh M. Metkar

The Impact of Additive Manufacturing on Indian GDP 173
J. Avinash, K. UdayKiran and K. Srujita

Optimization of the Print Quality by Controlling the Process Parameters on 3D Printing Machine 187
R. Devicharan and Raghav Garg

A Review on Current State of Art of Bioprinting 195
Devarsh Vyas and Divya Udyawar

A Turnkey Manufacturing Solution for Customized Insoles Using Material Extrusion Process 203
Jia-Chang Wang, Hitesh Dommati and Jung Cheng

Parameter Optimization for Polyamide in Selective Laser Sintering Based on Mechanical Behavior 217
B. Karthick Raja, R. Jegan Pravin Raja, K. Karan, R. Soundararajan and P. Ashokavarthanan

Analysis of Adjacent Vertebrae Post Vertebroplasty 233
A. Kavitha, G. Sudhir, V. D. Deepak, M. Pavithra and V. Vallabhi

Design and Processing of Functionally Graded Material: Review and Current Status of Research 243
Manu Srivastava, Sandeep Rathee, Sachin Maheshwari and T. K. Kundra

Development of Electrical Discharge Machining (EDM) Electrode Using Fused Deposition Modeling (FDM) 257
Piyush Saxena and R. M. Metkar

State of the Art of Powder Bed Fusion Additive Manufacturing: A Review 269
R. Verma and G. Kaushal

Distortion in Metal Additive Manufactured Parts 281
Hemnath Anandan Kumar and Senthilkumaran Kumaraguru

Laser Metal Deposition of Titanium Parts with Increased Productivity 297
Vishnuu Jothi Prakash, Mauritz Möller, Julian Weber and Claus Emmelmann

About the Editors

Mr. L. Jyothish Kumar is the Founder CEO of Rapitech Solutions Inc., Bangalore, and the Founder President of the Additive Manufacturing Society of India (AMSI), Bangalore. He received his bachelor's degree in Mechanical Engineering from the National Institute of Engineering, Mysore, and his master's in Rapid Product Development from De Montfort University, UK. He is currently pursuing Ph.D. research in Aerospace Applications of Additive Manufacturing Technologies. Prior to that, he served in various areas of the Mechanical Engineering Industry, such as Quality Assurance, Marketing, Sales, and Product Development in India and abroad. He has specialized experience in Quality Management Systems and Rapid Product Development. He is currently the Managing Editor of the Additive Manufacturing Technology Magazine, the only journal for *3D Printing and Additive Manufacturing Technologies* in India.

Prof. Pulak M. Pandey is currently serving as a Professor in the Department of Mechanical Engineering at the Indian Institute of Technology (IIT) Delhi. After completing his B.Tech. from HBTI Kanpur in 1993, he went on to do his master's and Ph.D. from IIT Kanpur, where his Ph.D. was in the area of Additive Manufacturing/3D Printing. In IIT Delhi, he diversified his research areas in the field of micro- and nano-finishing and micro-deposition and also continued working in the area of 3D Printing. He supervised 21 Ph.D.s and more than 33 M.Tech. theses in the last 10 years and also filed 13 Indian patent applications. He has approximately 119 international journal papers and 44 international/national refereed conference papers to his credit. These papers have been cited for more than 3056 times with h-index as 26. He received Highly Commended Paper Award by Rapid Prototyping Journal for the paper "Fabrication of three dimensional open porous regular structure of PA 2200 for enhanced strength of scaffold using selective laser sintering" published in 2017. Many of the B.Tech. and M.Tech. projects he has supervised have received awards and accolades, and his students have won the Gandhian Young Technological Innovation (GYTI) Award in 2013, 2015, and 2017. He is the recipient of the Outstanding Young Faculty Fellowship

(IIT Delhi) sponsored by Kusuma Trust Gibraltar and J. M. Mahajan Outstanding Teacher Award of IIT Delhi.

Prof. David Ian Wimpenny is currently the Chief Technologist at The Manufacturing Technology Centre (MTC), Coventry, UK. He joined MTC as a Technology Manager in 2011 and worked as a Full-Time Technologist of the Component Technology Group at MTC. He is the Chairman of the Additive Manufacturing and 3D Printing Forum for HVM Catapult. His past roles include being Head of the Research at De Montfort University, Leicester, UK, from 2009 to 2011, and Director of the Additive Manufacturing Technology Group in the Department of Engineering and Technology, De Montfort University, from 2001 to 2011. He is also a Member of the Additive Manufacturing Special Interest Group (AM-SIG), which was established by the Technology Strategy Board to develop a road map for the UK AM sector. His major activities are in the areas of Additive Manufacturing, Rapid Product Development, Laser Printing, Surface Engineering, and Manufacturing Production Tooling. He has published more than 60 papers in international/national journals and presented papers at seminars and international conferences. He holds three patents for Rapid Prototyping, Reverse Engineering, and Computer-Aided Design. He also serves on the review committee of several reputed international journals like *Additive Manufacturing*, *Rapid Prototyping*, *Materials Processing Technology*. He has two books to his credit and was a Co-Editor of the Rapid Prototyping Case Book, Professional Engineering Publications.

Finite Element Analysis of Melt Pool Characteristics in Selective Laser Spot Melting on a Powder Layer

Akash Aggarwal and Arvind Kumar

Abstract In this work, volume contraction of powder layer and convective flow in the melt pool during laser spot melting of Ti–6Al–4V powder layer are investigated using a transient two-dimensional finite element model. An algorithm, coupled with the finite element model, accounting for volume contraction due to melting of porous powder to a denser liquid is proposed, which is thereafter used to understand the role of natural and Marangoni convection on the melt pool behaviour. Results for the melt pool characteristics, such as melt pool geometry, melt pool fluid flow dynamics and thermal behaviour are presented.

Keywords Selective laser melting · Ti–6Al–4V · Powder layer
Marangoni convection · Natural convection

Nomenclature

C_p	Specific heat capacity (J kg^{-1} K^{-1})
g	Acceleration due to gravity (m s^{-2})
K	Thermal conductivity (W m^{-1} K^{-1})
L	Latent heat of fusion (J kg^{-1})
T	Temperature (K)
$\vec{u}$	Velocity vector (m s^{-1})
φ	Porosity
P_{laser}	Laser power (W)
v_g	Velocity magnitude of Gaussian profile (m s^{-1})
k_{powder}	Thermal conductivity of powder layer (W m^{-1} K^{-1})
k_{solid}	Thermal conductivity of substrate (W m^{-1} K^{-1})
R	Gaussian beam spot size (m)
h_c	Heat convection coefficient (W m^2 K^{-1})

A. Aggarwal · A. Kumar (✉)
Department of Mechanical Engineering, Indian Institute of Technology Kanpur, Kanpur 208016, India
e-mail: arvindkr@iitk.ac.in

L. J. Kumar et al. (eds.), *3D Printing and Additive Manufacturing Technologies*,
https://doi.org/10.1007/978-981-13-0305-0_1

β_T Coefficient of thermal expansion (K^{-1})
$T_{solidus}$ Solidus temperature (K)
$T_{liquidus}$ Liquidus temperature (K)

Greek Symbols

μ Dynamic viscosity (kg m^{-1} s^{-1})
ρ Density (kg m^{-3})
γ Surface tension (N m^{-1})
σ Stefan–Boltzmann constant (W m^2 K^{-4})

Subscripts

solidus Solidus temperature
liquidus Liquidus temperature

1 Introduction

Selective Laser Melting (SLM) is a type of additive manufacturing process that begins with placing a layer of powder onto a solid substrate. A laser beam traverses the surface of the powder layer along a predefined profile. The intense heat generated by the laser beam causes powder particles to melt and to form a melt pool which solidifies as a fused layer of material. This mechanism operates over the entire layer according to the need and is repeated for each subsequent layer until the desired product is attained [1]. SLM can manufacture the final geometry with part density similar to that of the bulk solid. SLM has a wide range of potential applications in aerospace, tooling, automotive and prototyping industries. SLM is also used in manufacturing of orthopaedic and dental bio-implants, where porous titanium structures are used [2] and also in stainless steel sandwich structures where remarkable progress has been achieved [3].

SLM process involves interaction with very high energy beam and high temperature gradients causing the problem of thermal distortion, balling and the formation of cracks in the product [4]. Temperature gradients within the melt pool cause a surface tension driven fluid flow from regions with low surface tension to regions with high surface tension, known as Marangoni convection. It is very challenging to study the process experimentally as it involves high energy beam interaction, rapid melting and rapid solidification [5]. Therefore, alternative strategies, such as computational modelling and simulations are an emerging area to study this process. Numerical modelling becomes an effective tool to understand the effect of numerous process parameters and to obtain optimized conditions for the SLM process [6].

In the past some works were reported to investigate the SLM process. Loh et al. [7] developed a 3D finite element heat transfer model to study the SLM process. They have divided the powder layer in two sub layers and when the temperature of the powder layer reaches the melting temperature then the top sub-layer was removed. Verhaeghe et al. [8] presented a model to study various aspects of the SLM process using an enthalpy formulation. Their model accounted for shrinkage and laser light penetration. They investigated the significance of evaporation for various process parameters. Antony et al. [9] presented a numerical model to investigate the effect of process parameters during laser melting of grade 316L stainless steel powder on an AISI 316L substrate using a pulsed Nd-YAG laser. Effect of laser power, scanning speed and size of the beam on the melt pool dimensions was studied and it was found that the laser power and the scanning speed of the laser beam have a significant effect on the track characteristics. Kruth et al. [10] studied the surface morphology of SLM produced parts. They reported that with the help of laser remelting one could achieve higher density and better surface quality of SLM produced parts but it will take longer production time. Dou et al. [11] experimentally studied and measured the effect of surface tension on Al-alloy melt and found that the effect significantly depends on the operating temperature during the process. Zhang et al. [12] developed a three dimensional model for laser sintering of metal powders and investigated impact of the intensity of the moving heat source, the traverse speed, and the thickness of the powder layer on the thermal behaviour.

The primary aim of this paper is to develop a model which includes volume contraction of powder layer coupled with other attendant transport phenomena in the SLM process. A transient two-dimensional finite element heat transfer model, incorporating the volume contraction algorithm along with buoyancy and surface tension driven flow, and melting phase change, is proposed. Using the model laser-powder interaction during selective laser spot melting on a Ti–6Al–4V powder layer is simulated.

2 Model Description

Figure 1 shows the computational domain considered. The laser beam is stationary focusing on a spot of the powder layer. Both the powder layer and the substrate are of Ti–6Al–4V material. Its material properties are taken from references [13–15] and are given in Table 1. Powder layer was considered as a homogeneous layer with uniform material properties.

To account for porosity, the material properties were obtained by suitable relationships. Melt flow was considered as laminar and the fluid was assumed to be an incompressible Newtonian fluid. The laser beam with Gaussian distribution of

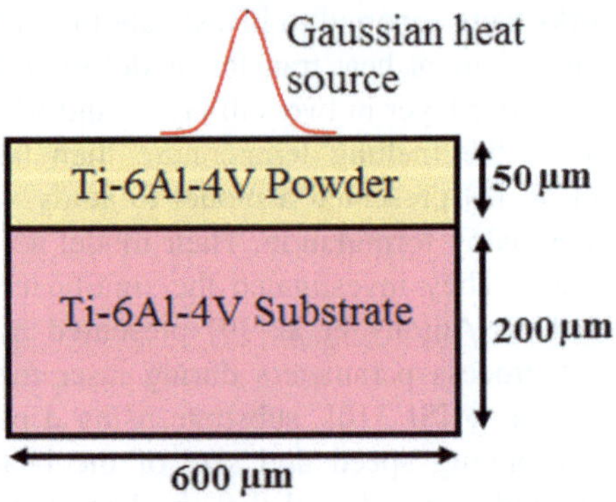

Fig. 1 Schematic of Ti–6Al–4V powder layer and the substrate

Table 1 Thermophysical properties of Ti–6Al–4V

Liquidus temperature (K)	1923
Solidus temperature (K)	1876
Solid specific heat (J kg^{-1} K^{-1})	$483.04 + 0.215T \quad T \leq 1268$ $412.7 + 0.1801T \quad 1268 < T \leq 1923$
Liquid specific heat (J kg^{-1} K^{-1})	831
Thermal conductivity (W m^{-1} K^{-1})	$1.25 + 0.015T \quad T \leq 1268$ $3.15 + 0.012T \quad 1268 < T \leq 1923$ $-12.75 + 0.024T \quad T > 1923$
Solid density (kg m^{-3})	$4420 - 0.154\ (T - 298)$
Liquid density (kg m^{-3})	4122
Latent heat of fusion (kJ K^{-1})	286
Dynamic viscosity (N m^{-1} s^{-1})	3×10^{-3}
Thermal expansion coefficient (K^{-1})	1.1×10^{5}
Surface tension coefficient (N m^{-1} K^{-1})	-0.28×10^{-3}

Table 2 Parameters for simulation

Parameter	Value
Laser spot size (μm)	80
Ambient temperature (K)	293.15
Laser power (W)	40
Powder layer thickness (μm)	50
Absorptivity	0.4
Porosity (φ)	0.35

heat was applied to the top surface of the powder layer for 200 μs, and during this period melt pool thermal and fluid flow behaviour was analysed. Parameters considered in the simulations are listed in Table 2.

2.1 *Powder Bed Properties*

Accounting the power porosity in the powder layer is crucial in predicting accurate results for powder bed melting [16]. The gas-filled pores and powder particle size controls the thermal conductivity of the powder bed [17]. Thummler and Oberacker [18] proposed a relationship for the thermal conductivity of a powder bed given as

$$k_{\text{powder}} = k_{\text{solid}}(1 - \varphi), \tag{1}$$

where k_{powder} and k_{solid} are the thermal conductivity of the powder bed and the bulk solid, respectively. Similarly, effective density of the powder bed can be determined by [19]

$$\rho_{\text{powder}} = \rho_{\text{solid}}(1 - \varphi), \tag{2}$$

where ρ_{powder} and ρ_{solid} are the density of the powder bed and the bulk solid, respectively.

2.2 *Governing Transport Equations*

2.2.1 Phase Change in the Powder Layer

Only one phase change from solid powder particles to molten metal is considered. Vaporization is not taken into account as the evaporation temperature for Ti–6Al–4V is very high and it was not reached in the present simulations.

Mass Conservation:

The continuity equation during the phase change in the powder layer is given by

$$\frac{\partial \rho}{\partial t} + \nabla.(\rho \vec{u}) = 0 \tag{3}$$

Energy Conservation:

The governing heat transfer equation during the phase change in the powder layer is given by

$$\frac{\partial(\rho C_{\text{p}} T)}{\partial t} + \vec{u}.\nabla(\rho C_{\text{p}} T) = \nabla.(K \nabla T) \tag{4}$$

During phase change, the values of ρ, C_p and k were determined by the following equations

$$\rho = \theta\rho_{\text{phase1}} + (1-\theta)\rho_{\text{phase2}} \tag{5}$$

$$C_p = \frac{1}{\rho}\left(\theta\rho_{\text{phase1}}C_{p,1} + (1-\theta)\rho_{\text{phase2}}C_{p,2}\right) + L\frac{\partial\alpha_m}{\partial T} \tag{6}$$

$$k = \theta k_{\text{phase1}} + (1-\theta)k_{\text{phase2}}, \tag{7}$$

where phase1 represents the solid phase, i.e., powder bed with modified properties), phase2 represents the liquid phase, θ is a linear function, varying between 1 and 0 and representing the fraction of phase before transition.

Momentum Conservation:

The governing momentum conservation equation is given by

$$\frac{\partial(\rho\vec{u})}{\partial t} + \vec{u}.\nabla(\rho\vec{u}) = -\nabla p + \nabla.(\mu\left(\nabla\vec{u} + (\nabla\vec{u})^{T}\right) + \vec{F} \tag{8}$$

$$\vec{F} = \overrightarrow{F_S} + \overrightarrow{F_N} \tag{9}$$

The source term $\overrightarrow{F_S}$ in Eq. (9) is defined in Eq. (10). It aids to bring down the velocity of the fluid at the liquid-solid phase transition interface and makes the fluid motion in the unmelted powder zone as zero [19, 20].

$$\overrightarrow{F_S} = \frac{(1-\beta)^2}{\beta^3 + b}C\vec{u} \tag{10}$$

The constant C in Eq. (10) represents mushy zone constant and a value of 150,000 kg m^{-3} s^{-1} is considered in the current model [20]. The term b is a constant with small value to prevent division by zero, and β represents the liquid fraction given by [21, 22]

$$\beta = \begin{pmatrix} 0 & T < T_{\text{solidus}} \\ \frac{T - T_{\text{solidus}}}{T_{\text{liquidus}} - T_{\text{solidus}}} & T_{\text{solidus}} \leq T < T_{\text{liquidus}} \\ 1 & T \geq T_{\text{liquidus}} \end{pmatrix} \tag{11}$$

where T_{solidus} and T_{liquidus} represent solidus and liquidus temperature, respectively.

Natural Convection:

The buoyant flow is included in this model with the help of Boussinesq approximation. The source term $\overrightarrow{F_N}$ in Eq. (9) is given as

$$\overrightarrow{F_N} = \rho_{\text{liquid}} \vec{g} \beta_T (T - T_{\text{ref}}) \tag{12}$$

where ρ_{liquid}, β_T and T_{ref} are density of liquid, coefficient of thermal expansion and reference temperature, respectively.

Marangoni Convection:

The following equation describes the forces that the Marangoni convection induces on the interface at the free surface [23, 24]

$$\mu\left(\frac{\partial u}{\partial y}\right) = -\left(\frac{\partial \gamma}{\partial T}\right)\left(\frac{\partial T}{\partial x}\right), \tag{13}$$

where μ and γ are viscosity and surface tension, respectively.

2.2.2 Substrate

Energy Conservation

The heat transfer equation in the substrate is given by

$$\frac{\partial(\rho_{\text{solid}} C_p T)}{\partial t} = \nabla.(k_{\text{solid}} \nabla T), \tag{14}$$

where ρ_{solid}, k_{solid} and C_p are density, thermal conductivity and specific heat capacity of the solid substrate, respectively.

2.3 *Boundary Conditions*

The heat energy from the laser beam can be approximated by a Gaussian distribution and the heat flux (W m^{-2}) on the powder bed is given by the following expression:

$$q = \frac{2AP}{\pi R^2} \exp\left(-\frac{2r^2}{R^2}\right), \tag{15}$$

where q is the input heat flux, A is the absorptivity, P is the power of laser beam and R is the radial distance in which energy density equals to e^{-2} times that at the centre of the laser spot.

Energy balance at the top surface leads to the following boundary equation

$$k\frac{\partial T}{\partial n} = q - h_c(T - T_\infty) - \varepsilon\sigma\left(T^4 - T_\infty^4\right) \tag{16}$$

Terms on the right-hand side are heat energy from the laser beam, convective heat loss and radiation heat loss to the surrounding, respectively. In Eq. (16) h_c is the convective heat transfer coefficient, ε is the emissivity and σ is the Stefan-Boltzmann constant.

2.4 Volume Contraction of the Powder Layer

When powder layer is irradiated with a laser beam, the powder particles start to melt and thereby the pores initially present between them are removed. This causes the volume of the powder layer to contract in size as shown in Fig. 2 [9]. The laser heat energy profile is considered as Gaussian, meaning that maximum intensity of the heat source will be at the centre of the beam and it decreases as one move away from the centre. Taking an analogy from the case of laser drilling, during which keyhole formed is of Gaussian shape, it is considered that when a laser beam is focused on a powder layer the volume contraction that occurs after powder particles melt will also be of Gaussian profile. The maximum melting will occur at the centre as the heat source intensity is maximum there, accordingly, it can be assumed that the volume contraction will be maximum at the centre and will decrease while moving away from the centre.

To incorporate this strategy, the deformed geometry physics in COMSOL Multiphysics^{TM} software has been used. This enables the motion of boundary of the computational domain with time. This is implemented by solving PDEs for the mesh displacements. The top interface on which the laser beam is focused moves with a normal velocity v_g, which is of Gaussian profile. This leads to time-dependent formation of a cavity as soon as melt pool starts to form.

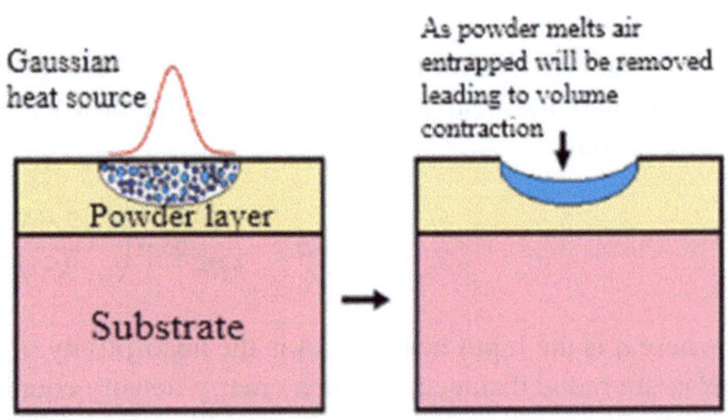

Fig. 2 Volume contraction in the powder layer

3 Results

In simulations, a laser power of 40 W was applied for a period of 200 μs to the top surface of the powder layer. Gaussian beam spot radius was taken to be 40 μm. During the heating by the laser, the formation of melt pool is analysed. The model was implemented using COMSOL Multiphysics™ software.

Figure 3 shows the temperature distribution at t = 200 μs in the entire domain containing the powder layer and the substrate. As mentioned earlier, the heat source is applied until 200 μs. The melt pool is confined by the melt pool boundary as shown in the figure. In the figure the substrate, the unmelted powder layer, the melt pool and the volume contraction caused by powder bed melting can be clearly seen. It can also be observed that there is not much temperature rise in the portion of the substrate which is away from the melt pool boundary. This is due to low thermal conductivity of Ti–6Al–4V and shorter heat source interaction time (200 μs). It may be noted that in subsequent figures the results will be shown in region comprising only the melt pool and the nearby region.

Figures 4 and 5 show the evolution of the melt pool and the flow, and the temperature distribution at different times. In Fig. 4 velocity vectors are shown. In the figure the melt pool and flow in the melt pool, the unmelted powder layer, and the volume contraction caused by powder bed melting can be clearly seen. It can be observed that the maximum velocity occurs along the free surface and the molten liquid tend to flow outward from the centre of melt pool. This is because of the temperature gradients within the melt pool that causes a surface tension driven flow, often called thermo-capillary fluid flow (or Marangoni convection), from a region with low surface tension to regions with high surface tension. Due to this Marangoni convection there is a recirculating fluid flow. The magnitude of the velocity in the melt pool is in the range of 4–8 m s^{-1} which is quite substantial. The high velocity in the melt pool makes the molten liquid to flow away from the centre resulting in a shallow melt pool. From results shown at different time, it can be

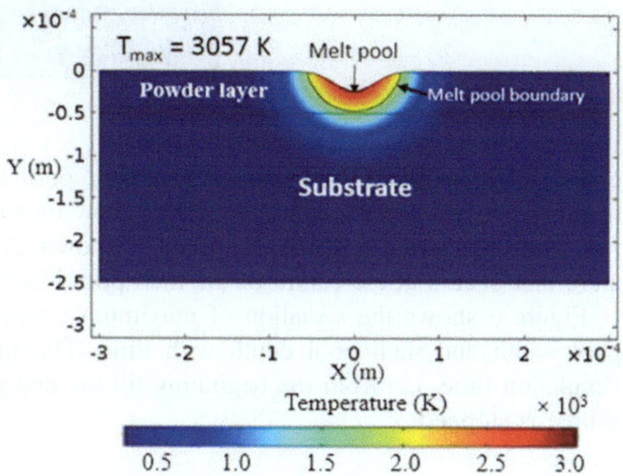

Fig. 3 Temperature map in the complete domain at t = 200 μs containing the powder layer and the substrate

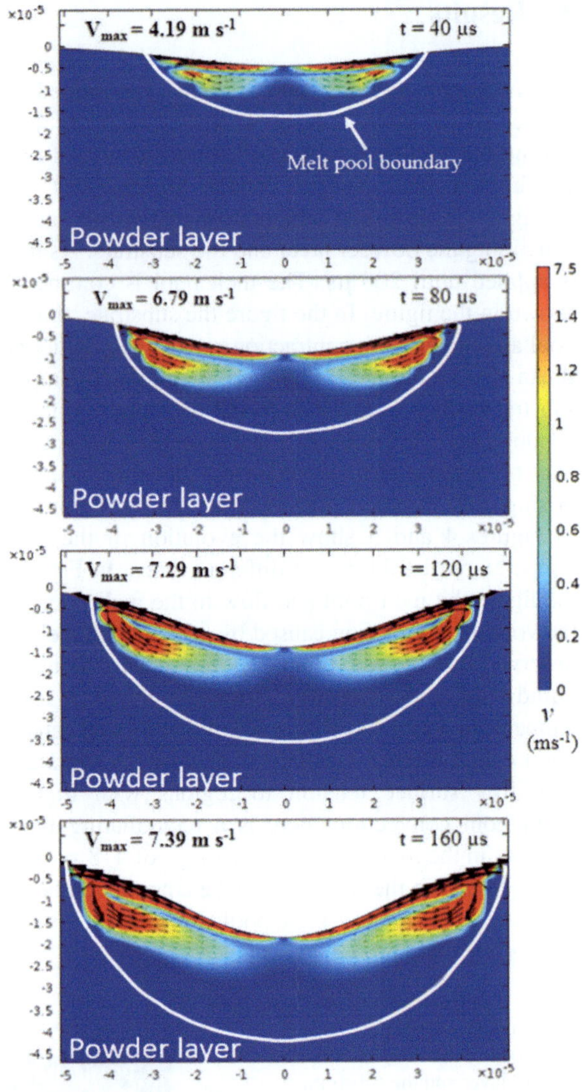

Fig. 4 Evolution of the melt pool and the flow, coloured region shows the velocity magnitude

clearly observed that the depth of the melt pool is lower than its width. The magnitude of velocity in the recirculating flow increases with progress in time. In Fig. 5 the temperature distribution map is shown. As seen the maximum temperature that occurs at the centre of the melt pool increases with progress in time.

Figure 6 shows the variation of maximum temperature in the melt pool, melt pool width and melt pool depth with time. The time axis shows the complete simulation time, i.e. from the beginning till the end time of 200 μs when the heat source is stopped.

Fig. 5 Evolution of the melt pool temperature

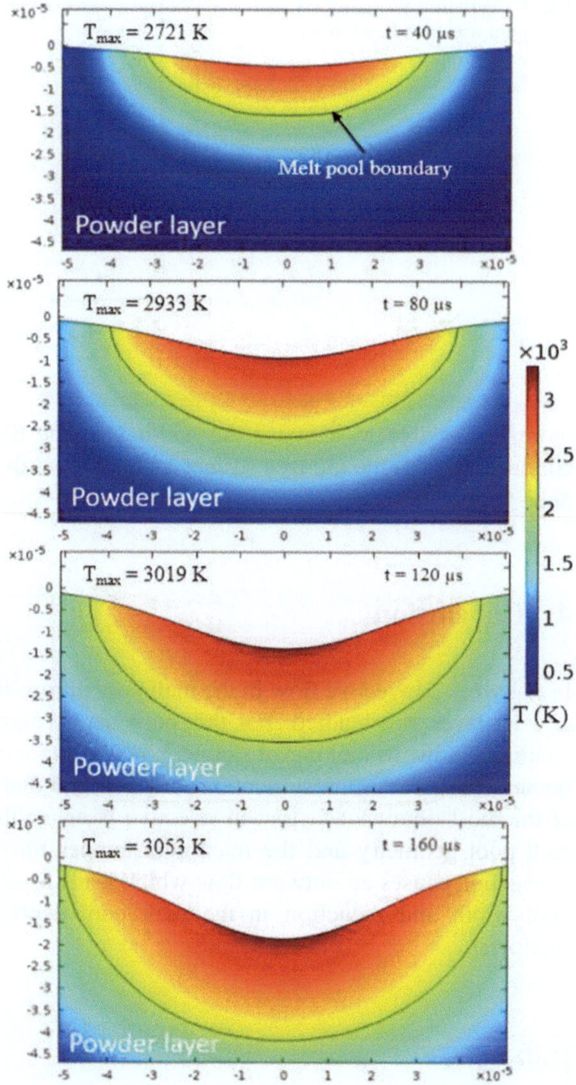

It can be seen that initially the maximum temperature in the melt pool rises rapidly, but it reaches a quasi-steady state after 130 μs. This is because of the melt pool mixing caused by the convective flow whose intensity increases with time.

The same behaviour is also exhibited by the melt pool depth. On the other hand, the behaviour of the melt pool width is different, it continuously increases with time. This is due to surface tension driven fluid flow, which causes hot molten metal to flow from centre to sideways leading to increase in melt pool width. This flow is

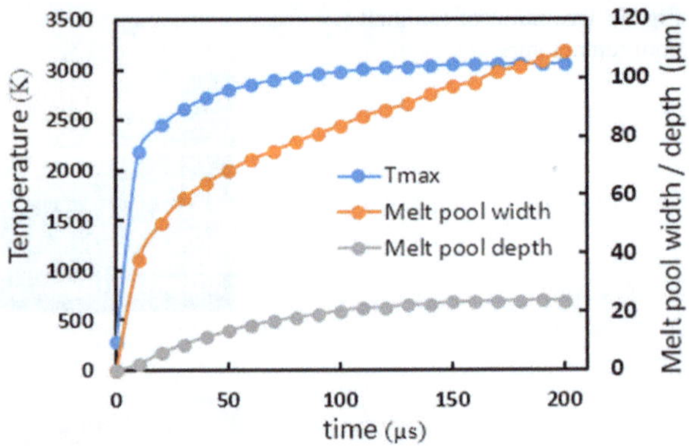

Fig. 6 Variation of T_{max}, melt pool width and melt pool depth with time

more responsible to increase the width than the depth of the melt pool as it is confined close to the top free surface and not towards the deeper portion of the melt pool.

4 Conclusion

In this work convective flow in the melt pool during selective laser spot melting of a powder layer was studied. For this purpose, an algorithm accounting for volume contraction due to melting of porous powder into a denser liquid is developed and implemented. From the simulation, it was found that Marangoni convection is one of the most dominant factors in selective laser melting process that influences the melt pool geometry and the maximum temperature in the melt pool. Marangoni convection causes an outward flow which leads to substantial widening of the melt pool width and reduction in the melt pool depth, thus making the melt pool shallower.

References

1. E.C. Santos, K. Osakada, M. Shiomi, Y. Kitamura, F. Abe, Microstructure and mechanical properties of pure titanium models fabricated by selective laser melting. Proc. Inst. Mech. Eng., Part C: J. Mech. Eng. Sci. **218**(7), 711–719 (2010)
2. L. Mullen, R.C. Stamp, W.K. Brooks, E. Jones, C.J. Sutcliffe, Selective laser melting: a regular unit cell approach for the manufacture of porous, titanium, bone in-growth constructs, suitable for orthopedic applications. J. Biomed. Mater. Res. B Appl. Biomater. **89**(2), 325–334 (2009)
3. Y. Shen, S. McKown, S. Tsopanos, C.J. Sutcliffe, R.A.W. Mines, W.J. Cantwell, The mechanical properties of sandwich structures based on metal lattice architectures. J. Sandwich Struct. Mater. (2009)

4. D. Gu, Y. Shen, Balling phenomena during direct laser sintering of multi-component Cu-based metal powder. J. Alloy. Compd. **432**(1), 163–166 (2007)
5. S.M. Thompson, L. Bian, N. Shamsaei, A. Yadollahi, An overview of Direct Laser Deposition for additive manufacturing. Part I: Transp. Phenom. Model. Diagn. Add. Manuf. **8**, 36–62 (2015)
6. N. Shamsaei, A. Yadollahi, L. Bian, S.M. Thompson, An overview of Direct Laser Deposition for additive manufacturing. Part II: Mech. Behav. Process Parameter Optim. Control Add. Manufact. **8**, 12–35 (2015)
7. L.E. Loh, C.K. Chua, W.Y. Yeong, J. Song, M. Mapar, S.L. Sing, Z.H. Liu, D.Q. Zhang, Numerical investigation and an effective modelling on the Selective Laser Melting (SLM) process with aluminium alloy 6061. Int. J. Heat Mass Transf. **80**, 288–300 (2015)
8. F. Verhaeghe, T. Craeghs, J. Heulens, L. Pandelaers, A pragmatic model for selective laser melting with evaporation. Acta Mater. **57**(20), 6006–6012 (2009)
9. K. Antony, N. Arivazhagan, K. Senthilkumaran, Numerical and experimental investigations on laser melting of stainless steel 316L metal powders. J. Manuf. Process. **16**(3), 345–355 (2014)
10. E. Yasa, J. Deckers, J.P. Kruth, The investigation of the influence of laser re-melting on density, surface quality and microstructure of selective laser melting parts. Rapid Prototyp. J. **17**, 312–327 (2011)
11. D. Lei, Y. ZhangFu, L. JianQiang, L. Jing, W. XiaoQiang, Surface tension of molten Al-Si alloy at temperatures ranging from 923 to 1123 K. Chin. Sci. Bull. **53**(17), 2593–2598 (2008)
12. T. Chen, Y. Zhang, Numerical simulation of two-dimensional melting and resolidification of a two-component metal powder layer in selective laser sintering process. Numer. Heat Transf. Part A: Appl. **46**(7), 633–649 (2004)
13. Z. Fan, F. Liou, *Numerical Modeling of the Additive Manufacturing (AM) Processes of Titanium Alloy, Titanium Alloys–Towards Achieving Enhanced Properties for Diversified Applications* (2012)
14. G. Welsch, R. Boyer, E.W. Collings, *Materials Properties Handbook: Titanium Alloys* (ASM International, Materials Park, Ohio, 1998)
15. K.C. Mills, *Recommended Values of Thermophysical Properties for Selected Commercial Alloys* (Woodhead Publishing, 2002)
16. Y. Li, D. Gu, Thermal behavior during selective laser melting of commercially pure titanium powder: numerical simulation and experimental study. Add. Manuf. **1**, 99–109 (2014)
17. M. Rombouts, L. Froyen, A.V. Gusarov, E.H. Bentefour, C. Glorieux, Light extinction in metallic powder beds: correlation with powder structure. J. Appl. Phys. **98**(1), 013533 (2005)
18. F. Thummler, R. Oberacker, *An Introduction to Powder Metallurgy* (Institute of Materials, 1993)
19. A. Kumar, P. Dutta, S. Sundarraj, M.J. Walker, Remelting of solid and its effect on macrosegregation during solidification. Numer. Heat Transf. Part A: Appl. **51**(1), 59–83 (2007)
20. N. Pathak, A. Kumar, A. Yadav, P. Dutta, Effects of mould filling on evolution of the solid–liquid interface during solidification. Appl. Therm. Eng. **29**(17), 3669–3678 (2009)
21. A. Kumar, S. Gu, H. Tabbara, S. Kamnis, Study of impingement of hollow ZrO_2 droplets onto a substrate. Surf. Coat. Technol. **220**, 164–169 (2013)
22. A. Kumar, S. Gu, S. Kamnis, Simulation of impact of a hollow droplet on a flat surface. Appl. Phys. A **109**(1), 101–109 (2012)
23. V.K. Arghode, A. Kumar, S. Sundarraj, P. Dutta, Computational modeling of GMAW process for joining dissimilar aluminum alloys. Numer. Heat Transf. Part A: Appl. **53**(4), 432–455 (2008)
24. D. Dai, D. Gu, Thermal behavior and densification mechanism during selective laser melting of copper matrix composites: simulation and experiments. Mater. Des. **55**, 482–491 (2014)

Thermal Transport Phenomena in Multi-layer Deposition Using Arc Welding Process

Anshul Yadav, Aniruddha Ghosh and Arvind Kumar

Abstract The repair of steel plate using welding can be optimized by multi-layer deposition. In this numerical study, semi-automatic arc welding is used to deposit single-track multi-layers of mild steel on a same material mild steel plate. Three-dimensional transient numerical simulations of the transport phenomena involved in the melt pool are performed. The model considers heat transfer, convective and radiative losses, phase change, re-melting, solidification, and buoyancy and Marangoni convection driven fluid flow in the melt pool. The model predicts the temperature and velocity fields, and the evolution of melt pool shape and size.

Keywords Additive manufacturing · Fluid dynamics · Multi-layer deposition Phase change · Re-melting

Nomenclature

C_p	Specific heat of work piece (J kg^{-1} K^{-1})
f_l	Volume fraction of liquid phase
$\mathbf{g}$	Acceleration due to gravity (m s^{-2})
h	Convective heat transfer coefficient (W m^{-2} K^{-1})
k	Thermal conductivity of the work piece (W m^{-1} K^{-1})
p	Pressure (N m^{-2})
q	Heat input (W m^{-2})
T	Temperature (°C)
T_∞	Ambient temperature (°C)

A. Yadav · A. Ghosh · A. Kumar (✉)
Department of Mechanical Engineering, Indian Institute of Technology Kanpur, Kanpur 208016, India
e-mail: arvindkr@iitk.ac.in

L. J. Kumar et al. (eds.), *3D Printing and Additive Manufacturing Technologies*,
https://doi.org/10.1007/978-981-13-0305-0_2

t Time (s)
V Welding speed (m s^{-1})
$\overrightarrow{u}$ Continuum velocity vector (m s^{-1})

Greek Symbols

α_m Mass fraction
β_T Coefficient of thermal expansion (K^{-1})
γ Surface tension (N m^{-1})
ε Emissivity of the work piece
σ Stefan-Boltzmann constant (W m K^{-4})
μ Dynamic viscosity (N s m^{-2})
ρ Density of the work piece (kg m^{-3})

Subscripts

∞ Ambient
l Liquid state
s Solid state

1 Introduction

In recent years, additive manufacturing (AM) processes with high-energy heat sources have been developed and applied to fabricate multifunctional, complex-shaped, or custom-designed components. Different types of mechanical components can be manufactured with such a layer-based fabrication technology through a computer-controlled machine. AM is used to build component layer-by-layer material deposition. Similar to AM, welding process is also capable to generate layer-by-layer deposition of material.

Chiumenti et al. [1] described the formulations adopted for the numerical simulation of the shaped metal deposition (SMD) process and the experimental work to calibrate and validate the proposed model. Lan et al. [2] performed a multi-pass submerged arc welding (SAW) on HSLA steels using multi-micro alloyed electrodes, and three different heat input processes to investigate the microstructure evolution and corresponding mechanical properties of weldments.

The heat input to various layers governs residual stress distribution in layer-wise deposition. Residual stress could be decreased using multi-layer deposition and higher heat input. Heinze et al. [3] experimentally determined and calculated residual stresses using both two-dimensional and three-dimensional numerical models. Jiang et al. [4] estimated residual stress and deformation in the repair weld of a stainless steel clad plate by using finite-element methods. They also studied the effects of heat input and number of layers on residual stresses and deformation.

Cho et al. [5] analyzed the effect of torch angle and current polarities on the convective heat transfer in single wire SAW. They concluded that the variation of arc forces, the direction of droplet flight with polarity, and the torch angle significantly affect the molten pool flow and the resultant weld beads. Duggan et al. [6] presented a novel meso-scale (mm–cm) numerical model of multi-pass welding of stainless steel based on the front tracking formulation.

During a typical butt weld, there are multiple cycles of melting and solidification that occur throughout the weld pool. These thermal effects of the current deposition of liquid filler metal are being super-imposed onto those of the previous pass result in thermal cycles. Lee and Farson [7] conducted a three-dimensional numerical simulation of transport phenomena using volume of fluid (VOF) method for multi-layer single-track laser additive manufacturing (LAM). They predicted thermal and flow fields, transient variation of the melt pool fluid boundary shape and re-melting, and solidified build geometry during deposition of successive LAM layers which was in good agreement with the experimental results. Mishra et al. [8] developed transport phenomena-based mathematical model to study liquation cracking in weldments during fusion welding. Traidia et al. [9] proposed a hybrid two-dimensional and three-dimensional model for the numerical simulation of gas tungsten arc welding (GTAW). They predicted the temperature field as well as the shape of the solidified weld joint for different operating parameters, with relatively good accuracy and reasonable computational cost.

Boumerzoug et al. [10] studied the effect of arc welding on microstructures and mechanical properties of industrial low carbon steel. Nami et al. [11] studied thermal response of thick plate weldments under multi-layer and multi-block welding. The addition of welded material that normally occurs during the welding was also simulated in their work using the element rebirth technique. The numerical model they developed was able to predict temperature and velocity distribution, evolution of the melt pool shape and size, re-melting in the previous layer, heat affected zone, and solidified layer built-up. Zhao et al. [12] conducted a three-dimensional transient heat transfer numerical simulation with variable material properties to investigate thermal field, thermal cycling, thermal gradient and the effects of depositing directions on the thermal process of single-pass multi-layer rapid prototyping. Roberts et al. [13] used a new simulation technique called element birth and death, for modeling the three-dimensional thermal field in multiple layers in additive powder bed based process. Ding et al. [14] studied a three-dimensional thermo-elastic–plastic transient model and a model based on an advanced steady-state thermal analysis. Gosh et al. [15] studied the effect of electrode's tilt angle on transient temperature distribution, HAZ width and microstructure of various zones.

In this work, a numerical study of arc welding is performed to deposit a single-track three layers of mild steel on a mild steel plate. Three-dimensional transient numerical simulations of the transport phenomena involved in the melt pool are performed and thermal and velocity fields in the weld plate are numerically predicted. The model accounts for heat transfer, temperature dependent properties of the material, phase change, re-melting, re-solidification, and fluid flow in the melt pool.

2 Model Description

Figure 1 shows the schematic diagram of the computational domain. Figure 2 shows the geometrical size of computational domain considered. Since, the physical geometry undertaken is symmetric, a three-dimensional symmetric computational domain is considered. The process parameters and material properties are given in Table 1. Both the filler and the substrate are of same material (mild steel). The boundary conditions for temperature and velocity are as follows: heat flux was considered at top surface, bottom surface was insulated, and convective and radiative losses were accounted for the rest of the surfaces including the melt pool region.

The following assumptions are made in constructing the mathematical model:

1. The flow in the weld pool is laminar and incompressible.
2. The Boussinesq approximation is applied to simulate buoyancy-induced convection in the melt pool.
3. Gaussian distribution is used to describe the arc heat flux.

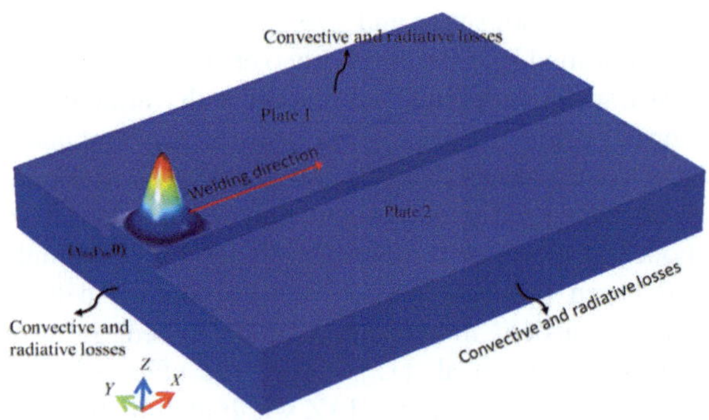

Fig. 1 Schematic of the computational domain

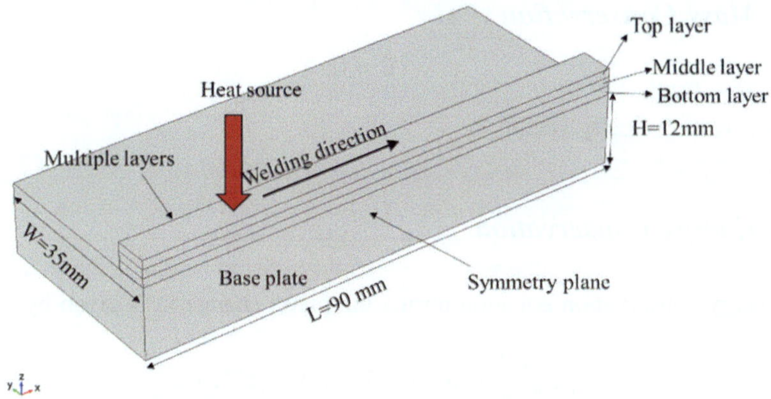

Fig. 2 Schematic of the computational domain showing geometry

Table 1 Material properties of the workpiece [16]

Ambient temperature (K)	300
Power (kJ m^{-1})	1750
Arc travel speed (mm s^{-1})	5
Liquidus temperature (K)	1802
Solidus temperature (K)	1770
Heat of fusion (kJ kg^{-1})	240
Solid specific heat (J kg^{-1} K^{-1})	750
Liquid specific heat (J kg^{-1} K^{-1})	840
Solid thermal conductivity (W m^{-1} K^{-1})	39.4
Liquid thermal conductivity (W m^{-1} K^{-1})	36.5
Solid and liquid density (kg m^{-3})	7530
Thermal expansion coefficient of liquid (K^{-1})	11.9×10^{-6}
Viscosity (kg m^{-1} s^{-1})	6.3×10^{-3}
Surface tension (N m^{-1})	1.65
Convective coefficient (W m^{-2} K^{-1})	20
Surface emissivity	0.75

3 Governing Transport Equations

In the present model, a fixed-grid continuum formulation based on the classical mixture theory [17] is adopted for modeling melting and solidification phase change in the meltpool. It may be noted that in the present formulation solid phase and liquid phase densities are same, and hence solidification shrinkage is not considered.

3.1 Mass Conservation

$$\nabla \cdot \vec{u} = 0, \tag{1}$$

where $\vec{u}$ is the velocity vector.

3.2 Energy Conservation

The energy conservation equation during the phase change in is given by

$$\rho C_\mathrm{p} \frac{\partial T}{\partial t} + \rho C_\mathrm{p} \vec{u} \cdot \nabla T = \nabla \cdot (k \nabla T), \tag{2}$$

where ρ is density, C_p is specific heat capacity, $\vec{u}$ is the continuum velocity, k is thermal conductivity and T is the temperature.

During phase change, the values of ρ, k and C_p were determined by the following equations:

$$\rho = f_\mathrm{l}\rho_\mathrm{l} + (1 - f_\mathrm{l})\rho_\mathrm{s} \tag{3a}$$

$$k = f_\mathrm{l}k_\mathrm{l} + (1 - f_\mathrm{l})k_\mathrm{s} \tag{3b}$$

$$C_\mathrm{p} = \frac{1}{\rho}\left(f_\mathrm{l}\rho_\mathrm{l}C_\mathrm{p,l} + (1 - f_\mathrm{l})\rho_\mathrm{s}C_\mathrm{p,s}\right) + L\frac{\partial \alpha_\mathrm{m}}{\partial T} \tag{3c}$$

where ρ_s represents the density of solid material, ρ_l represents the density of liquid material, ρ_s represents the density of solid material, $C_\mathrm{p,l}$ represents the specific heat of liquid material, $C_\mathrm{p,s}$ represents the specific heat of solid material, f_l is a linear function, varying between 1 and 0 and representing the fraction of liquid phase before transition. Its value is equal to 1 before solidus temperature and to 0 after liquidus temperature. L represents the latent heat of fusion and α_m is the mass fraction given as

$$\alpha_\mathrm{m} = \frac{1}{2}\frac{f_\mathrm{l}\rho_\mathrm{l} - (1 - f_\mathrm{l})\rho_\mathrm{s}}{\rho}. \tag{4}$$

3.3 Momentum Conservation

The governing momentum conservation equation is given by

$$\rho\frac{\partial \vec{u}}{\partial t} + (\rho\vec{u} \cdot \nabla)\vec{u} = -\nabla p + \nabla \cdot \left(\mu\left(\nabla\vec{u} + (\nabla\vec{u})^T\right) + \vec{F}\right. \tag{5}$$

$$\vec{F} = \overrightarrow{F_S} + \overrightarrow{F_N}, \tag{5a}$$

where μ is viscosity and p is pressure.

The source term $\overrightarrow{F_S}$ in (5a) aids to bring down the velocity of the fluid at the liquid–solid phase transition interface and is defined as

$$\overrightarrow{F_S} = -\frac{(1-f_l)^2}{f_l^3 + d} C\vec{u} \tag{6}$$

The constant C represents mushy zone constant. The term b is a constant with small value to prevent division by zero.

The buoyant flow is included in this model with the help of Boussinesq approximation. The source term $\overrightarrow{F_N}$ in (5a) is given as

$$\overrightarrow{F_N} = \rho\vec{g}\beta_T(T - T_{ref}), \tag{7}$$

where β_T and T_{ref} are thermal expansion coefficient of liquid and reference temperature, respectively.

3.4 Boundary Conditions

The heat source energy can be approximated by a Gaussian distribution and the heat flux (q) on the top surface is given by the following expression:

$$q = \frac{2AP}{\pi W_0^2} \exp\left(-\frac{2r^2}{W_0^2}\right), \tag{8}$$

where P is the arc power of heat source, A is the absorptivity, W_0 is the radial distance in which energy density equals to e^{-2} times that at the center of the heat source spot and r is the radial distance in meter.

Energy balance at the top surfaces leads to the following boundary equation:

$$k\frac{\partial T}{\partial n} = q - h_c(T - T_\infty) - \varepsilon\sigma\left(T^4 - T_\infty^4\right), \tag{9}$$

where h_c is the convective heat transfer coefficient, ε is the emissivity, σ is the Stefan-Boltzmann constant and T_∞ is the ambient temperature.

In (9), terms on the right-hand side are heat energy from the heat source, convective heat loss and radiation heat loss to the surrounding, respectively.

On the lateral surfaces, only heat loss due to convection is considered as losses due to radiation is negligible [16]. Therefore, energy balance at the side surfaces leads to the following boundary equation:

$$k\frac{\partial T}{\partial n} = -h_c(T - T_\infty) \tag{10}$$

The following equations describes the forces that the Marangoni convection induces on the interface at the free surface.

$$\tau_{xz} = -\mu\left(\frac{\partial u}{\partial z}\right) = \gamma\left(\frac{\partial T}{\partial x}\right) \tag{11}$$

$$\tau_{yz} = -\mu\left(\frac{\partial v}{\partial z}\right) = \gamma\left(\frac{\partial T}{\partial y}\right), \tag{12}$$

where γ is the surface tension coefficient.

4 Results and Discussions

The governing equations are solved using finite-element analysis software COMSOL Multiphysics™. In this study, multi-layer single-track semi-automatic arc welding experiment is carried out to join two mild steel plates with size 12 mm (thickness) × 35 mm (width) × 90 mm (length) in a butt joint configuration. The arc travel speed is 5 mm/s for each layer and the power of 1750 kJ m^{-1} is applied. Heat source is applied for 18 s in each layer. To obtain a better understanding of the deposition process, numerical simulations were also performed for this process. With the help of simulations, the temperature plots and the velocity distribution for each layer is studied. Also, the formation of melt pool is analyzed during the multi-layer deposition.

The results are presented in two parts: first, the temperature distribution is studied, in the second part study on velocity field, and meltpool shape and size is carried out. Physical phenomena, such as solidification and re-melting, phase change, natural convection, Marangoni convection, and convective and radiative losses have been included in the model.

4.1 Temperature Distribution

Figures 3, 4, and 5 show the temperature distribution plots and variation of the melt pool shape for bottom, mid and top layer, respectively for various time instants in the entire domain. The black contour shows the boundary of the melt pool. The maximum temperature rises up to 2732 K which is higher than the liquidus temperature of steel which is about 1802 K.

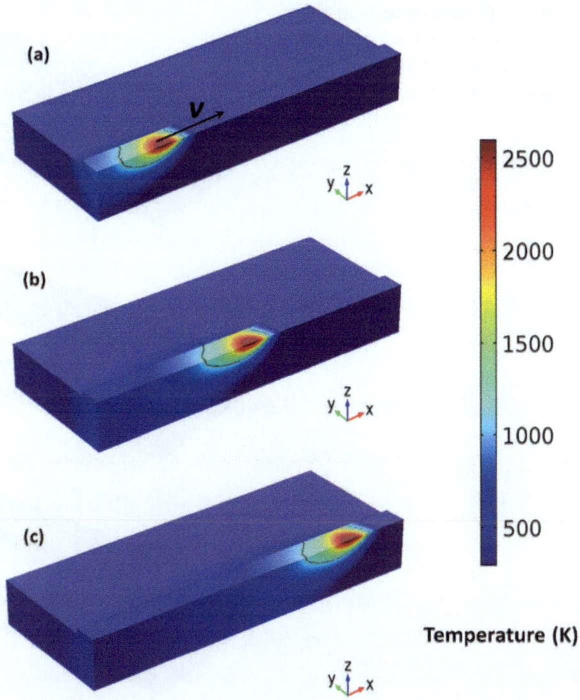

Fig. 3 Temperature map for bottom layer and variation of the melt pool fluid boundary shape at time. **a** t = 4 s, **b** t = 9 s, and **c** t = 14 s

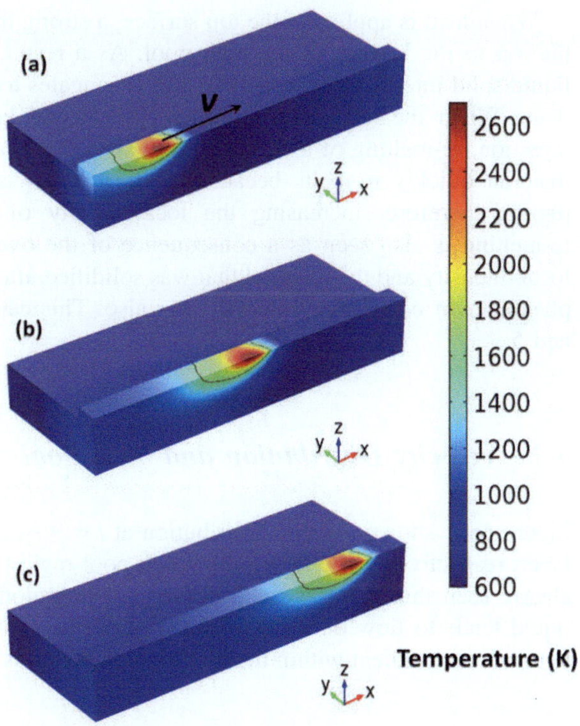

Fig. 4 Temperature map for the mid layer and variation of the melt pool fluid boundary shape at time. **a** t = 4 s, **b** t = 9 s, and **c** t = 14 s

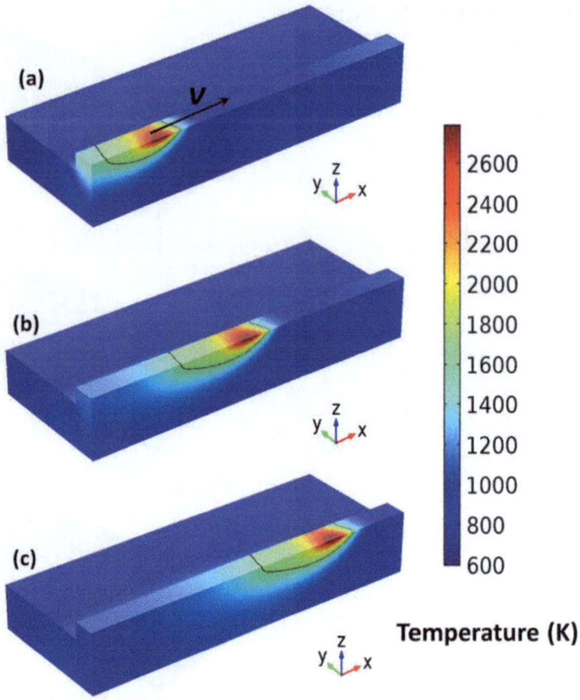

Fig. 5 Temperature map for the top layer and variation of the melt pool fluid boundary shape at time. **a** $t = 4$ s, **b** $t = 9$ s, and **c** $t = 14$ s

When heat is applied at the top surface, a strong thermal gradient develops from the top to the bottom of the melt pool. As a result, solidification starts from the liquid/solid interface at the bottom and propagates towards the top surface of weld plate. When the heat source reaches the second layer and travels along the x-direction, re-melting of the already deposited layer occurs. At this point, solidified material quickly re-melts because the heat beam is spending more time in this region, therefore, increasing the local supply of heat to the material. Here, re-melting is also seen as a consequence of the overlap between the heat source local intensity and the material that was solidified after the first layer. Similarly, this phenomenon occurs for the next layer also. This can be seen clearly in Figs. 3, 4 and 5.

4.2 *Velocity Distribution and Melt Pool Shape*

Figure 6a–c shows velocity distribution at $t = 9$ s for the bottom, middle, and top layer, respectively. The flow in the melt pool region is assumed laminar. It can be clearly seen that the maximum velocity occurs along the surface and the molten liquid tends to flow outward from the center of melt pool. This is because of the temperature gradient within the melt pool that causes a surface tension driven flow,

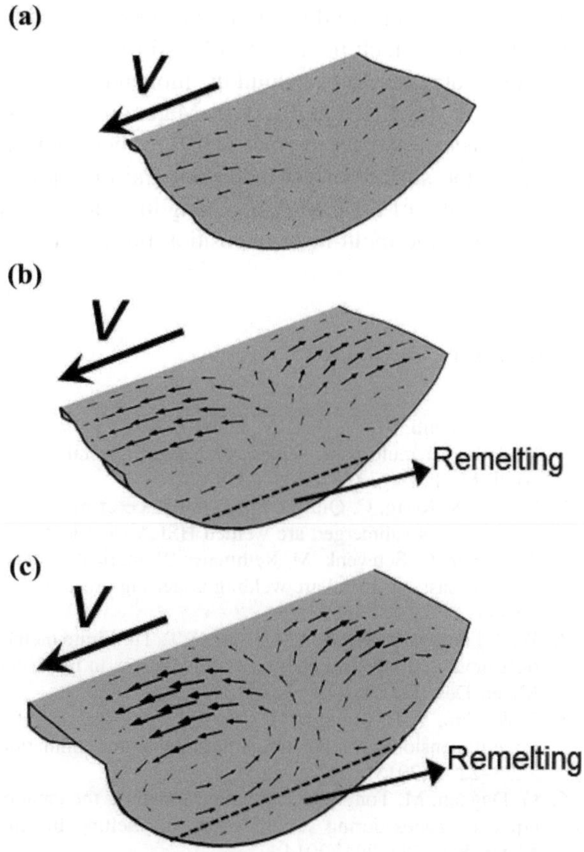

Fig. 6 Velocity distribution in the melt pool for different layers at time $t = 9$ s. **a** bottom layer, **b** mid layer, and **c** top layer

often called Marangoni convection, from a region with low surface tension to regions with high surface tension. Due to this Marangoni convection there is a recirculating fluid flow as shown in Fig. 6. The magnitude of velocity in the melt pool is in the range of 0.2–0.8 m/s which is quite common [18]. This high velocity in the melt pool makes the molten liquid to flow away from the center resulting in a shallow melt pool. Re-melting is also present for the middle and the top layer. This has been marked in Fig. 6b, c. A proportional scale is used to size the velocity arrows; therefore, the lengths of the vectors do not represent the magnitude of the velocity.

5 Conclusions

In this work, transport phenomena in the melt pool during single-track multi-layer deposition using arc welding were investigated. A three-dimensional transient model is developed incorporating heat transfer, convective and radiative losses, phase change, re-melting, solidification and, buoyancy and Marangoni convection driven fluid flow in the melt pool. It was found that with increase of the deposition height, the high temperature area behind the molten pool becomes larger gradually as the arc is moving in the direction of x-axis. Marangoni convection strongly governs the melt pool geometry, and its thermal and flow behavior. The maximum velocity occurs along the top surface of the melt pool and the molten liquid tends to flow outward from the center of the melt pool. Using the current model it was possible to simulate re-melting in the multi-layer deposition process and the flow therein.

References

1. M. Chiumenti, M. Cervera, A. Salmi, C.A. Saracibar, N. Dialami, K. Matsui, Finite element modeling of multi-pass welding and shaped metal deposition processes. Comput. Methods Appl. Mech. Eng. **199**, 2343–2359 (2010)
2. L. Lan, X. Kong, C. Qiu, D. Zhao, Influence of microstructural aspects on impact toughness of multi-pass submerged arc welded HSLA steel joints. Mater. Des. **90**, 488–498 (2016)
3. C. Heinze, C. Schwenk, M. Rethmeier, Numerical calculation of residual stress development of multi-pass gas metal arc welding under high restraint conditions. Mater. Des. **35**, 201–209 (2011)
4. W.C. Jiang, B.Y. Wang, J.M. Gong, S.T. Tu, Finite element analysis of the effect of welding heat input and layer number on residual stress in repair welds for a stainless steel clad plate. Mater. Des. **32**, 2851–2857 (2010)
5. D.W. Cho, W.H. Song, M.H. Cho, S.J. Na, Analysis of submerged arc welding process by three-dimensional computational fluid dynamics simulations. J. Mater. Process. Technol. **213**, 2278–2291 (2013)
6. G. Duggan, M. Tong, D.J. Browne, Modelling the creation and destruction of columnar and equiaxed zones during solidification and melting in multi-pass welding of steel. Comput. Mater. Sci., 285–294 (2014)
7. Y. Lee, D.F. Farson, Simulation of transport phenomena and melt pool shape for multiple layer additive manufacturing. J. Laser Appl. **28**, 012006s (2015)
8. S. Mishra, S. Chakraborty, T. DebRoy, Probing liquation cracking and solidification through modeling of momentum, heat, and solute transport during welding of aluminum alloys. J. Appl. Phys. **97**, 094912s (2005)
9. A. Traidia, F. Roger, E. Guyot, J. Schroeder, G. Lubineau, Hybrid 2D–3D modelling of GTA welding with filler wire addition. Int. J. Heat Mass Transf. **55**, 3946–3963 (2012)
10. Z. Boumerzoug, C. Derfouf, T. Baudin, Effect of welding on microstructure and mechanical properties of an industrial low carbon steel. Engineering **2**(7), 502–506 (2010)
11. M.R. Nami, M.H. Kadivar, K. Jafarpur, Three-dimensional thermal response of thick plate weldments: effect of layer-wise and piece-wise welding. Modell. Simul. Mater. Sci. Eng. **12** (4), 731–743 (2004)
12. H. Zhao, G. Zhang, Z. Yin, L. Wu, A 3D dynamic analysis of thermal behavior during single-pass multi-layer weld-based rapid prototyping. J. Mater. Process. Technol. **211**(3), 488–495 (2010)
13. I.A. Roberts, C.J. Wang, R. Esterlein, M. Stanford, A three-dimensional finite element analysis of the temperature field during laser melting of metal powders in additive layer manufacturing. Int. J. Mach. Tools Manuf. **49**, 916–923 (2009)
14. J. Ding, P. Colegrove, J. Mehnen, S. Ganguly, Thermo-mechanical analysis of wire and arc additive layer manufacturing process on large multi-layer parts. Comput. Mater. Sci. **50**, 3315–3322 (2011)

15. A. Ghosh, A. Yadav, A. Kumar, Modelling and experimental validation of moving tilted volumetric heat source in gas metal arc welding process. J. Mater. Process. Technol. **239**, 52–65 (2017)
16. A. Yadav, A. Ghosh, A. Kumar, Experimental and numerical study of thermal field and weld bead characteristics in submerged arc welded plate. J. Mater. Process. Technol. **248**, 262–274 (2017)
17. N. Pathak, A. Kumar, A. Yadav, P. Dutta, Effects of mould filling on evolution of the solid–liquid interface during solidification. Appl. Therm. Eng. **29**(17), 3669–3678 (2009)
18. A. Bahrami, D.K. Aidun, D.T. Valentine, Interaction of gravity forces in spot GTA weld pool. Weld. J. **93**, 139–144 (2014)

Comparison of Bonding Strength of Ti–6Al–4V Alloy Deposit and Substrate Processed by Laser Metal Deposition

L. Jyothish Kumar and C. G. Krishnadas Nair

Abstract Laser metal deposition (LMD) is a metal additive manufacturing technique where in a metal component is built through layer-by-layer approach. Laser metal deposition is broadly used in fabrication and repairing of complex components for aerospace applications. The objective of this research paper is to focus on the bonding strength of Ti–6Al–4V alloy deposit fabricated under optimal process parameters and the substrate. Three-point bending test has been carried out on Ti–6Al–4V deposit and substrate shown that the bending strength of Ti–6Al–4V deposit is closer to the Ti–6Al–4V substrate. This result helps in restoration of complex aero engine parts with reduced lead-time and cost.

Keywords Laser metal deposition · Process parameters · Ti–6Al–4V alloy
Three-point bending test · Bonding strength

1 Introduction

Laser Metal Deposition (LMD) also termed as direct metal deposition (DMD)—is a laser solid forming process, which has a capability to form complex three-dimensional objects by adding material in the form layers [1]. The key industrial applications for LMD technology are freeform fabrication or repairing of high-value parts. In these instances, titanium alloys play an important role, such as in freeform fabrication or repairing of critical Ti–6Al–4V aero engine components. Ren et al. [2] studied the microstructural and mechanical properties of graded Ti–6Al–4V material processed by LMD technology. The results revealed that with the increase of solute elements and microstructure refinement the microhardness value increases, which is an indication of good metallurgical bonding and high

L. Jyothish Kumar (✉) · C. G. Krishnadas Nair
Jain University, Bangalore, India
e-mail: jyothish@amsi.org.in

L. J. Kumar et al. (eds.), *3D Printing and Additive Manufacturing Technologies*,
https://doi.org/10.1007/978-981-13-0305-0_3

strength of the deposit. Arif et al. [3] investigated the mechanical properties of the HVOF Inconel 625 coating using three-point bending test. The results of three-point bending test revealed that the compression stress level developed on the tensile surface of the substrate material and in the deposit surpasses the welding limit. Because of mechanical properties differences between the deposit and substrate, the deposition failure is resulted with succeeding bending. Yao et al. [4] investigated the interfacial element diffusion and fractured surface morphology under three-point bending test. It was found that increased laser power resulted in increased deposition efficiency with good interface bonding at lower laser scan speed. Liu et al. [5] investigated the microstructural formation and bonding strength of multi-layer deposition of NiCoCr alloy. It was observed that three-point bending test resulted in the enhanced bonding strength with the multi-layer deposition method.

The deposition fine microstructure and increased hardness results in good bonding strength of the deposit and the bonding strength is significant in repairing complex aero engine Ti–6Al–4V components. In view of the above, the current study focuses on understanding the bonding strength Ti–6Al–4 alloy deposit and substrate processed by laser metal deposition process.

2 Experimental Methods

TRUMPF LASER CELL LMD system with 5200 W CO_2 laser power was employed to deposit Ti–6Al–4V powder on Ti–6Al–4V substrate with a single powder feeder system. Ti–6Al–4V alloy was obtained in powder form from M/s. LPW U.K. The powder was in spherical shape with 44–106 μm (140–350 mesh) with a standard alloy composition of 0.01 C, 0.063 O, 0.02 N, 0.004 H, 0.21 Fe, 6.40 Al, 4.0 V, and balance Ti.

As per the dimensions given in Table 1 two flat test specimens were fabricated. The optimal process parameters used for deposition are summarized in Table 2. Two Ti–6Al–4V deposits and substrates test specimens dimensions were maintained using wire EDM as per the requirement of the three-point bending. Ti–6Al–4V substrate was chosen for bending test for comparison of the bending strength of the deposit and substrate. Three-point bending test specimens of Inconel Ti–6Al–4V deposit and substrate are shown in Fig. 1a, b.

Table 1 Bending test specimen dimensions [7]

L (mm)	$b = w$ (mm)	d_s (mm)	d_d (mm)	$P(N)$
44	18	5	0.6	10,000

Table 2 Optimized process parameters

Expt. No.	Laser power (W)	Powder flow rate (g/min)	Laser scan speed (mm/min)	Hardness (HV)
1	2650	5	700	461.22

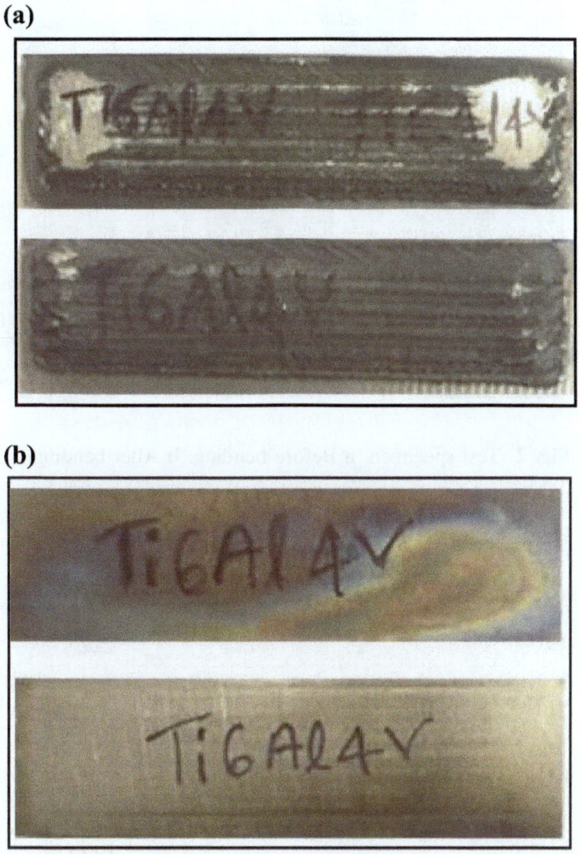

Fig. 1 Ti–6Al–4V three-point bending test specimens. **a** Deposit. **b** Substrate

2.1 *Three-Point Bending Test of Ti–6Al–4V Deposit and Substrate*

Three-point bending test was performed on Ti–6Al–4V deposit and substrates using Mecmesin–Multitest 10-xt UTM machine. Figure 2a, b reveals the test specimen before and after bending test. A load of 10 kN load was applied in each experiment and maintained constant. The bending stress, bending strain and Young's modulus or Modulus of elasticity was calculated using the following equations [4]. The load versus displacement curve was plotted for Ti–6Al–4V deposit and substrate after bending test.

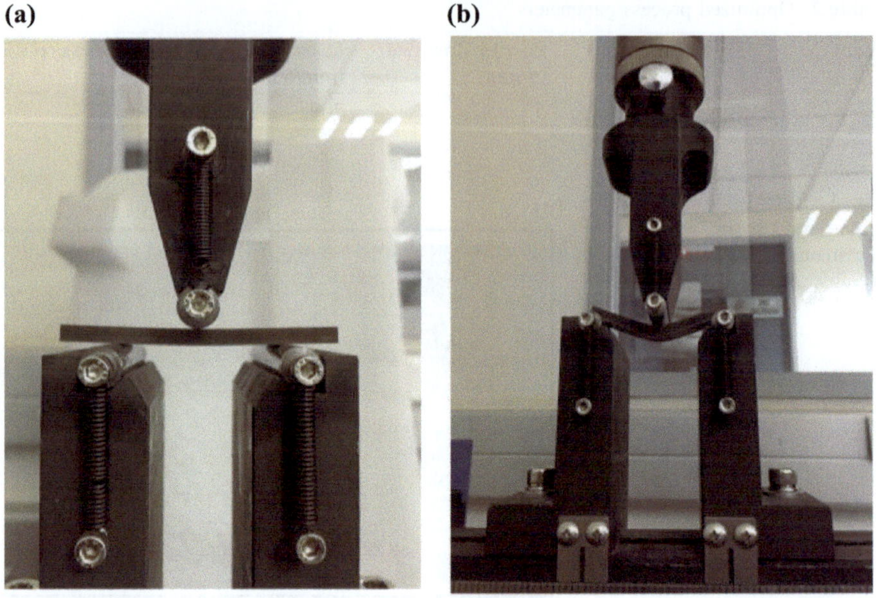

Fig. 2 Test specimen. **a** Before bending. **b** After bending

Bending stress

$$(\sigma b) = \frac{3PL}{2bd^2}, \tag{1}$$

where L is the span length in mm, P is the load in N and d is the depth or thickness and b is the width of the specimen.

Bending Strain

$$(\varepsilon b) = \frac{6Dd}{L^2}, \tag{2}$$

where L is the span length in mm, d is the thickness of the specimen in mm and D is the deflection of the beam in mm.

Modulus of elasticity

$$(Eb) = \frac{L^3 m}{4bd^3}, \tag{3}$$

where L is the span length in mm, d is the thickness of the specimen in mm and b is the width of the specimen in mm and m is the slope of the tangent to the straight-line position of the load displacement.

Scanning Electron Microscope (SEM) and Energy Dispersive X-ray Spectroscopy (EDS) experiment was performed to study the type of cracks established on the bent surface using Carl Zeiss 55 SEM and EDS.

3 Results and Discussion

3.1 Three-Point Bending Test of Ti–6Al–4V Deposit and Substrate

Figure 3 reveals the cracks developed in Ti–6Al–4V deposit and substrate after three-point bending test. Table 3 shows the load, displacement and calculated bending strength, bending strain and Young's modulus of Ti–6Al–4V deposit and substrate. The load versus displacement graph shown in Fig. 4 shows the average displacement of Ti–6Al–4V substrate is more compared to the Ti–6Al–4V deposit, which is attributed to the higher plasticity of the substrate. However the calculated average bending strength and bending strain of the deposit and substrate is marginal, which is due to the lower plastic deformation of the deposit. Whereas the Young's modulus of the Ti–6Al–4V deposit is decreased compared to the substrate.

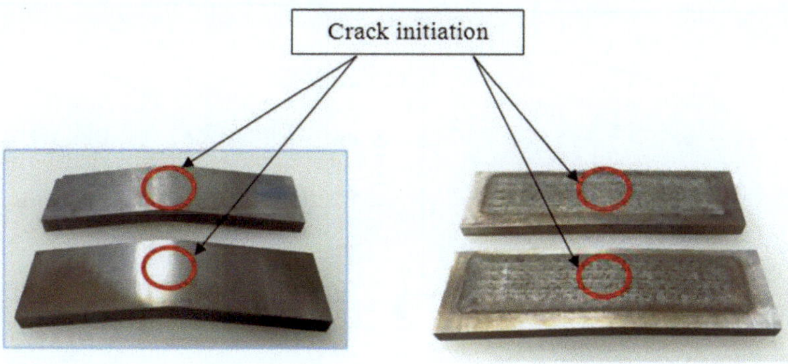

Fig. 3 Crack initiation in Ti–6Al–4V deposit and substrate after three-point bending test

Table 3 Bending test properties of Ti–6Al–4V specimens

Specimen	Load (N)	Displacement (mm)	Bending strength (MPa)	Bending strain	Young's modulus (GPa)
Substrate 1	9867	8.99	1447.16	0.13	30.30
Substrate 2	9809	7.61	1438.65	0.11	30.06
Average	9838	8.30	1442.90	0.12	30.18
Deposit 1	9367	4.64	1095.20	0.08	20.71
Deposit 2	9305	4.31	1087.95	0.07	22.84
Average	9336	4.475	1091.57	0.075	21.57

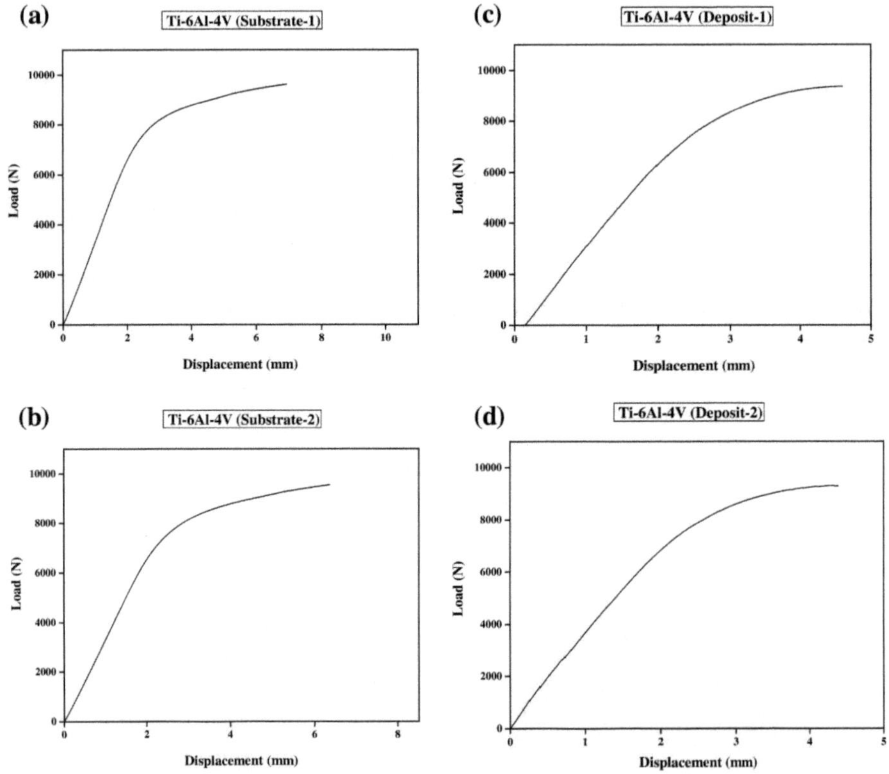

Fig. 4 Load versus displacement curves of Ti–6Al–4V. **a** Substrate 1. **b** Substrate 2. **c** Deposit 1. **d** Deposit 2

3.2 Scanning Electron Micrograph (SEM) and Energy Dispersive X-ray Spectroscope Studies (EDS)

Figure 5 shows the SEM micrograph of Ti–6Al–4V deposit before and after bending test. It shows the dense deposition area before bending test and transgranular cracks after bending test. These developed cracks are attributed to the transgranular failure of the deposit, which mainly due to the dendrite structure formation during the solidification. This is in line with the findings Miranda et al. [6].

Figure 6 reveals the EDS analysis of Ti–6Al–4V deposit area. No significant elemental changes were observed in the deposit however it reveals the presence of key alloying elements such as carbon, oxygen, aluminium, titanium, vanadium, chromium, and iron.

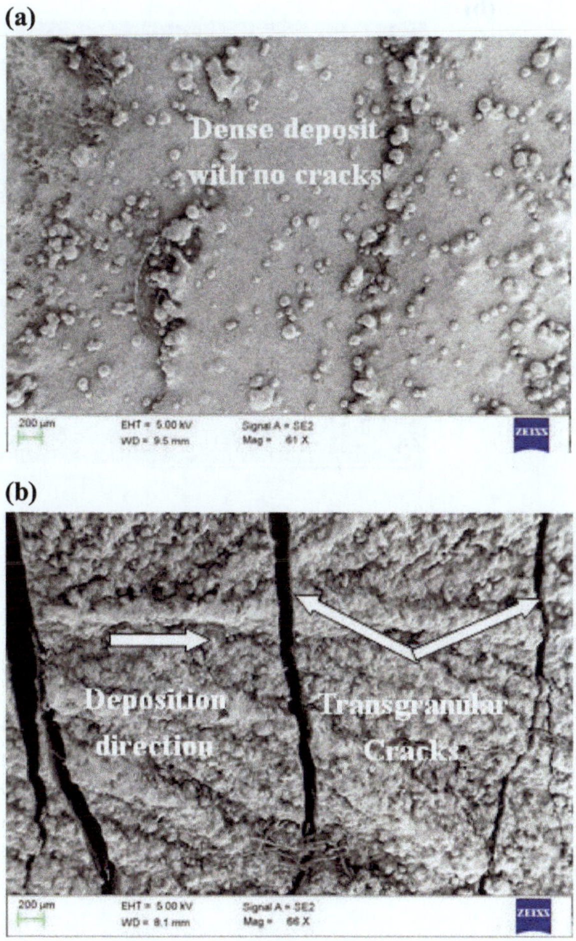

Fig. 5 SEM micrograph of the Ti–6Al–4V deposit surface. **a** Before bending. **b** After bending

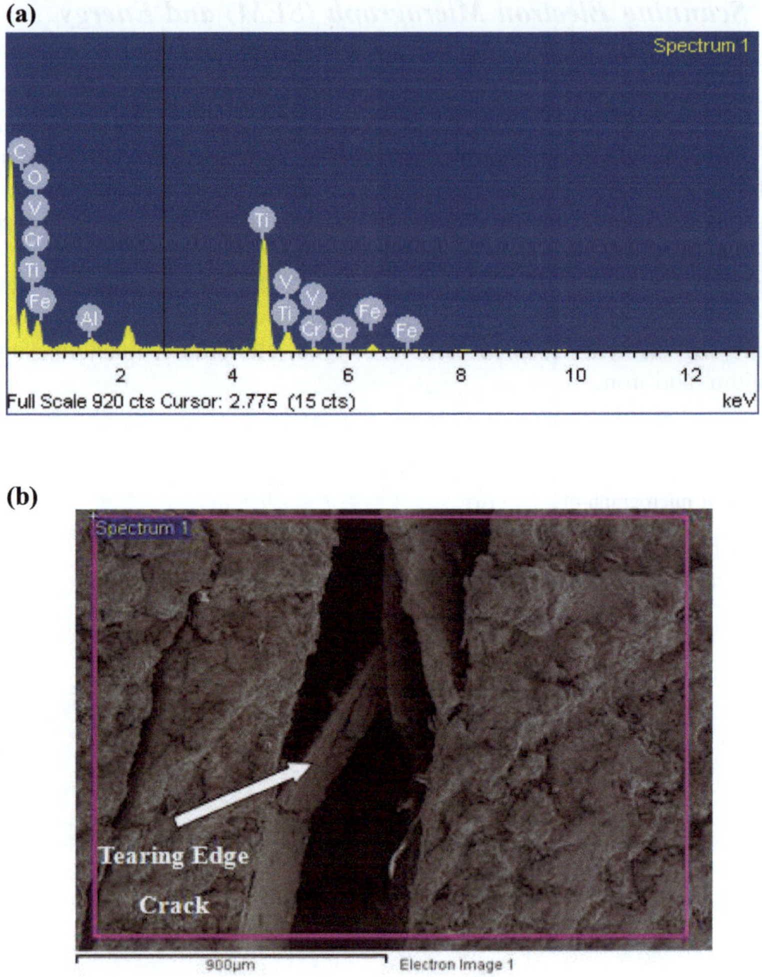

Fig. 6 EDS of Ti–6Al–4V deposit in the crack zone

4 Conclusion

The aim of the research work was to study the bonding strength of laser metal deposited Ti–6Al–4V alloy and substrate. Three-point bending test was performed on Ti–6Al–4V deposit and substrate. SEM and EDS analysis was performed on deposit before and after bending test to study the nature of crack and main alloying elements in the crack area. The following conclusions can be drawn from this study:

(1) The calculated average bending strength and bending strain of Ti–6Al–4V deposit and substrate closer, which is attributed to the lower plastic deformation of the deposit.
(2) The developed transgranular cracks in the deposit are mainly due to the formation of dendrite structures during solidification process.
(3) The obtained marginal bonding strength results and the methodology could be applied to repair critical aero-engine components.

Acknowledgements The authors thank Dr. C. K. Srinivasa, Former Head, Additive Manufacturing Technology Centre, Central Manufacturing Technology Institute, Bangalore and Dr. C. S. Ramesh, Dean, Dayanand Sagar, University, Bangalore for their technical guidance and encouragement.

References

1. J. Yu, M. Rombouts, G. Maes, F. Motmans, Material properties of Ti-6Al-4V parts produced by laser metal deposition. Phys. Procedia **39**, 416–424 (2012)
2. H.S. Ren, D. Liu, H.B. Tang, X.J. Tian, Y.Y. Zhu, H.M. Wang, Microstructure and mechanical properties of a graded structural material. Mater. Sci. Eng. A **611**, 362–369 (2014)
3. A.F.M. Arif, B.S. Yilbas, Three-point bend testing of HVOF Inconel 625 coating, FEM simulation and experimental investigation. Surf. Coat. Technol. **201**, 1873–1879 (2006)
4. J. Yao, Z. Li, B. Li, L. Yang, Characteristics and bonding behavior of stellite 6 alloy coating processed with supersonic laser deposition. J. Alloy. Compd. **661**, 526–534 (2016)
5. H. Liu, J. Hao, Z. Han, Gang Yu, X. He, H. Yang, Microstructural evolution and bonding characteristic in multi-layer laser cladding of NiCoCr alloy on compacted graphite cast iron. J. Mater. Process. Technol. **232**, 153–164 (2016)
6. R.M. Miranda, E. Assuncao, R.J.C. Silva, J.P. Oliveria, L. Quintino, Fiber laser welding of NiTi to Ti-6Al-4V. Int. J. Adv. Manuf. Technol. (2015)

Study on Rayleigh–Bénard Convection in Laser Melting Process

Kurian Antony and T. R. Rakeshnath

Abstract The Rayleigh–Bénard is also known as Gibbs–Marangoni effect. In this effect the mass transfer occurs along an interface between liquid and air due to its surface tension gradient. Rayleigh–Bénard convection effect plays a vital role in getting smooth continuos track in laser melting process. The Rayleigh–Bénard convection enhances the heat transfer phenomena throughout the powder bed. The type of Rayleigh–Bénard convection flow identified during simulation is source flow. Rayleigh–Bénard convection flow which shows the pattern of clockwise flow is defined as source flow. Overall convection flow found in this studies are of clockwise flow. This study mainly focuses its attention on the effect of Rayleigh–Bénard convection flow in laser melting process.

Keywords Additive manufacturing · Rayleigh–Bénard convection
Navier–Strokes equation

1 Introduction

The laser melting process is an additive manufacturing process technique involves many unique physical phenomena mainly consist of heat, mass, and momentum transfer [1]. These physical phenomena have direct effect on the quality of laser melted parts which is highly dependence on the factor of transient melting temperature. Moreover, the temperature field along the powder bed varies rapidly with higher laser scan speed, which will further leads to phase transformation of solid to liquid to solid within a very short span of time. If the laser processing parameters are

K. Antony (✉) · T. R. Rakeshnath
Department of Mechanical Engineering, Muthoot Institute of Technology & Science, Kochi 682308, Kerala, India
e-mail: kuriantony@mgits.ac.in

L. J. Kumar et al. (eds.), *3D Printing and Additive Manufacturing Technologies*,
https://doi.org/10.1007/978-981-13-0305-0_4

not properly selected, balling effect, or distortion might occur. Moreover the rapid phase change is usually accompanied by a greater interfacial force [2]. The temperature distribution and energy density plays a vital role in the smoothness of each laser-melted layer [3].

The major parameters that affect the laser melting process such as wetting, energy density, capillary forces, absorptivity, heat conduction, melting/solidification, radiation, Rayleigh–Bénard convection effect, distortion, curling, balling.

Due to Gaussian laser beam, localized melting of powder occurs, due to this the temperature gradient will be formed. The temperature gradient mainly affects on the centre and edge of the molten pool. Concentration differences and temperature gradient at solid–liquid interfaces of the molten pool causes surface tension difference which finally results in Rayleigh–Bénard convection flow [4, 5]. The Rayleigh–Bénard convection flow can be defined into two, the thermal Rayleigh–Bénard convection flow and the solutal flow [6, 7]. Based on the [6] experimental studies by various researchers, the thermal Rayleigh–Bénard convection flow will lead to a clockwise flow pattern which is defined as source flow effect, whereas the solutal flow effect also known as converging flow moves the opposite direction; the source flow thermal Rayleigh–Bénard convection is weaker than the solutal flow effect. Rayleigh–Bénard convection effect plays a major role in laser melting process, since this effect have lesser been studied, this work gives a better discussion about the Rayleigh–Bénard convection flow happening during laser melting process of commercially pure Titanium powder.

2 Numerical Model

Direct laser melting experimental investigations are extremely difficult to be carried out due to the higher working temperature. Hence for visualizing the Rayleigh–Bénard convection flow, numerical analysis is highly supportive in conducting these studies.

The pressure distribution and velocity field is described by the Navier–Strokes equation [8].

$$-\eta\nabla^2 u + \rho u.\nabla u + \nabla p = F. \tag{1}$$

$$\nabla u = 0 \tag{2}$$

The first term in Eq. (1) [8] indicates the momentum transfer occurred during laser melting process followed by the second and third term which is viscous force, convective force, and pressure force respectively. The source term is denoted by F term which represents external forces per unit volume N/m^3. By coupling the fluid flow to an energy balance equation, the heating of the fluid has been considered..

$$\nabla.(-k\nabla T + \rho C_p T u) = Q \tag{3}$$

The expression in Eq. (3) within the closed brackets represents the heat flux vector, which consist of a conductive and a convective expression. Equation (3) expresses the following properties and variables:

- The thermal conductivity is k (W/m K)
- The temperature is T (K)
- The heat capacity is C_p (J/kg K)
- Source term Q (W/m^3)

Boussinesq approximation [8]

$$F = \begin{pmatrix} F_x \\ F_y \end{pmatrix} = \left(\alpha g \rho \left(T^0 - \frac{\Delta T}{2}\right)\right), \tag{4}$$

where acceleration due to gravity m/s^2 is denoted by g, thermal expansion coefficient (1/K) represents by α and ΔT denotes the temperature difference. The Boussinesq approximation is incorporated to determine the effect of temperature on the velocity field. In this approximation, the buoyancy force which will be formed is due to the variations in temperature. Moreover this force is also known as Archimedes force that lifts the fluids.

The Rayleigh–Bénard convection effect at the fluid interface is expressed by the below equation

$$\frac{\partial u}{\partial y} = \gamma \frac{\partial t}{\partial x} \tag{5}$$

Equation (5) describes the shear stress applied is directly proportional to the temperature gradient on the surface.

Assumptions considered for modeling the Rayleigh–Bénard convection effect.

1. For the simulation of Rayleigh–Bénard convection. The dimension of 0.1×0.1 mm dimension has been selected.
2. Top and Bottom area no slip boundary condition is applied.
3. Slip on the side walls.
4. Bottom area is thermally insulated.
5. At the top surface Gaussian heat flux is applied.
6. Temperature-dependent material properties given to the model.
7. Boussinesq approximation included
8. Phase change also considered.

3 Results and Discussions

The simulation of Rayleigh–Bénard convection flow has been performed by using the parameters in Table 1. The multi-physics simulation of temperature distribution as well as the isothermal contours has been represented in Fig. 1a–f.

Temperature distribution during Rayleigh–Bénard convection clearly shows that the melt pool is wider and deeper in nature as represented in Fig. 1a–f. The arrows representing the counter clockwise direction is the convection and it is a free convection. The result Fig. 1e, f shows that at higher energy density, the melt pool is comparatively larger and this could be due to higher convection effect. The overall flow Fig. 1a–f identified during the simulation is counter clockwise flow. Further, the thermal and solutal gradients could affect the Rayleigh–Bénard convection flow on the free surface. However, the flow in such a molten stainless steel pool is very difficult to be determined quantitatively under the present experimental conditions. Due to higher temperature in the melt pool the surface tension tends to be lesser. Hence the cooler metal with higher surface tension pulls the liquid metal apart from the center of the liquid pool, whereas the molten metal is hotter and its surface tension is lower. The Rayleigh–Bénard convection flow influences the flow of the fluid in the powder bed and the distribution of temperature along the molten metal pool. The study shows that the flow of molten metal mainly depends on the thermal diffusion, surface tension gradient temperature gradient and viscosity.

Table 1 Physical properties of Cp Titanium

Parameters	Value	Region
Density (kg/m^3) Density (g/cm^3)	4500 $4.10–9.90 \times 10^{-4}\ (T–T_m)$	Solid Liquid
Thermal conductivity (W/m K)	21	Solid
Specific heat (J/kg K)	545	Solid
Effective thermal conductivity (W/m K)	0.07467	Solid
Viscosity (kg/m s)	0.05	
Thermal expansion coefficient (1/K)	9×10^6	

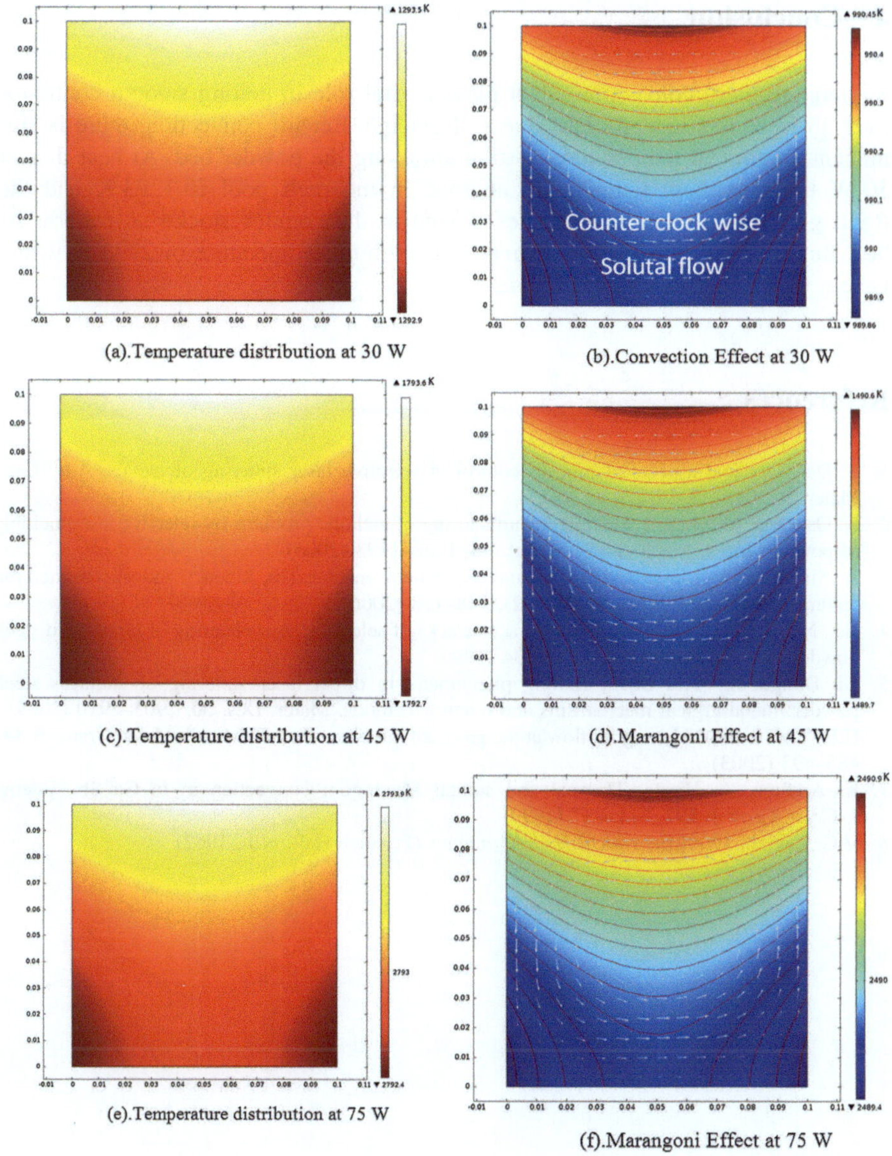

(a).Temperature distribution at 30 W

(b).Convection Effect at 30 W

(c).Temperature distribution at 45 W

(d).Marangoni Effect at 45 W

(e).Temperature distribution at 75 W

(f).Marangoni Effect at 75 W

Fig. 1 **a** Temperature distribution at 30 W. **b** Convection Effect at 30 W. **c** Temperature distribution at 45 W. **d** Marangoni effect at 45 W. **e** Temperature distribution at 75 W

4 Conclusion

Rayleigh–Bénard convection effect plays a vital role in getting smooth continuos track in laser melting process. Due to Rayleigh–Bénard convection, there is significant amount of temperature distribution along the powder bed. At heat flux of 30 W the maximum temperature attained in the melt pool is 1293 K and the Rayleigh–Bénard convection eddies moved in the counter clockwise pattern. At heat flux 75 W the melt pool attained a maximum temperature of 2793 K which gave a wider and broader melt zone.

References

1. S. Das, Physical aspects of process control in selective laser sintering of metals. Adv. Eng. Mater. **5**, 701–711 (2003)
2. K. Osakada, M. Shiomi, Flexible manufacturing of metallic products by selective laser melting of powder. Int. J. Mach. Tools Manuf. **46**, 1188–1193 (2006)
3. A. Simchi, Direct laser sintering of metal powders: mechanism, kinetics and microstructural features. Mater. Sci. Eng., A **428**(1–2), 148–158 (2006)
4. H.J. Niu, I.T.H. Chang, Instability of scan tracks of selective laser sintering of high speed steel powder. Scripta Mater. **41**, 1229–1234 (1999)
5. Gu Dongdong, Yifu Shen, Balling phenomena in direct laser sintering of stainless steel powder: metallurgical mechanisms and control methods. Mater. Des. **30**, 2903–2910 (2009)
6. H.B. Yin, T. Emi, Marangoni flow at the gas/melt interface of steel. Metall. Mater. Trans. B **34**, 483–493 (2003)
7. K. Arafune, A. Hirata, Thermal and solutal Marangoni convection in In–Ga–Sb system. J. Cryst. Growth **197**, 811–817 (1999)
8. V.G. Levich, *Physicochemical Hydrodynamics* (Prentice-Hall, N.J., 1962)

Enhancing Surface Finish of Fused Deposition Modelling Parts

M. S. Khan and J. P. Dash

Abstract Rapid prototyping is a manufacturing process in which a computer-aided design (CAD) model is used to fabricate a physical model without the use of fixtures, tools, and human intervention. The prototype is made by deposition of material in layers. The major advantage of this manufacturing process is that it can fabricate complex part quickly with minimum loss of material. There are many rapid prototyping techniques available commercially. Fused deposition modelling (FDM) is one of the most widely acceptable methods in industry due to its simplicity of operation and ability to fabricate parts with locally controlled properties. However, the surface of the FDM parts shows a very low surface finish. In order to find out the effect of important factors that influence the surface roughness, two machining parameters such as layer thickness and orientation are considered in this paper. The specimens are fabricated with various combination of orientation and layer thickness in a FDM machine which uses Acrylonitrile Butadiene Styrene (ABS) plastics. Roughness of the top and side surface is measured with a stylus profilometer. The parameters are classified into 'cost component' which comprises number of layers, part building time, part and support materials and a 'quality component' which comprises roughness of top and side surface. The best orientation-layer thickness combination is found out statistically considering trade-off between considering the cost and quality components. In order to enhance surface finish, the above-selected specimen is subjected to chemical post treatment which shows a significant level of surface finish.

Keywords Rapid prototyping (RP) · Fused deposition modelling (FDM)
ABS plastics

M. S. Khan (✉) · J. P. Dash
KIIT University, Bhubaneswar, India
e-mail: mskhanfme@kiit.ac.in

J. P. Dash
e-mail: dashjyoti60@gmail.com

L. J. Kumar et al. (eds.), *3D Printing and Additive Manufacturing Technologies*,
https://doi.org/10.1007/978-981-13-0305-0_5

1 Introduction

Additive manufacturing is a type of non-conventional manufacturing system where physical models are fabricated automatically with the help of CAD (Computer-Aided Design) data. These processes are also known as RP (Rapid Prototyping) or 3D printing as it can build intricate parts rapidly using 2D CAD data. Minimum loss of material and absence of fixing devices are the major advantage of this manufacturing process which makes it less expensive and faster. A variety of applications such as design testing of prototypes and fabrication of small and complex working models specially in automotive, aerospace and other fields have been proposed. RP techniques are also used for making tools known as rapid tooling (RT), where patterns are made for investment casting and to create mould for sand casting in a faster way. Rapid Manufacturing (RM) is another evolution of RP technique in which automated fabrication of parts can be carried out directly as finished products or components from CAD data [1]. There are over 20 RP processes available today which are divided into three categories such as liquid-based-, power-based- and solid-based system according to the raw material used in the process [2, 3]. Out of these existing processes, the most widely used include Stereolithography (SLA), Fused Deposition Modelling (FDM), 3D Printing (3DP) and Laminated Object Manufacturing (LOM) [4, 5]. The major disadvantages of RP technique is that large parts are difficult to fabricate. Metal prototypes are still in an infant stage. Barring these limitations, rapid prototyping is a remarkable technology that is revolutionizing the manufacturing process.

Fused Deposition Modelling (FDM) is one of the widely accepted 3D printing technique due to its simplified operations. Usually thermoplastic materials are used in the form of filament. the filament is heated electrically to molten state and is extruded through a nozzle tip moving in *X-Y* direction. The nozzle is fitted to the extrusion head which is controlled by computer software that deposits material in a thin layer on the build platform. Low temperature is maintained on the platform in order to solidify the liquid thermoplastic. After the platform lowers to a distance of one layer thickness, the extrusion head deposits a second layer upon the first. Supports are built up along with the part by a weaker material. After completion of the model the support material is removed to get the finished part. The major advantage of FDM process is its simplicity, reliability and cost effective. However, it exhibits very poor surface finish of the fabricated part. This paper investigates the issue of surface roughness in FDM parts in relations to the various factors and the way of improving the surface roughness.

Basically, there are five steps in all types RP techniques as illustrated in Fig. 1 which can be divided into pre-processing, machine processing and post processing. Pre-processing include design of part in CAD as well as converting it into STL file format which has to be exported to RP machine for further processing. Two operations are carried out during the second stage (machine processing). First, the CAD models are sliced into required thickness of layers and then the physical 3D model is generated with layer by layer deposition of material. In the third stage,

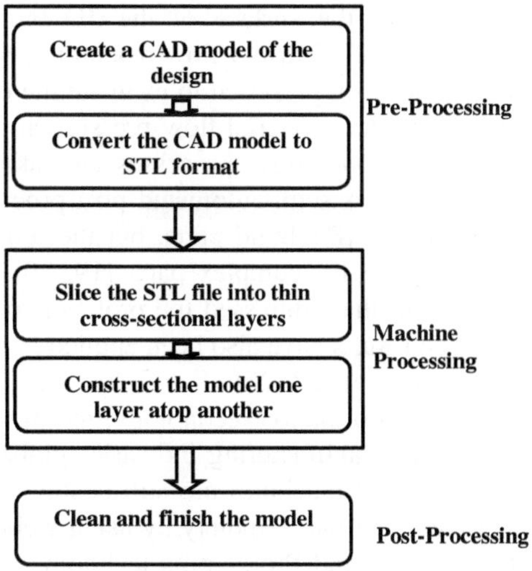

Fig. 1 Steps in RP

the finished part is removed from machine and is processed with various methods in order to enhance its mechanical and other properties.

2 Literature Review

Surface roughness of RP parts and its elimination is a major concern that have been addressed in various literatures [6, 7]. Some authors have demonstrated that an optimized building direction can reduce surface roughness as well as complexity of support structure for a given design [8]. In another study, two aspects of FDM technology such as roughness and manufacturing time optimization has been investigated using genetic algorithm [9]. Yet, in another study, along with manufacturing time and roughness, cost of the prototype has also been considered in relation to the part orientation and build direction [10]. In a study the author proposed on region-based adaptive slicing where different surface finish can be achieved in different surface area of the prototype and the errors can be determined for each [11, 12]. In another paper, various parameters of FDM process such as roughness, geometrical and dimensional features, are optimized using Taguchi design of experiment to determine the best parameter setting [13]. It has been proposed in the literature that the quality of the end product bears a functional relationship with the manufacturing process parameters in stereolithography parts [14]. Some authors used CNC milling machines for post processing part produced by RP technique, that usually contribute better surface finish [15]. However, if the part to be generated has complex surfaces with small details, this approach is a time consuming and costly due to machining setup, CNC code generation and the subtractive machining phase. All these are absent in additive manufacturing.

To improve the surface of the 3D build parts, experiments with abrasive jet deburring and abrasive flow-machining were carried out [16, 17]. An empirical model has been proposed to evaluate the dimensional errors and surface roughness in the RP part. Material flow pressure and the generation time in the FDM (fused deposition modelling) processes are taken into consideration for evaluating the model [18]. A semi-automated post-processing system was patented by Stratasys recently for ABS build parts, but the process is very costly as well as the curing time is more for complex parts [19, 20]. A good number of research have been carried out that addressed the problems of surface improvements of RP parts [21, 22]. The results have also been applied in many industries to in a diverse number of technical areas [23–25].

Using information technology, the 3D printing technique has been extensively utilized in manufacturing industries globally to connect and enhance the productivity. As RP machines are still costlier, small- and medium-scale industries can fabricate their part remotely by using web based 3D printing system with less cost. Therefore, Rapid Prototyping technology is now considered as the future of manufacturing industries [26, 27].

The literature above reveals that FDM is the widely used additive manufacturing process due to its compatibility with the number of material options, simplified process and less costly in comparison to other methods. However, the surface roughness problem is an important aspect of further study as FDM produces a poorly finished parts. It has also been noticed in the literature that FDM methods are least explored out of many RP techniques as far as studies on surface roughness are concerned. The present work emphasise on further study of surface roughness in the FDM part with regards to various criteria such as layer thickness, orientation, etc. Based on these guiding principles, the objective of the present work are (1) to study the effect of fabrication parameters on part quality, specially, surface roughness (2) to analyse the experimental results and (3) to select optimum parameter for overall improvement in part quality.

3 Experimental Method

In this work, an investigation is being done to link between the FDM process parameters and the surface roughness of prototypes. The experimental methodology consists of two stages, stage I—Prototype fabrication and stage II—Chemical processing that focus on FDM process and the post process (independent variables). At the beginning, FDM parts are being manufactured, the variables considered are part orientation and other machine parameters like, layer thickness, number of layers, build time, part material consumed, support material consumes, etc. As per the machine constraints, the tip size considered are T14 for part material and T16 for support material. The tip size refers to the diameter of the nozzle through which Acrylonitrile Butadiene Styrene (ABS) is extruded. The slice thickness affects the number of slices making up each prototype.

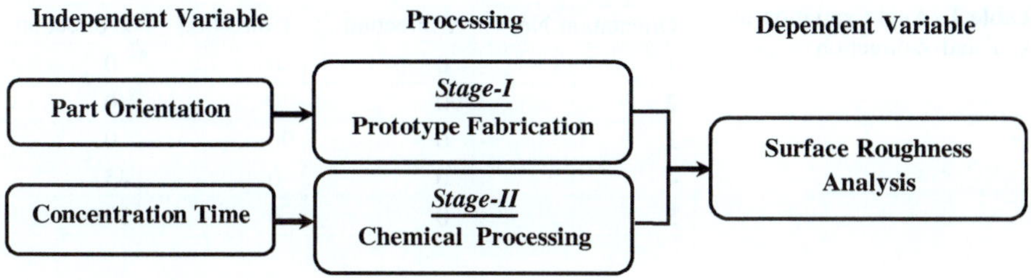

Fig. 2 Proposed experimental approach

In the second stage, post processing with a chemical method is carried out to improve the surface finish of the products. The purpose of chemical post processing is to reduce machining time, cost and finally the curing time also reduced a lot. However, dimensional loss occurs due to chemical reactions. This can be compensated by making slightly larger size of the part. Figure 2 shows the proposed methodology carried out in this paper.

3.1 Stage-I: Prototype Fabrication

The experiment has been carried out using the fabricated specimen shown in Fig. 3 with dimensions 20 mm × 20 mm × 10 mm square base prisms. This has been intentionally kept very simple in shape for ease in measurements. The various levels like part orientations and layer thickness are considered while fabricating the specimens. The various angles were also considered in *X*, *Y* and *Z*-directions. The angles were changed manually with 0°, 45° and 90° in various directions (Table 1). It was found that parts could not be completed for orientation $X = 0°$, $Y = 45°$, $Z = 0°$ and $X = 45°$, $Y = 0°$, $Z = 0°$. The layer thickness available in this machine is 0.1778, 0.2540 and 0.3302 mm. The other machine parameter such as build time,

Fig. 3 ABS plastic specimen

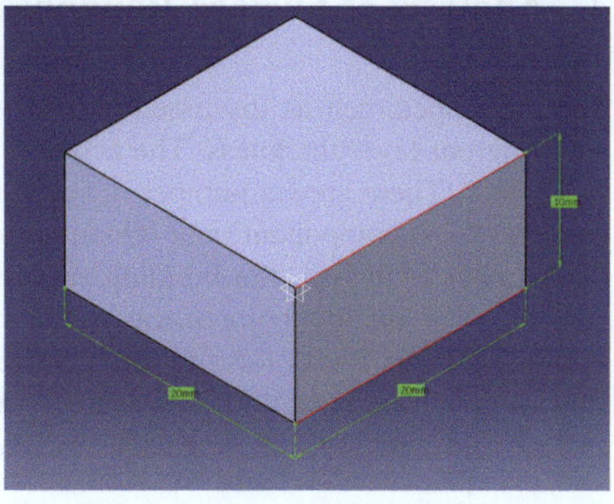

Table 1 Angle variation in *X*, *Y* and *Z*-direction

Orientation No.	*X*-direction	*Y*-direction	*Z*-direction
1	0	0	0
2	90	0	0
3	0	90	0
4	0	0	45
5	0	0	90

part material, support material and number of layers which can be considered as cost components of the prototype manufactured.

It is evident from the literature that the layer thickness of the build part and the deposition direction influence much on the surface roughness. Therefore, two different areas are considered, the top and the side surface. The results of the surface roughness were compared with the chemically treated parts. A contact system of surface tester (Taylor Hubson) has been use to measure surface roughness of the fabricated part. Fifteen (15) specimens were manufactured with various parameters and the average roughness (Ra) was evaluated on the top surface (Ra-T) and on one side surface (Ra-S) for each piece.

3.2 Stage-II: Chemical Post Processing

The present stage describes the procedure for chemical post processing of the specimen. A chemical bath is prepared with acetone (Acetone 85% and distilled water 15%) as ABS plastic is soluble in acetone and also due to its low cost, low toxicity and good diffusion property. The specimen is now immersed for one hour. After removing it from ethylene, it is dried and weighed and its dimensions are measured. The surface roughness of the part is again measured to find out any difference in roughness.

4 Analysis of Surface Roughness Results

The ABS specimen as discussed above were fabricated with various orientations and different layer thickness. The actual readings for all the three layers are shown in Table 2. There are six number of variables considered here under two categories namely, 'Cost component' and 'Quality component'. The cost component includes the number of layers, the building time and the material used for both part and support. The quality component of the reading consists of the average surface roughness value Ra for top and side surface of the build part. The surface roughness Ra (average roughness) were measured with a surface roughness tester on both top surface and side surface. For each orientation and layer thickness the number of

Table 2 Experimental readings of three layer thickness

Layers	Orientations	Cost component				Quality Component	
		No. of layers	Build time (min)	Part material (cm^3)	Support material (cm^3)	Ra-Top (μm)	Ra-Top (μm)
$L_1 = 0.1778$	1	72	22	3.060	0.611	5.540	2.320
	2	128	27	2.927	0.424	7.340	6.285
	3	128	27	2.936	0.421	6.260	7.195
	4	72	18	2.873	0.587	5.030	2.480
	5	72	22	3.060	0.609	4.195	3.750
$L_2 = 0.2540$	1	50	14	3.127	0.605	2.590	1.220
	2	89	17	2.982	0.423	9.290	5.925
	3	89	17	2.974	0.423	7.560	4.970
	4	50	12	2.964	0.573	3.420	0.970
	5	50	14	3.129	0.606	8.875	3.100
$L_3 = 0.3302$	1	39	10	3.054	0.634	3.960	0.958
	2	69	13	3.197	0.438	12.400	1.445
	3	69	13	3.198	0.437	9.670	2.680
	4	39	9	2.928	0.627	4.205	2.110
	5	39	10	3.033	0.636	11.250	2.060

Table 3 Normalized value of experimental data (layer thickness in mm)

Layers	Orientations	Cost component				Quality component	
		No. of layers	Build time (min)	Part material (cm^3)	Support material (cm^3)	Ra-Top (μm)	Ra-Top (μm)
$L_1 = 0.1778$	1	0.153	0.190	0.206	0.230	0.195	0.105
	2	0.272	0.233	0.197	0.160	0.259	0.285
	3	0.272	0.233	0.198	0.159	0.221	0.327
	4	0.153	0.155	0.193	0.221	0.177	0.113
	5	0.153	0.190	0.206	0.230	0.148	0.170
$L_2 = 0.2540$	1	0.153	0.189	0.206	0.230	0.082	0.075
	2	0.272	0.230	0.197	0.161	0.293	0.366
	3	0.272	0.230	0.196	0.161	0.238	0.307
	4	0.153	0.162	0.195	0.218	0.108	0.060
	5	0.153	0.189	0.206	0.230	0.280	0.192
$L_3 = 0.3302$	1	0.153	0.182	0.198	0.229	0.095	0.104
	2	0.272	0.237	0.207	0.158	0.299	0.156
	3	0.272	0.237	0.208	0.158	0.233	0.290
	4	0.153	0.164	0.190	0.226	0.101	0.228
	5	0.153	0.182	0.197	0.229	0.271	0.223

layers, building time, part and support material consumed were directly noted from the machine data. These machine parameters are considered as cost components of the model and the roughness values were considered as quality component. Since all the parameters described above have different units, the actual data are normalized in order to get uniform value between 0 and 1 which will help in analysing the data more precisely (Table 3).

The objective of this analysis is to find out the parameters those are minimum in case of cost component. However, in the later case the quality of the prototype must be maximum for which the surface roughness should be minimum. For layer thickness $L_1 = 0.1778$ mm, if the cost of the part is considered, the best orientation is number 4 as these have approximately lower value whereas the roughness on the top surface is minimum for orientation 5 and for roughness on the side the best value comes for orientation number 1 (refer Figs. 4 and 5).

For layer thickness $L_2 = 0.2540$ mm, the orientation number 4 shows the minimum values for almost all parameters in cost components which implies cost is minimized for the orientation number 4. As far as quality component is concerned, roughness for top surface shows minimum value for orientation number 1 and roughness for side surface shows minimum value for orientation number 4 (refer Figs. 6 and 7).

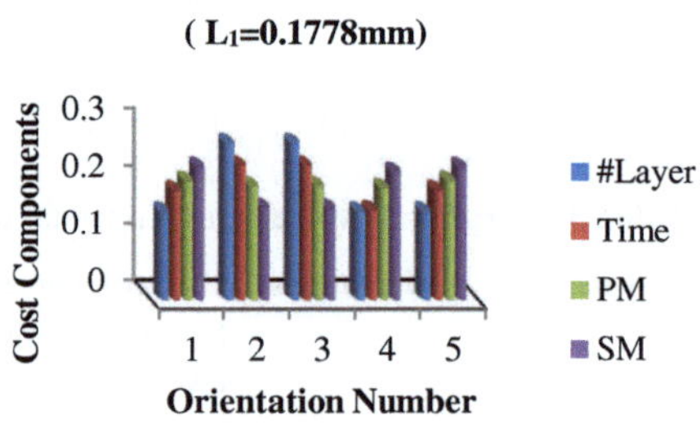

Fig. 4 Parameters related to cost

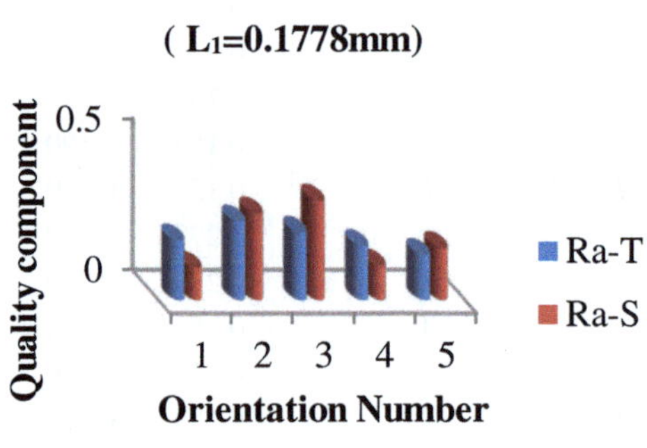

Fig. 5 Parameters related to quality

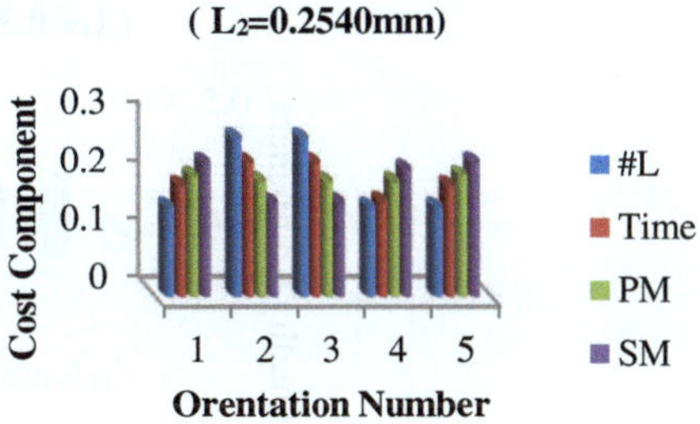

Fig. 6 Parameters related to cost

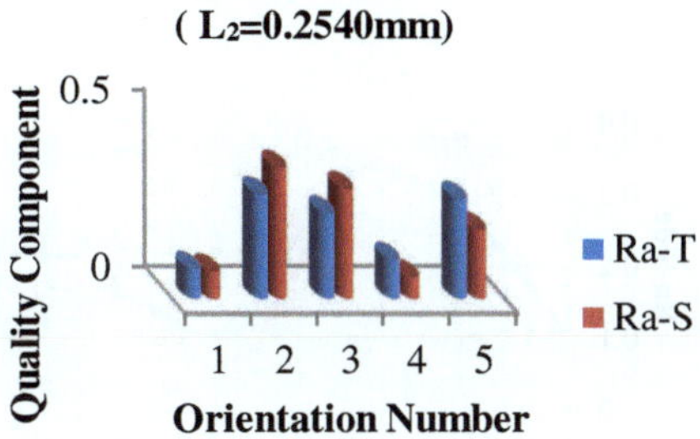

Fig. 7 Parameters related to quality

Finally, for layer thickness L_3 = 0.3302 mm, all the parameters in cost component shows minimum value for orientation 4, except support material which gives a slightly higher value. The minimum value of roughness for top surface is found to be almost same for orientation number 1 and 4, whereas side surface roughness is minimum for orientation number 1 (refer Figs. 8 and 9).

It is observed from the above that the parameters responsible for cost components are almost minimum and hence cost effective for orientation number 4 in layer 2. However, for roughness (quality component) the scenario is different. The minimum value of Ra for top surface and side surface varies between orientation number 1, 4 and 5. In order to get a clear picture regarding the roughness, the values for Ra (Top and side) considering the three layer thickness are plotted (Table 4 and Fig. 10). It is observed from Fig. 10 that layer 2 and layer 3 have low roughness value for orientation number 1 and 4 for Ra-top, where as only layer 2 with orientation number 4 has minimum roughness value for Ra-side. We can presume

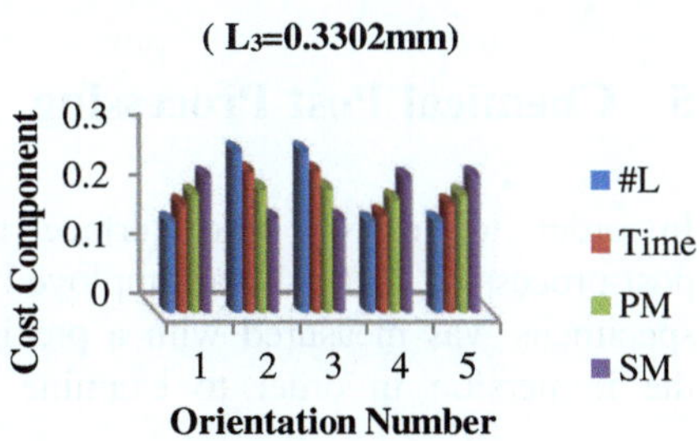

Fig. 8 Parameters related to cost

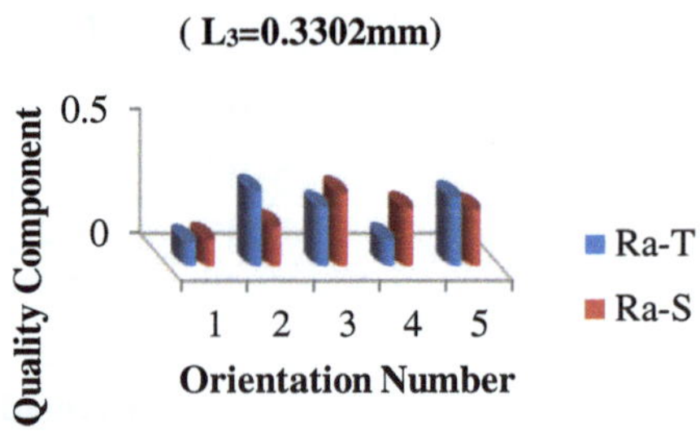

Fig. 9 Parameters related to quality

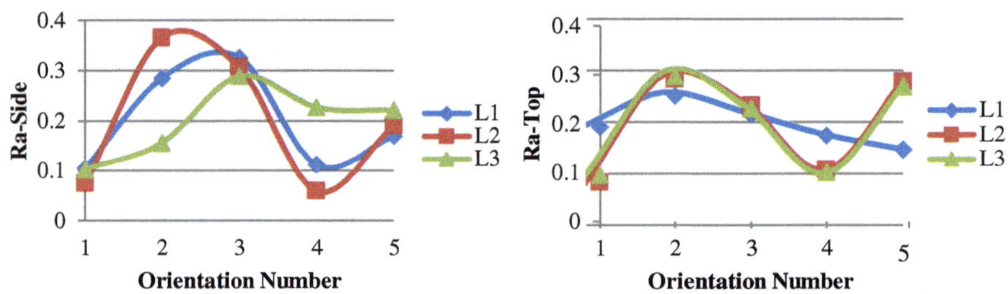

Fig. 10 Layer thickness versus orientation numbers for Ra-Side and Ra-Top

that 'Layer 2 with orientation 4 can be a better option as far as surface roughness or the quality component of prototype is concerned. Further, looking into Table 4, the minimum values of Ra-Top for L_2 and L_3 are 0.082 and 0.095 for orientation number 1 and the next higher values 0.108 and 0.101 for orientation number 4 in L_2 and L_3 layer. It can be observed that the difference in the values is very less. Similarly, in case of roughness Ra-side, L_2 value is minimum for orientation 4 (0.060) and minimum value for L_3 is 0.104 for orientation number 1.

From the above discussions, it can be concluded that 'layer 2 with orientation 4' is the best combination of parameters as far as cost and quality of the build part is concerned. Hence, it is proposed that Layer thickness L_2= 0.2540 mm with orientation number 4 (X = 0°, Y = 0° and Z = 45°) can be considered as the highest quality of the prototype with minimum cost.

5 Chemical Post Processing

In order to improve the surface roughness of the specimen, a chemical post-processing method was employed as discussed earlier. The dimensions of the specimens was measured with a precision calliper and weighed before and after the immersion in order to examine the variations due to the chemical bath.

Table 4 Normalized value for Ra-Side and Ra-Top for all three layers

Orientation	Ra-Side (μm)			Ra-Top (μm)		
	L_1	L_2	L_3	L_1	L_2	L_3
1	0.105	0.075	0.104	0.195	0.082	0.095
2	0.285	0.366	0.156	0.259	0.293	0.299
3	0.327	0.307	0.290	0.221	0.238	0.233
4	0.113	0.060	0.228	0.177	0.108	0.101
5	0.170	0.192	0.223	0.148	0.280	0.271

Table 5 Dimensions and roughness before and after chemical treatment

	Length (mm)	Breadth (mm)	Height (mm)	Weight (gm)	Ra-Top (μm)	Ra-Side (μm)
Before	20	20	10	2.5	0.107767	0.059932
After	19.82	19.75	9.91	2.92	0.085467	0.048795

After immersion the specimen were kept at room temperature for one hour. The linear dimension of the specimen before chemical treatment is 20 mm × 20 mm × 10 mm. After immersion, it is observed that the size variations take place in all sides (Table 5). The specimens shrank by approximately 1%, while they increased their average weight by less than 1%. This is because the chemical treatment removes some material but the subtracted ABS is balanced by the absorption of the solution. It is also observed from Table 5 that the roughness of the part after chemical treatment has improved significantly in both the cases that means at the top and side surfaces.

Thus, to have a better surface quality of ABS fabricated 3D models, the optimum layer thickness ($L_2 = 0.2540$ mm) and orientation number 4 ($X = 0$, $Y = 0$, $Z = 45$) can be employed followed by post processing chemical treatment with acetone.

6 Conclusions

This investigation provides information about impact of parameters on dimensional accuracy of FDM fabricated part using ABS plastic as raw material. Optimum combination of parameters that shows better surface quality has been found out. Post chemical treatment using acetone effectively enhanced the surface finish of the model. This leads to reduce the effective time needed to determine process conditions, thus saving in costs and wasted products. Thus, considering the cost and quality, the best combination of parameters emerged in this work is $L_2 = 0.25407$ mm with orientation number 4 ($X = 0$, $Y = 0$, $Z = 45$). Further, in order to improve the roughness quality, a chemical post processing was conducted with the above parameters which showed better results. However the dimension of the part is slightly decreased due to chemical reaction which can be compensated by slightly increasing the dimension before chemical reaction.

Some of the limitations of this work are that, the initial FDM Machine set up cost is high. Also, it is difficult to put any combination of orientation in all the three directions. Moreover, the roughness quality of the FDM fabricated part can only be enhanced cannot be overcome. This investigation can be extended by taking more orientations in all three directions. The chemical used for post chemical treatment can be changed with other ethylene solvents which may indicate better results.

References

1. N.A. Waterman, P. Dickens, Rapid product development in the USA, Europe and Japan. World Class Des. Manuf., I **3**, 27–36 (1994)
2. C.K. Chua, K.F. Leong, C.S. Lim, *Rapid Prototyping, Principles and Applications* (World Scientific Publishing Co. Pte. Ltd, Singapore, 2003)
3. N. Hopkinson, R.J.M. Hague, P.M. Dickens, *Rapid Manufacturing: An Industrial Revolution for the Digital Age* (Wiley, Chichester, England, 2006)
4. E. Grenda, *Rapid Tooling & Metal Parts by Additive Manufacturing* (2007). http://www.additive3d.com/tl_c.htm
5. F.W. Liou, *Rapid Prototyping and Engineering Applications: A Toolbox for Prototyping Development* (Taylor & Francis, 2008)
6. R. Ippolito, L. Iuliano, A. Gatto, Benchmarking of rapid prototyping techniques in terms of dimensional accuracy and surface finish. Ann. CIRP **44**(1), 157–160 (1995)
7. A. Armillotta, Assessment of surface quality on textured FDM proto-types. Rapid Prototyping J. **12**(1), 35–41 (2006)
8. P. Lan, S. Chou, L. Chent, D. Gemmilt, Determining fabrication orientations for rapid prototyping with stereolithography apparatus. Comput. Aided Des. **29**(1), 53–62 (1997)
9. K. Thrimurthulu, P. Pandey, N. Reddy, Optimum part deposition orientation in fused deposition modelling. Int. J. Mach. Tools Manuf. **44**, 585–594 (2004)
10. D.K. Ahn, H.C. Kim, S.H. Lee, Fabrication direction optimization to minimize post-machining in layered manufacturing. Int. J. Mach. Tools Manuf. **47**(3–4), 593–606 (2007)
11. K. Mani, P. Kulkarni, D. Dutta, Region-based adaptive slicing. Comput. Aided Des. **31**(5), 317–333 (1999)
12. M. Kumar, A.R. Choudhury, Adaptive slicing with cubic patch approximation. Rapid Prototyping J. **8**(4), 224–232 (2002)
13. J. Zhou, D. Herscovici, C. Chen, Parametric process optimization to improve the accuracy of rapid prototyped stereolithography parts. Int. J. Mach. Tools Manuf. **40**, 363–379 (2000)
14. Y.Y. Chiu, Y.S. Liao, S.C. Lee, Slicing strategies to obtain accuracy of feature relation in rapidly prototyped parts. Int. J. Mach. Tools Manuf. **44**, 797–806 (2004)
15. P. Kulkarni, D. Dutta, On the integration of layered manufacturing and material removal process. Int. J. Mach. Sci. Eng. **122**, 100–108 (2000)
16. R.E. Williams, V.L. Melton, Abrasive flow finishing of stereolithography prototypes. Rapid Prototyping J. **4**, 56–67 (1998)
17. K.F. Leong, C.K. Chua, G.S. Chua, C.H. Tan, Abrasive jet deburring of jewellery models built by stereolithography apparatus. J. Mater. Process. Technol. **83**, 36–47 (1998)
18. P.M. Pandey, N.V. Reddy, D.G. Sanjay, Improvement of surface finish by staircase machining in fused deposition modelling. J. Mater. Process. Technol. **132**, 323–331 (2003)

19. http://www.stratasys.com/uploadedFiles/North_America/Products/Smoothing_Station/Stratasys_FinishingStations.pdf. Last accessed on 2 Jan 2009
20. D. Espalin, F. Medina, R. Wicker, Vapor smoothing, a method for improving FDM-manufactured part surface finish. Int. Rep. of the W.M. Keck Center for 3D Innovation, University of Texas at El Paso, provided by Mrs. C. Gerling, Fortus Marketing Association, Stratasys Inc. (2009)
21. T. Tolio, D. Ceglarek, H.A. ElMaraghy, A. Fischer, S.J. Hu, L. Laperrière, S.T. Newman, J. Váncza, SPECIES-co-evolution of products processes and production systems. CIRP Ann.-Manuf. Technol. **59**, 672–693 (2010)
22. S.H. Masood, Advances in fused deposition modelling. Compr. Mater. Process. **10**, 69–91 (2014)
23. T. Wohler, *Additive Manufacturing State of the Industry Annual Worldwide Progress Report* (Wohler associates, Colorado, USA, 2010)
24. C.K. Chua, S.H. The, R.K.L. Gay, Rapid prototyping and virtual manufacturing in product design and manufacturing. Int. J. Adv. Manuf. Technol. **15**, 597–603 (1999)
25. Q. Liu, M.C. Leu, S.M. Schmitt, Rapid prototyping in dentistry: technology and application. Int. J. Adv. Manuf. Technol. **29**, 317–335 (2005)
26. V. Raja, S. Zhang, J. Garside, C. Ryall, D. Wimpenny, Rapid and cost-effective manufacturing of high integrity aerospace components. Int. J. Adv. Manuf. Technol. **27**, 759–773 (2005)
27. B. Dong, G. Qi, X. Gu, X. Wei, Web service oriented manufacturing resource applications for networked product development. Adv. Eng. Inform. **22**, 282–295 (2008)

Development and Analysis of Accurate and Adaptive FDM Post-finishing Approach

Mohammad Taufik and Prashant K. Jain

Abstract This paper presents a novel post-processing tool, namely selective melting (SM) tool for thermally assisted finishing (TAF) of Fused Deposition Modelling (FDM) build parts. During FDM process layer upon layer, part fabrication leads to an inherent surface constraint, i.e. generation of stepped surfaces that cannot be avoided while fabrication of parts. Hence, these stepped surfaces need to be removed to achieve desired surface quality of FDM build parts for different applications. Therefore, post-finishing has been performed by means of selective melting of stepped surfaces. This paper studied coupled parameters during pre- and post-processing operation and their effect and estimation of randomness in peak to valley height variation on surface profile. Post-finishing of the FDM modelled surfaces is performed by a significant amount of feed rate variation along the tool path which results in temperature variation over the part surface. Feed rate regulation is used in this research to vary the depth of melting as the part surface changes its orientation along the tool path. Proposed selective melting (SM) tool adaptively melts the requisite or extra material to provide adaptive finishing tool path planning for improving the surface finish conditions. The effectiveness of the proposed technique was analysed and studied. The results showed that the proposed approach played a significant role in improving the surface finish conditions of FDM build parts.

Keywords Additive manufacturing · Fused deposition modelling (FDM)
Staircase effect · Surface roughness · Post-finishing

M. Taufik · P. K. Jain (✉)
Mechanical Engineering Discipline, PDPM Indian Institute of Information Technology, Design and Manufacturing Jabalpur, Jabalpur, Madhya Pradesh, India
e-mail: pkjain@iiitdmj.ac.in

M. Taufik
e-mail: mohammad.taufik@iiitdmj.ac.in

L. J. Kumar et al. (eds.), *3D Printing and Additive Manufacturing Technologies*,
https://doi.org/10.1007/978-981-13-0305-0_6

1 Introduction

Additive Manufacturing refers to various processes by which CAD model information is used to build a component from successive adding material in the form of layers [1–3]. Nowadays with the advancement in the materials, AM is being used in different application areas [4–8]. Surface finish in additive manufacturing (AM) is highly variable due to variable staircase formation during layer upon layer part fabrication process [9–13]. Post-finishing operations have an important role in improving the surface finish of AM build parts [14]. Part fabrication through Additive Manufacturing (AM) process followed by finishing on the different setup is known as post-finishing operation. A general technique for surface finish improvement is typically done through the use of milling machine with customized or standard mill cutters, which machine the staircase of an AM build parts to reduce the roughness of the surface during post-finishing operation. Since post-finishing operation are performed after fixturing the FDM parts on a different setup, this induces location error due to which staircases are not removed properly [15]. This results in a non-uniform surface generation in finished parts and is the main disadvantage of such post-finishing techniques as shown in Fig. 1.

As a result of this, additive manufacturing and material removal process were used on the same setup as an integrated finishing operation for solving the issue of location error. In integrated finishing operation, not only the lead time reduces but also location errors problem minimized. Kulkarni and Dutta [15] developed an integrated approach of additive manufacturing and material removal process to control the location error effect. They developed a new CNC cutter path algorithm as a solution of poor surface quality through which they performed machining of RP built part with ball end milling tool on the same setup and completely eliminates the location errors problem during staircase machining. Frank et al. [16] described the new methodology using CNC machining for layer manufacturing on the same setup. In their approach, process factors for the prototype fabrication were estimated according to slicing data. Similarly, Santos et al. [17] utilized layer-based slicing concept for milling complex shapes and thin sections. For better dimensional accuracy Hu and Lee [18] introduced part decomposition algorithm. Developed algorithm was able to find out total layers in a LM part, and accordingly provided extra layers for machining purpose. This algorithm also considered the maximum allowable sheet thickness and tool accessibility into account. Pandey et al. [19] described a virtual hybrid fused deposition modelling system in terms of integrated

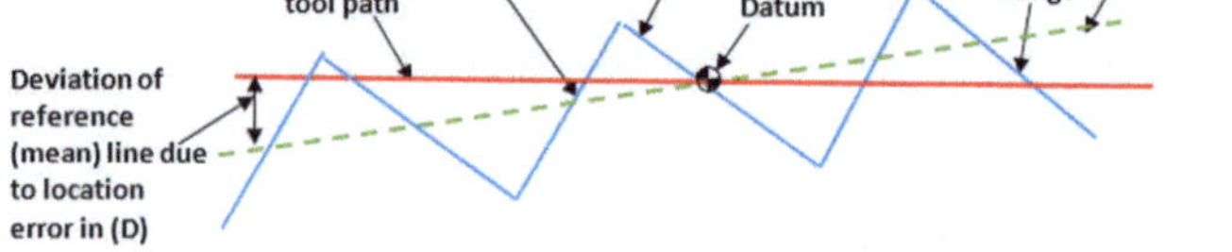

Fig. 1 The finishing inaccuracies due to location error

approach of additive manufacturing and material removal process. To reduce surface roughness of FDM formed prototypes hot cutter machining (HCM) is recommended. It was proposed to machine build edges with HCM based on the adaptively sliced data of a CAD model. With a view to access complicated features of a part they suggested a process planning system to pick the number of layers to be deposited and afterward machined. However, the integrated finishing operation is very efficient in removal of cusps due to minimization of location error. But, it is hard to finish the part on the same setup with most of the available AM machines in the market. For example, in commercial AM/3D printing systems, the part is additively manufactured, and they do not provide the material removal processes on the same setup. To finish the FDM build part, it needs to be removed from the build machine platform. To have the integrated finishing operation technique, there is a need for the installation of material removal hardware setup with AM systems. Therefore, to improve the surface finish of AM build parts, the different method of post-processing are explored/developed. The surface roughness value achieved by using the chemical dipping process (Galantucci et al. [20]) shows an enhancement in surface finish of FDM parts. Similar findings were observed for the cold vapour treatment technique used by Garg et al. [21]. The major constraint of these techniques is to understanding the behaviour of finishing. The use of barrel finishing process to reduce roughness of parts is reported by Singh and Singh [22]. They improved the surface quality of FDM part in a very slow rate of finishing operation. Therefore, more attention should be paid to the development of efficient post-treatment process.

To overcome the above disadvantages, in this research work, a new technique of post-finishing is proposed. This technique design and developed a selective melting tool; first to eliminate the effect of the location error which is induced by fixturing the part for finishing process on the different setup. To selectively and adaptively heating of build edge profile, proposed post-finishing operation provides forced closed follower mechanism and heating arrangements to the finishing tool. Then it provides non-rotational linear feed rate to the heated tool to fills the neighbouring valley through the softened material which is piled up near the border of a heated tool. A surface roughness measurement subsequently analyses the finished surface through the use of contact type surface tester. The data analysis technique estimates the change in roughness parameter value (ΔR) by taking values of roughness parameters before and after thermally assisted finishing operation.

2 A New Methodological Framework for Post-Finishing Operation

In this research study, a new methodological framework for post-finishing operation is proposed. A new three-axis CNC-assisted selective melting tool is design and developed to finish the FDM build part. The design of the tool is based on force closed translating follower so that tool can accurately and continuously trace the surface of FDM build parts during thermally assisted post-finishing operation. The theory of force closed translating follower mechanism for thermally assisted post-finishing of the FDM build parts is described in the following sections.

First, the post-finishing procedure is presented without application of force closed translating follower tool design. Manufacturing the part on the AM machine followed by milling on the CNC machine is shown in Fig. 1. This type of post-finishing operation, as discussed in the existing literature, does not perform accurately as a result of the location error due to this location error the tool path deviates from the reference (mean line). As shown in Fig. 2b mill tool is not able to remove cusps properly due to this deviation and the mill tool crosses the cusp

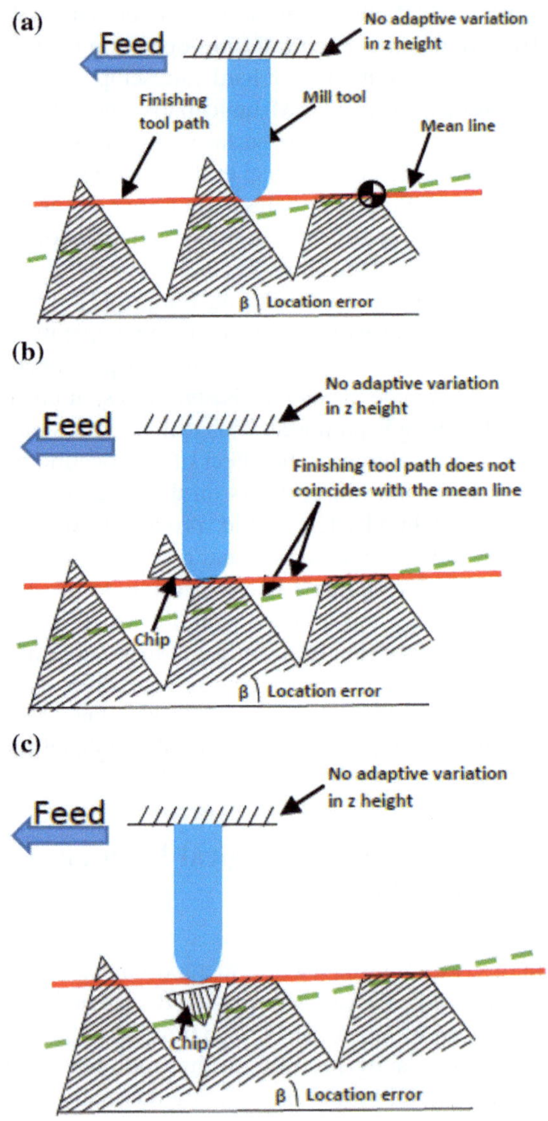

Fig. 2 Post-finishing of AM build surface profile through milling operation

without cutting it accurately as shown in Fig. 2c. Thus, in these types of post-finishing processes, once the deviation in tool path and mean line coexists, finishing inaccuracies are induced. Moreover, there is no facility available in the existing post-finishing operation to compensate this deviation.

To control the deviation during post-finishing operation, a new post-finishing tool is proposed. Figure 3 shows the working principle of proposed selective melting tool for the post-finishing purpose. In the present technique, the selective melting tool design consisting of ball nose tool tip for continuous contact, barrel for the translating movement and self-weight for force closed mechanism of the layer follower. The temperature of the selective melting tool is controlled with the digital thermocouple along with variance. With the proposed design of the selective melting tool, the deviation due to location error is adaptively controlled by assigning the proper feed rate value.

Feed rate variation is provided to the selective melting tool to obtain the depth of melting up to the mean line and then provide movement to soften material to fills the neighbouring valley along the mean line. Therefore, in the proposed TAF process depth of melting is working as a dependent parameter of feed rate. Thus, at the optimal value of feed rate selection, the finishing tool path coincides with the mean line as shown in Fig. 3b. At this feed rate value, the half of the height of roughness peak is selectively melted and then provides movement to melted material to fills the neighbouring valley along the mean line as shown in Fig. 3c. For the finishing of successive profile peaks, the variation in z-height ($D_1 > D_2 > D_3$) due to location error can be adaptively controlled by functions of the translating mechanism of the layer follower along z-height. Therefore, by implementing proposed concept, it is possible that half of the profile peaks accurately finished along the mean line. As a result of this accuracy, it is feasible to reduce the surface roughness of an AM part to the smallest possible value (i.e., theoretically zero) by providing the optimal value of the feed rate. Thus, the proper value of feed rate for the selective melting of the average amount of peak to valley height is required during proposed technique otherwise, non-optimized value of feed rate will lead to a different depth of melting (d_{mf}) that not follow the mean line, and hence, it results finishing in sub-optimization sense.

This can be theoretically expressed by schematization of 'sharp edges' of a rectangular layer profile as shown in Fig. 4, before and after TAF process as following:

$$R_{a(\mathrm{TAF})} = 1000\frac{t\sin\alpha}{4}\left(\frac{\cos^3\alpha}{\sin\alpha} + \cos\alpha\sin\alpha\right) - 1000\frac{\mathrm{d}_{mf}^2\sin\alpha(\cot\alpha + \tan\alpha)}{t} \quad (1)$$

This mathematical formulation modelled the thermally assisted finishing process of planar surfaces at various orientations at a given feed rate f. In the above mathematical formulation, the dotted triangle Ω represents the initial surface topography of an AM prototype: the cusp height $t\cos\alpha$ correspond to the layer thickness t and build angle α, respectively. The base length of this profile is equal to $t/\sin\alpha$, and it represents the width of the triangular profile's peak. The centerline

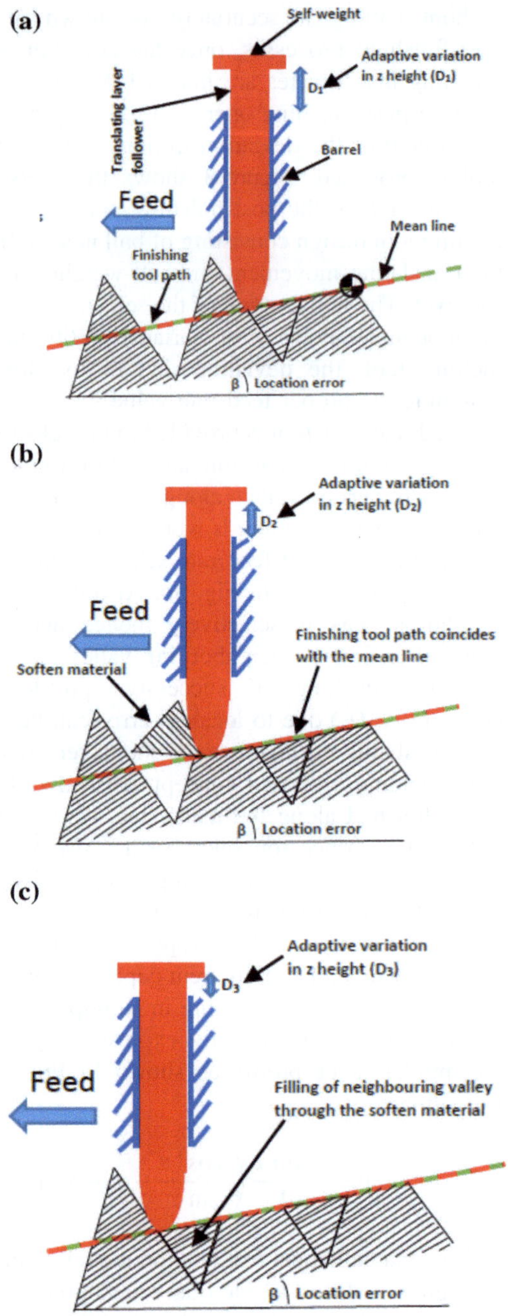

Fig. 3 A new methodological framework for post-finishing of AM build surface profile with the proposed selective melting tool

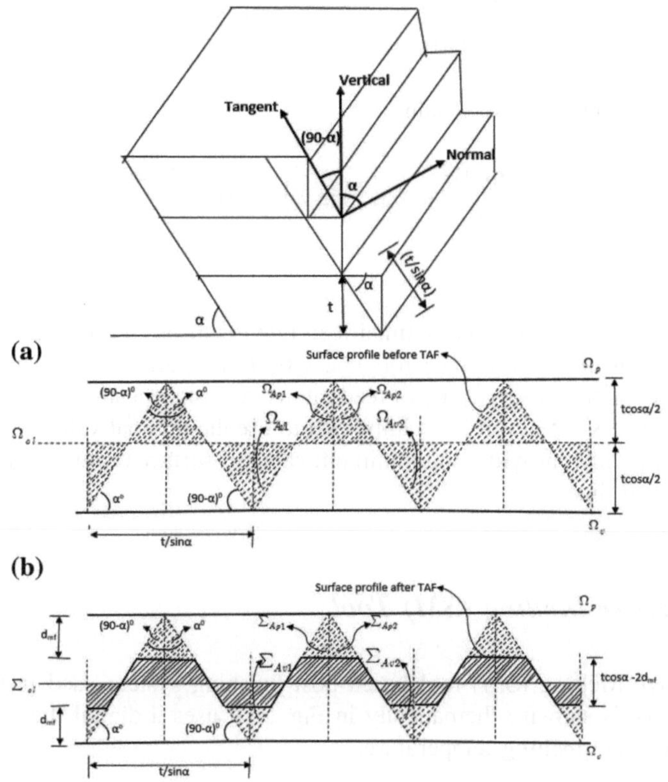

Fig. 4 Theoretical roughness peak for an additive manufactured part in the lower images: **a** before TAF **b** after TAF

(Ω_{cl}) is traced a parallel with the lines between two boundaries Ω_p and Ω_v respectively. Closed area of a peak and valley profile is enclosed by the centerline and the two respective lines Ω_p and Ω_v. The solid trapezium Σ is related to the surface topography of a finished surface. It is supposed that at the given feed rate f (at this feed rate tool not follow the mean line) the rubbery state ABS fills their adjacent valleys equally on both sides to the same depth of finishing (d_{mf}) as shown in Fig. 4. Therefore, the maximum peak to valley height t cosα of the surface topography decreases by the value 2 d_{mf}, which corresponds to the twice of the depth of finishing. After TAF, the closed areas Ω_{av} and Ω_{ap} should be changed in Σ_{av} and Σ_{ap} respectively.

As per the above discussion, if the optimal value of feed rate is provided then the depth of melting can be achieved up to the mean line. As a result of this, half of cusp height will exactly melt, and this exact magnitude of melted roughness peak, fills their adjacent valleys accurately along the mean line. Therefore, the average of peak to valley height during thermally assisted post-finishing approach is given as dependent parameter of optimal value of feed rate f and given by

$$d_{mf\,(\text{at optimal value of feed rate})} = \frac{t\cos\alpha}{2} \quad (2)$$

substituting (2) into (1), we obtain

$$R_{a(\mathrm{TAF})} = 1000\frac{t\sin\alpha}{4}\left(\frac{\cos^3\alpha}{\sin\alpha} + \cos\alpha\sin\alpha\right) - 1000\frac{\left(\frac{t\cos\alpha}{2}\right)^2\sin\alpha(\cot\alpha + \tan\alpha)}{t}$$

$$R_{a(\mathrm{TAF})} - 0$$

Therefore, by providing the optimal feed rate depth of melting can be achieved up to the mean line and surface roughness becomes zero. Thus, in the selective melting process, the proper value of feed rate is very significant in the reduction of surface roughness. Therefore, it is important to use the optimal value of feed during TAF process, which provides the minimum value of surface roughness by finishing along the mean line.

2.1 Selective Melting (SM) Tool

The Selective Melting (SM) tool based post-finishing system used in the present research work is shown schematically in Fig. 5. It uses a digital thermocouple to maintain constant melting temperature.

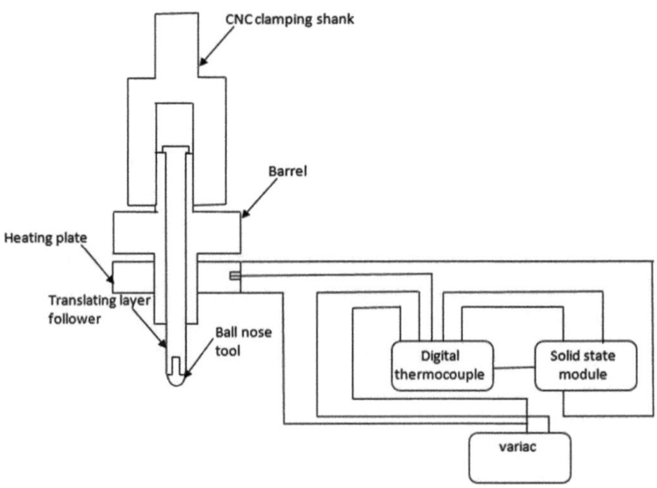

Fig. 5 Schematic diagram for test arrangement in selective melting tool

A barrel is provided to integrate the CNC clamping shank and heating plate along with translating layer follower. A tool is having ball nose shape provided at the bottom of the translating layer follower. A digital thermocouple maintains the heating plate temperature at 220 °C. This heat is transferred in translating layer follower by providing 30 min equilibrium time.

2.2 *Thermally Assisted Finishing (TAF) and Surface Roughness Measurement*

In the proposed thermally assisted finishing (TAF) process, the FDM build part was fixturing on the three axes OKUMA CNC milling machine platform, and the shank of the selective melting tool was clamping by the milling chuck/collets. Surface finishing and surface roughness measurement were then subsequently performed. The mathematical equation and the digital implementation of the arithmetic mean of the depth parameter and individual heights (asperities) are, respectively, given by Gadelmawla et al. [23]

$$R_a = \frac{1}{l}\int_0^l y(x)\,\mathrm{d}x \tag{3}$$

$$R_a = \frac{1}{n}\sum_{i=1}^{n} |y_i| \tag{4}$$

To maintain repeatability, each roughness reading was performed thrice. This was done using Rugosurf 10G surface tester, and it provides the profile height data of FDM build parts in micrometre. For all samples, sampling length was about 2.5 mm and cut-off lengths of 0.8 mm.

3 Results and Discussion

Thermally assisted finishing-based operations were performed on planer surface with 900 build orientation. To improve the surface finish of the FDM built part, a number of tests were conducted to choose the most suitable parameter values. Post-finishing of the FDM modelled surfaces is performed by a significant amount of feed rate variation along the tool path which results in temperature variation over the part surface. The effectiveness of the proposed technique was analysed and studied. The range of 400–900 feed rate in mm/min were selected. The significant variations in the change in surface roughness are associated with this rage. These results were then plotted between the change in surface roughness and feed rate as shown in Fig. 6.

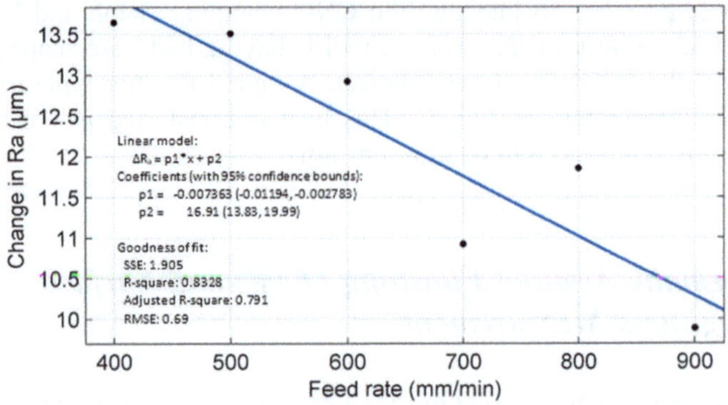

Fig. 6 Correlation between the change in surface roughness and the feed for selective melting

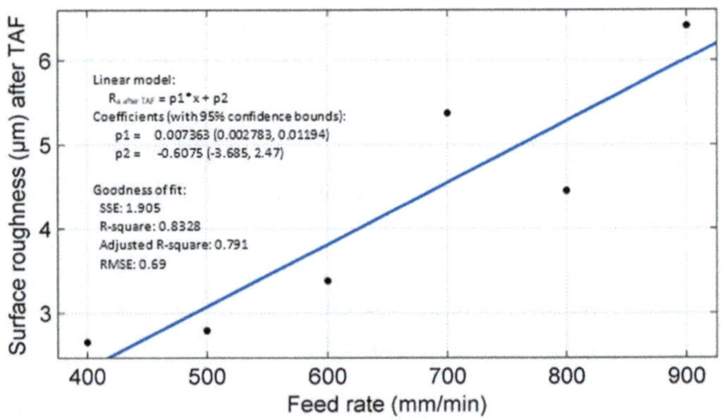

Fig. 7 Correlation between the surface roughness and feed rate after TAF process

It can be observed in Fig. 6 that a linear relationship is present between the change in surface roughness and feed rate.

$$\Delta R_a = -0.007363 * f + 16.91 \quad (5)$$

Similar correlation curve was obtained for surface roughness after TAF process. This is shown in Fig. 7. For Figs. 6 and 7, the linear relationship is very comparative. Nonetheless, the slope relies upon the type of the data representation technique.

$$R_{a\,(\text{after TAF})} = 0.007363 * f - 0.6075 \quad (6)$$

In the mathematical equations, a quite good correlation between the value of *R*-square (=0.8328) has been obtained. In a comparison between Figs. 6 and 7, one can recognize that optimal value of feed rate is 400 mm/min. The results showed that the proposed approach played a significant role in improving the surface finish conditions of FDM build parts. This may be because with 400 mm/min tool feed rate, stepped profile of FDM build part properly removed. Accordingly, the surface roughness and feed regression curve for FDM build part during post-finishing is unique and is also related to the finishing capability of the developed tool.

3.1 Surface Profiles Characterization Analysis

To determine the accuracy and adaptiveness of the developed post-finishing technique, surface profiles characterization analysis were performed before and after the TAF process. The surface profiles characterization data for FDM build part (R_a = 16.302 μm) before TAF process was measured using a Rugosurf 10G surface and presented in Fig. 8.

It was observed from the surface profiles characterization data before TAF process, the variation of the surface profile of the FDM build part is more pronounced, compared to a more uniform pattern of the FDM surface profile (R_a = 2.892 μm) after TAF process as shown in Fig. 9. Thus, the proposed post-finishing process gives more uniform surface profile than the original FDM surface. This improvement is the primary source of location error reduction that must be carefully examined. To show the minimization of location error effect, profile data which is above and below to the mean line is observed. By characterizing the profile data normal to the mean line, profile height variation between the original profile and the post-treated profile is found very low (R_a = 2.892 μm) as shown in Figs. 8 and 9. This is because the proper filling of staircases on the different set up are successfully performed by selectively melting of stepped profile at optimal value of feed rate(400 mm/min).

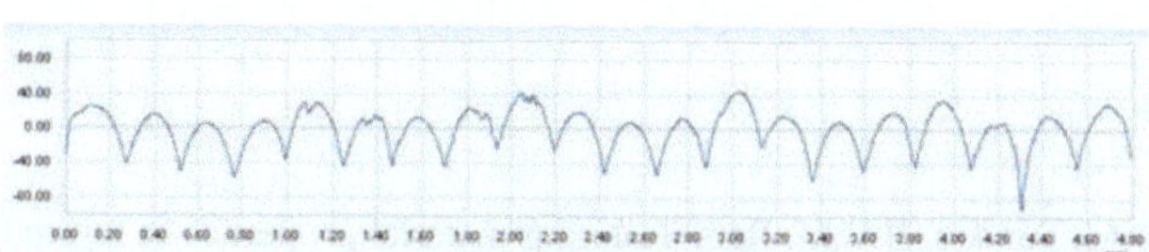

(a) primary profile

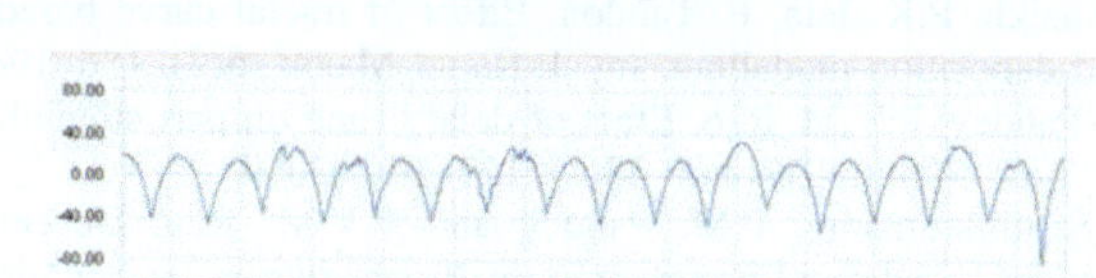

(b) Roughness profile

Fig. 8 Surface profiles characterization for original profile of FDM build part (R_a = 16.302 μm)

Fig. 9 Surface profiles characterization of TAF process profile (R_a = 2.892 μm, Feed rate = 400 mm/min)

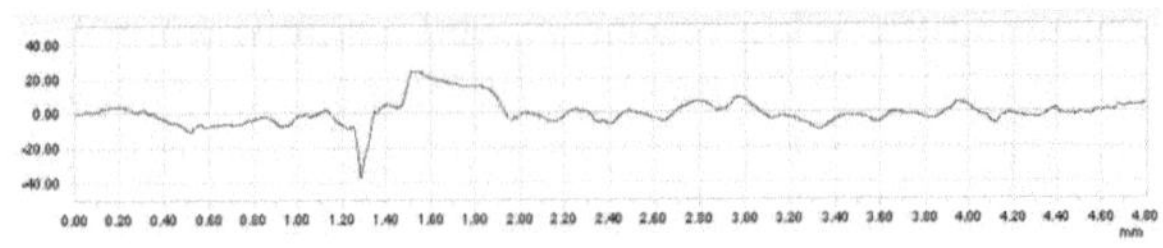

(a) primary profile

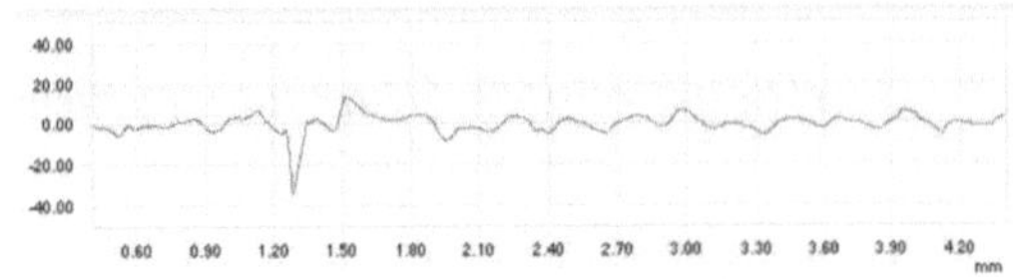

(b) Roughness profile

4 Conclusions

In this paper, a new method of post-finishing of FDM builds part has been developed. This process achieves a better part surface quality of FDM parts through the filling of the staircase by selectively melting. By developing a novel CNC-assisted selective melting tool post-finishing of FDM build parts is performed at the different feed rate. Mathematical modelling procedure is established for the thermally assisted finishing process. Using the regression analysis relationship is set between the surface roughness and feed rate. The results of the thermally assisted finishing show a significant improvement (82.26%) in the surface roughness of FDM build part which is post-treated on the OKUMA milling machine through the selective melting procedure. The results of the surface profiles characterization analysis show that stepped are properly removed. Furthermore, using this method, location error during selective melting reduces considerably.

References

1. P.K. Jain, P.M. Pandey, P.V.M. Rao, Tailoring material properties in layered manufacturing. Mater. Des. **31**(7), 3490–3498 (2010)
2. N. Kumar, S. Shaikh, P.K. Jain, P. Tandon, Effect of fractal curve based toolpath on part strength in fused deposition modelling. Int. J. Rapid Manuf. **5**(2), 186–198 (2015)
3. P.K. Jain, P.M. Pandey, P.V.M. Rao, Efect of delafy time on part strength in selective laser sintering. Int. J. Adv. Manuf. Technol. **43**(1), 117–126 (2009)
4. P.K. Jain, K. Senthilkumaran, P.M. Pandey and P.V.M. Rao, December. advances in materials for powder based rapid prototyping, in *Proceeding of International Conference on Recent Advances in Materials and Processing*, Coimbatore, India, 2006
5. V. Francis, P.K. Jain, Advances in nanocomposite materials for additive manufacturing. Int. J. Rapid Manuf. **5**(3/4), 215–233 (2015)
6. P.K. Jain, P.M. Pandey, P.V.M. Rao, Selective laser sintering of clay-reinforced polyamide. Polym. Compos. **31**(4), 732–743 (2010)

7. V. Francis and P.K. Jain, Experimental investigations on fused deposition modelling of polymer layered silicate nanocomposite. Virtual Phys. prototyping, https://doi.org/10.1080/17452759.2016.1172431 (in press)
8. P.K. Jain, P.M. Pandey, P.V.M. Rao, Experimental investigations for improving part strength in selective laser sintering. Virtual Phys. Prototyping **3**(3), 177–188 (2008)
9. M. Taufik, P.K. Jain, Volumetric error control in layered manufacturing, in *ASME 2014 International Design Engineering Technical Conferences and Computers and Information in Engineering Conference* (pp. V004T06A017–V004T06A017). American Society of Mechanical Engineers, August 2014
10. S.K. Singhal, P.K. Jain, P.M. Pandey, Adaptive slicing for SLS prototyping. Comp. Aided Des. Appl. **5**(1–4), 412–423 (2008)
11. S. Shaikh, N. Kumar, P.K. Jain, P. Tandon, Hilbert curve based toolpath for FDM process. in *CAD/CAM, Robotics and Factories of the Future* (Springer India, 2016), pp. 751–759
12. M. Taufik, P.K. Jain, A study of build edge profile for prediction of surface roughness in fused deposition modeling. ASME J. Manuf. Sci. Eng. **138**(6), 061002 (2016)
13. S.K. Singhal, P.K. Jain, P.M. Pandey, A.K. Nagpal, Optimum part deposition orientation for multiple objectives in SL and SLS prototyping. Int. J. Prod. Res. **47**(22), 6375–6396 (2009)
14. M. Taufik, P.K. Jain, Role of build orientation in layered manufacturing: a review. Int. J. Manuf. Technol. Manage. **27**(1–3), 47–73 (2013)
15. P. Kulkarni, D. Dutta, On the integration of layered manufacturing and material removal processes. J. Manuf. Sci. Eng. **122**(1), 100–108 (2000)
16. M.C. Frank, R.A. Wysk, S.B. Joshi, Rapid planning for CNC milling—a new approach for rapid prototyping. J Manuf Syst **23**(3), 242–255 (2004)
17. D.M.C. Santos, A.E.M. Pertence, H.B. Campos, P.R. Cetlin, The development of 3D models through rapid prototyping concepts. J. Mater. Process. Technol. **169**(1), 1–4 (2005)
18. Z. Hu, K. Lee, Concave edge-based part decomposition for hybrid rapid prototyping. Int. J. Mach. Tools Manuf. **45**, 35–42 (2005)
19. P.M. Pandey, N.V. Reddy, S.G. Dhande, Virtual hybrid-FDM system to enhance surface finish. Virtual Phys. Prototyp **1**(2), 101–116 (2006)
20. L.M. Galantucci, F. Lavecchia, G. Percoco, Quantitative analysis of a chemical treatment to reduce roughness of parts fabricated using fused deposition modeling. CIRP. Ann. Manuf. Technol. **59**(1), 247–250 (2010)
21. A. Garg, A. Bhattacharya, A. Batish, on surface finish and dimensional accuracy of FDM parts after cold vapour treatment. Mater. Manuf. Process. (just-accepted)
22. R. Singh, M. Singh, Surface roughness improvement of cast components in vacuum moulding by intermediate barrel finishing of fused deposition modeling patterns, in *Proceedings of the Institution of Mechanical Engineers, Part E: Journal of Process Mechanical Engineering*, vol. 0(0), pp. 1–8 (just-accepted)
23. E.S. Gadelmawla, M.M. Koura, T.M.A. Maksoud, I.M. Elewa, H.S. Soliman, Roughness parameters. J. Mater. Process. Technol. **123**, 133–145 (2002)

Toolpath Generation for Additive Manufacturing Using CNC Milling Machine

Narendra Kumar, Prashant K. Jain, Puneet Tandon and Pulak M. Pandey

Abstract Additive Manufacturing (AM) systems have gained popularity, due to their capability of producing parts in layer-by-layer manner with any degree of complexity. Although in the last three decades, various commercial AM systems have been developed, however, high cost of the indigenous materials used in commercial machines is the major hindrance for wide applications. Therefore, an economically viable system needs to be developed by exploring the existing manufacturing systems, i.e., CNC milling, etc., as AM technique. To perform AM operation on CNC milling machine, a toolpath is to be required which could be traced by the deposition tool to fabricate the part by adding material in layer-by-layer fashion. In this regard, the present paper proposes a raster and perimeter based toolpath for performing additive manufacturing on existing CNC milling machine. MATLAB platform has been used for algorithm implementation and development of toolpath. Graphical User Interface (GUI) has also been developed in MATLAB to provide the easy access of all parameters related to the toolpath generation, i.e., road gap, slice height, number of contours, etc. To check the feasibility of developed toolpath, experiments have been conducted on existing milling machine by developing and mounting a customized material deposition tool. 3D parts have been fabricated successfully using the developed toolpath. Results show that developed toolpath can be used for performing AM on CNC milling machine.

N. Kumar · P. K. Jain (✉) · P. Tandon
Mechanical Engineering Discipline, PDPM Indian Institute of Information Technology, Design and Manufacturing Jabalpur, Jabalpur, Madhya Pradesh, India
e-mail: pkjain@iiitdmj.ac.in

N. Kumar
e-mail: nyiiitj@gmail.com

P. Tandon
e-mail: ptandon@iiitdmj.ac.in

P. M. Pandey
Mechanical Engineering Department, Indian Institute of Technology Delhi, New Delhi, India
e-mail: pmpandey@mech.iitd.ac.in

L. J. Kumar et al. (eds.), *3D Printing and Additive Manufacturing Technologies*,
https://doi.org/10.1007/978-981-13-0305-0_7

Keywords Raster toolpath · Additive manufacturing · CNC milling
Fused deposition modeling · Hybrid manufacturing · Pellets · Screw extrusion

1 Introduction

Additive Manufacturing (AM) refers to the fabrication process of three-dimensional parts directly using the CAD model in layer-by-layer fashion [1, 2]. Varieties of AM systems have been developed and are available on the market with the capability of processing various materials for part fabrication [3, 4]. AM systems offer several advantages over the traditional manufacturing techniques such as relative time and cost saving in the complex part development [5, 6]. The high cost of the machine and used material are the main obstacles to the implementation of AM systems in small- and medium-scale industries since high-cost machines take a long period to payback. Therefore, researchers are finding alternative ways to reduce the cost of AM systems to increase the reach of these systems to the maximum number of people. CNC machines like milling, turning, etc., can be used for AM and readily available in the small and medium scale industries for the part fabrication. Use of existing CNC machines as an AM can be one of the alternatives to produce parts with less cost. In the past, some researchers developed the low-cost AM system by developing material deposition tool on the existing CNC milling machines and explored these machines as rapid prototyping and rapid tooling applications. Karunakaran et al. [7] developed hybrid deposition process using CNC milling machine and welding for part fabrication. They used Metal Inert Gas (MIG) welding for the material deposition on the CNC worktable and successfully fabricated metallic parts [7]. Bellini et al. [8] developed new setup, called mini extruder deposition (MED). They mounted this extruder system on a high-precision positioning system to fabricate parts [8]. Song et al. [9] also developed a process by combining welding and milling processes on a single platform. They developed a deposition tool using Gas Metal Arc Welding (GMAW) and then mounted on the CNC milling machine for the part fabrication [9]. Reddy et al. [10] also developed the extruder deposition systems using NC table for deposition of material. They successfully fabricated 3D components [10]. Kumar et al. [11] also developed a customized deposition tool for CNC milling to see the effect of fractal curve-based toolpath on the FDM produced parts. They found that CNC milling machine was able to extrude the material to fabricate 3D geometry with high strength through the developed deposition tool [11]. It is found from published literature that very less work reported in this area and hence more processes need to be developed using these machines to minimize cost of AM systems. As, it is known that CNC milling machines produce parts using subtractive manufacturing approach by removing material from the workpiece. Available softwares for CNC milling machines are only capable of generating toolpath for subtractive manufacturing, i.e., machining not for Additive Manufacturing (AM). Therefore, software is to be needed for toolpath generation for additive manufacturing on CNC milling machine. Raster and perimeter based toolpath is very common in AM machines for part fabrication. This paper presents the development of raster and perimeter based toolpath for part fabrication using CNC milling machine in additive manner. Developed toolpath is compatible with the CNC machine and could be utilized directly without making any modification. 3D parts have been fabricated successfully to show the feasibility of developed toolpath with CNC milling machine with the help developed Material Deposition Tool (MDT) system.

2 Methodology

The followed steps of part fabrication through CNC milling machine is shown in Fig. 1. The overall process can be divided into three stages such as STL file generation from CAD model, Toolpath generation through MATLAB and part fabrication using CNC milling machine.

2.1 CAD Model and STL Preparation

The CAD model of the object to be fabricated is prepared using any CAD package like CATIA, Unigraphics, SolidWorks, AutoCAD and Creo, etc. Then, this CAD model is exported as tessellated model, i.e., STL file format which is used as an input by almost all additive manufacturing systems for fabricating parts [12]. Also, in the proposed work, this file has been utilized as an input to generate the toolpath for part fabrication.

2.2 Importing STL in MATLAB

Error free STL file is used as an input to extract the contour coordinates through the developed code. After importing in MATLAB platform, all triangles of the STL file are visible in the figure window as shown in Fig. 2.

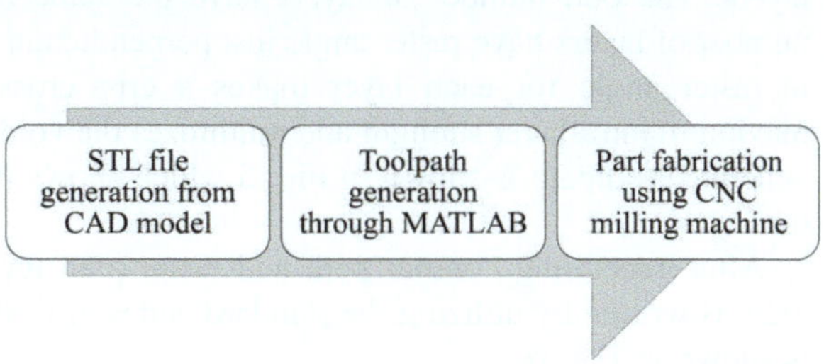

Fig. 1 Overview of part fabrication in AM manner on CNC milling machine

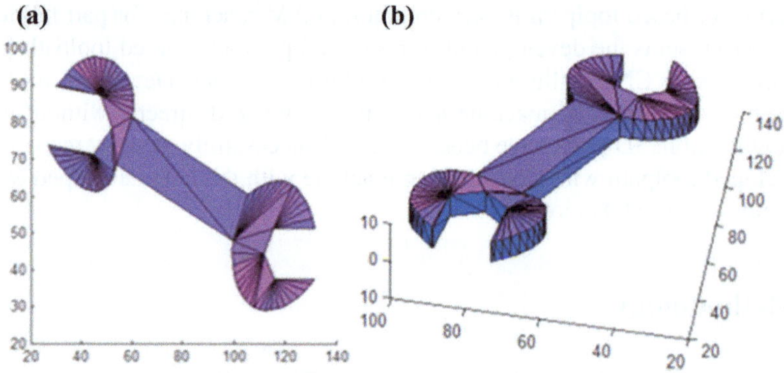

Fig. 2 Tessellated model of the part. **a** top view. **b** Isometric view

2.3 Slicing of the Tessellated Model

The whole tessellated model is then sliced into the number of layers using customized code of slicing. Slicing gives the contour information at various *Z* heights. The number of layers depends on the given value of the slice thickness and the model dimension in the *Z* direction. The slicing of the spanner is shown in Fig. 3a, b.

2.4 Raster/Perimeter Based Tool Path

Toolpath generation algorithm is implemented which takes slice file data as an input. A single slice is chosen at a time for toolpath generation. Initially, toolpath is generated for a slice along with the contour then the whole slice is filled with the rasters of different lengths as shown in Fig. 4. Then, further layers are chosen for toolpath generation, and this process continues until toolpath is generated for all layers. The odd number of layers have the same raster angle whereas the even number of layers have raster angle just perpendicular to the odd one. This variation in raster angle for each layer makes a criss-cross pattern which provides the maximum intra-layer strength and minimizes the voids between layers. Top view of generated toolpath is shown in Fig. 5 which shows the criss-cross pattern between the layers.

After generating contour path and raster path for each layer, the toolpath program is written by utilizing the standard codes of CNC milling. Generated program is shown in Fig. 6.

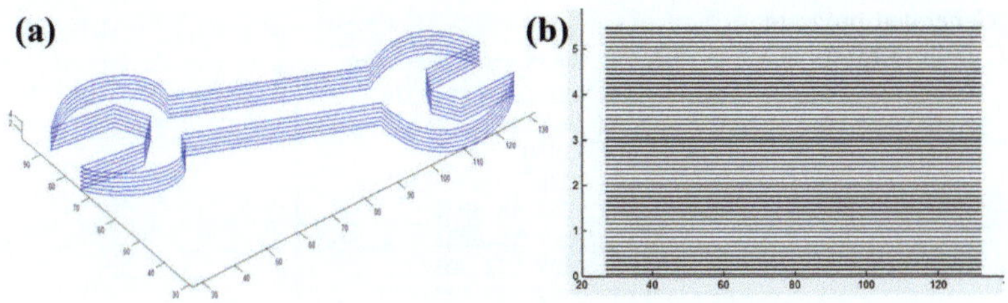

Fig. 3 **a** Extraction of contour information through slicing. **b** Y-Z view of slices

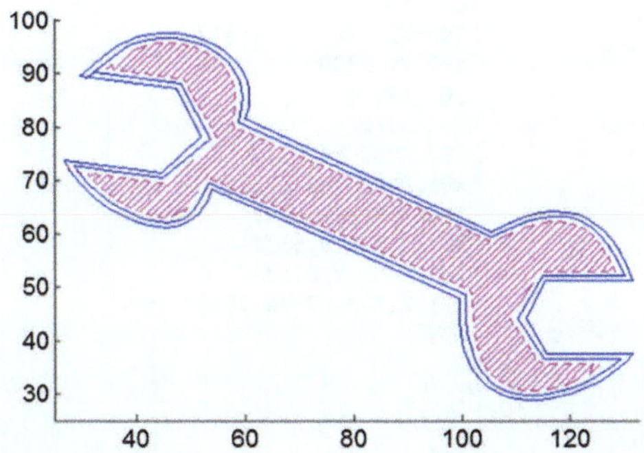

Fig. 4 Toolpath for the single layer

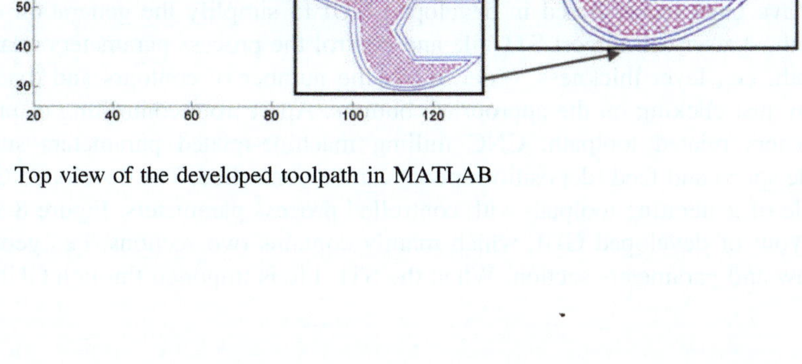

Fig. 5 Top view of the developed toolpath in MATLAB

Fig. 6 Generated program for CNC milling machine

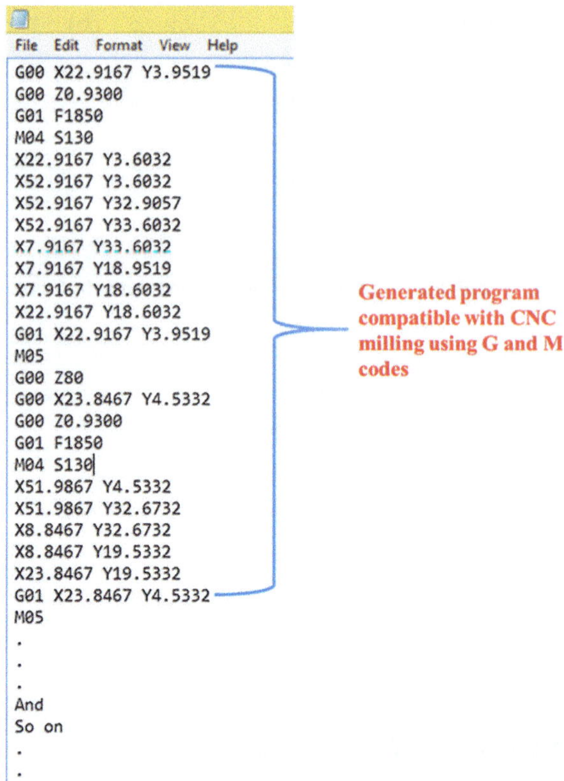

```
File Edit Format View Help
G00 X22.9167 Y3.9519
G00 Z0.9300
G01 F1850
M04 S130
X22.9167 Y3.6032
X52.9167 Y3.6032
X52.9167 Y32.9057
X52.9167 Y33.6032
X7.9167 Y33.6032
X7.9167 Y18.9519
X7.9167 Y18.6032
X22.9167 Y18.6032
G01 X22.9167 Y3.9519
M05
G00 Z80
G00 X23.8467 Y4.5332
G00 Z0.9300
G01 F1850
M04 S130
X51.9867 Y4.5332
X51.9867 Y32.6732
X8.8467 Y32.6732
X8.8467 Y19.5332
X23.8467 Y19.5332
G01 X23.8467 Y4.5332
M05
.
.
.
And
So on
.
.
```


All the steps have been followed for developing toolpath are described through a flow chart. Figure 7 depicts the typical flowchart of the implementation procedure.

3 Development of Graphical User Interface (GUI)

All the steps of the toolpath generation algorithm have been merged on the single platform by developing Graphic User Interface (GUI) in MATLAB. Various buttons have been incorporated in developed GUI to simplify the generation of the toolpath. A user can import STL file and control the process parameters related to toolpath, i.e., layer thickness, raster angle, the number of contours and road gap, etc., by just clicking on the appropriate buttons. Apart from controlling of process parameters related toolpath, CNC milling machine-related parameters such as spindle speed and feed (deposition speed) can be controlled. The developed GUI is capable of generating toolpath with controlled process parameters. Figure 8 shows the layout of developed GUI, which mainly contains two sections, i.e., geometry window and parameters section. When the STL file is imported through GUI, then

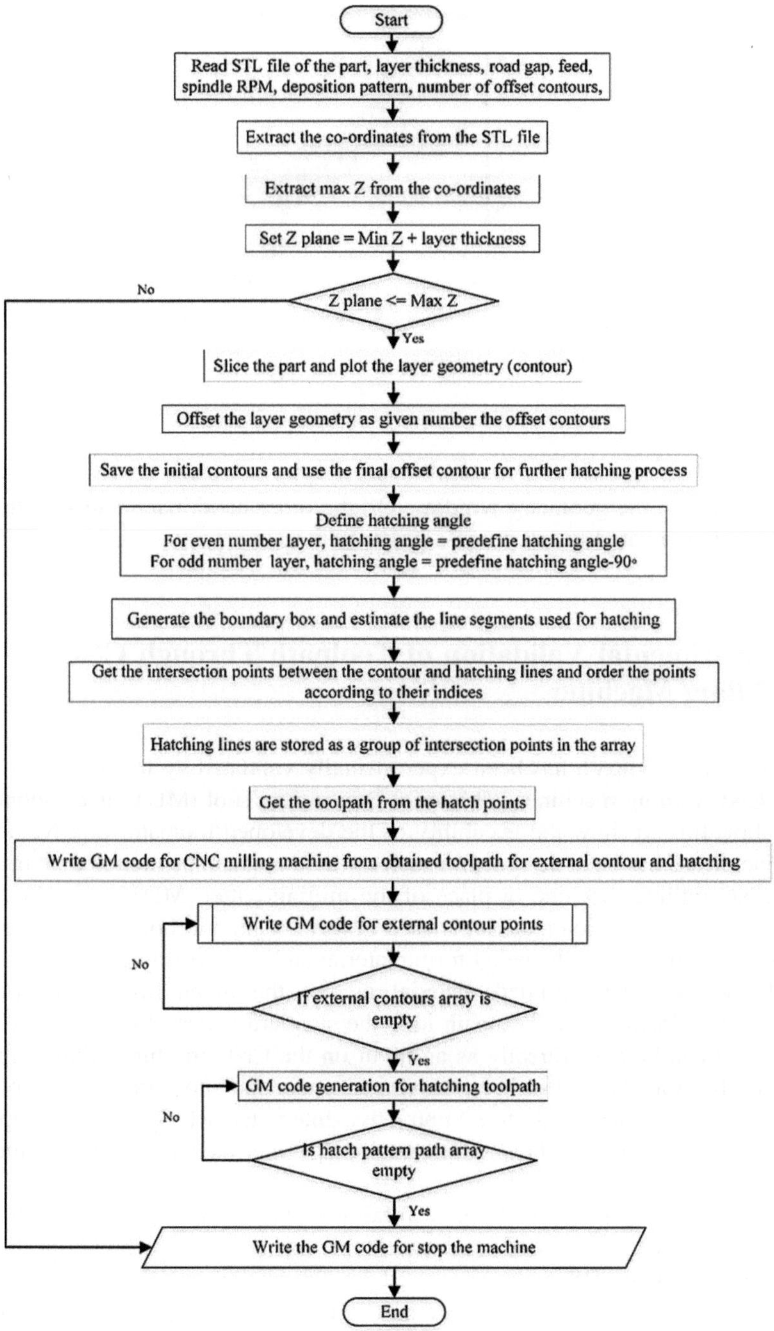

Fig. 7 Flow chart of the implementation procedure

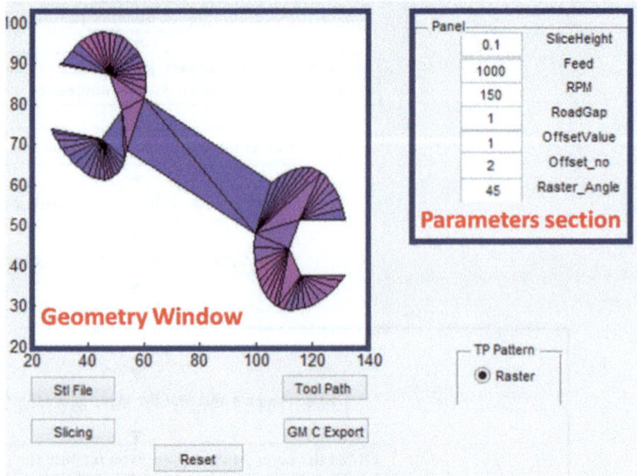

Fig. 8 Developed Graphical User Interface (GUI) for toolpath generation

it will appear in the geometry window. On the other hand, parameters section has all the options related to the process parameters.

4 Experimental Validation of Toolpath Through CNC Milling Machine

The developed toolpath has been experimentally validated by fabricating 3D parts using CNC milling machine. A Material Deposition Tool (MDT) is developed for part fabrication to show the feasibility of the developed toolpath for CNC milling machine. MDT has been designed in such a way that it can be mounted on the three axes CNC milling machine in place of the milling cutter. MDT is a pellet based extruder system in which material is used in form of granules or pellets. Screw and barrel assembly process the pellet form material and extrude it form the tiny nozzle. Nozzle deposits the extruded material as per the given instructions through developed toolpath. As the toolpath uses the standard codes of CNC, i.e., G & M codes so it can be used directly as an input on the CNC machine without making any modification. MDT has been developed under the sponsored project of DST titled "Development of Additive-Subtractive Integrated RP System for Improved Part Quality". Figure 9 shows the developed experimental set-up for the part fabrication.

Different geometric shapes have been fabricated using developed toolpath on the CNC milling machine. Fabricated prototype of spanner is shown in Fig. 10. Low-cost recycled ABS material is used for fabricating parts. Results show that developed toolpath is able to produce parts with acceptable quality. Parts could be produced with high quality after optimizing the deposition tool related parameters, i.e., nozzle deposition speed, road gap, build edge profile, etc. [13].

Fig. 9 Material deposition tool mounted on CNC milling machine

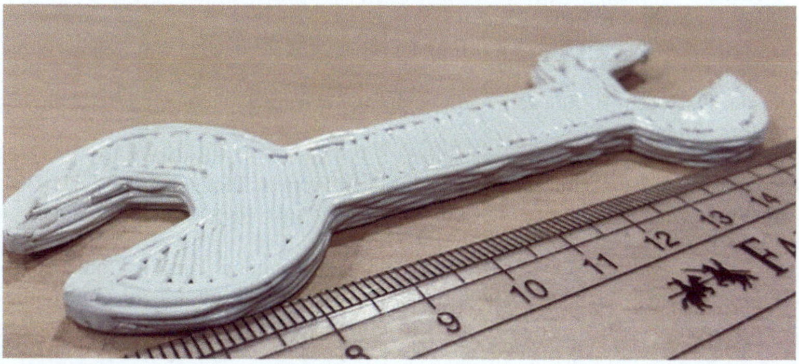

Fig. 10 Fabricated prototype of spanner

5 Conclusion

Raster and perimeter based toolpath is generally used by most of the AM machines. An algorithm has been presented and implemented successfully to generate the raster and perimeter based toolpath for performing additive manufacturing on the CNC milling machine. Developed toolpath has potential to fabricate 3D parts. Generated toolpath has been verified by fabricating the different 3D shapes using developed MDT which shows its feasibility with the CNC milling machine. In comparison to commercial AM machines, present approach shows that low-cost additive manufacturing of 3D parts could be performed using existing CNC milling machines. The major advantage of present approach is that it can be implemented in those industries where low cost CNC milling machines are available.

Acknowledgements This research was done under the sponsored project of DST titled "Development of Additive-Subtractive Integrated RP System for Improved Part Quality". Authors would like to thank their colleagues Durga Prasad Rakshit, Prafull Tripathi, Shashank Sharma, Saquib Shaikh, Mohammad Taufik, and Vishal Francis who provided their valuable suggestions throughout this research work.

References

1. M. Taufik, P.K. Jain, Role of build orientation in layered manufacturing: a review. Int. J. Manuf. Technol. Manag., **27**(1/2/3), 47 (2013)
2. S. Shaikh, N. Kumar, P.K. Jain, P. Tandon, Hilbert curve based toolpath for FDM process, in *CAD/CAM, Robotics and Factories of the Future* (Springer, India, 2016), pp. 751–759
3. X. Yan, P. Gu, A review of rapid prototyping technologies and systems. CAD Comput. Aided Des. **28**(4), 307–318 (1996)
4. V. Francis, P.K. Jain, Advances in nanocomposite materials for additive manufacturing. Int. J. Rapid Manuf. **5**, 215–233 (2015)
5. C.K. Chua, K.F. Leong, C.S. Lim, Rapid Prototyping: Principles and Applications. World Scientific (2010)
6. I. Gibson, D. Rosen, B. Stucker, *Additive Manufacturing Technologies*, 2nd edn. (Springer, New York, 2013)
7. K.P. Karunakaran, P.V. Shanmuganathan, S.J. Jadhav, P. Bhadauria, A. Pandey, Rapid prototyping of metallic parts and moulds. J. Mater. Process. Technol. **105**(3), 371–381 (2000)
8. A. Bellini, L. Shor, S.I. Guceri, New developments in fused deposition modeling of ceramics. Rapid Prototyp. J. **11**(4), 214–220 (2005)
9. Y.-A. Song, S. Park, D. Choi, H. Jee, 3D welding and milling: Part I—a direct approach for freeform fabrication of metallic prototypes. Int. J. Mach. Tools Manuf. **45**(9), 1057–1062 (2005)
10. B.V. Reddy, N.V. Reddy, A. Ghosh, Fused deposition modelling using direct extrusion. Virtual Phys. Prototyp. **2**(1), 51–60 (2007)
11. N. Kumar, S. Shaikh, P.K. Jain, P. Tandon, Effect of fractal curve based toolpath on part strength in fused deposition modelling. Int. J. Rapid Manuf. **5**(2), 186–198 (2015)
12. M. Taufik, P.K. Jain, Volumetric error control in layered manufacturing, in *19th Design for Manufacturing and the Life Cycle Conference; 8th International Conference on Micro- and Nanosystems*, vol. 4 (2014), p. V004T06A017
13. M. Taufik, P.K. Jain, A study of build edge profile for prediction of surface roughness in fused deposition modeling. J. Manuf. Sci. Eng. **138**(6), 061002 (2016)

Modelling of Heat Transfer in Powder Bed Based Additive Manufacturing Process Using Lattice Boltzmann Method

Priya Gupta, Anshul Yadav, Arvind Kumar and Niraj Sinha

Abstract One of the most promising additive manufacturing techniques is selective laser melting process. It is a complex process, which involves physical phenomena, such as absorption of the laser beam in the powder bed, melting and re-solidification, diffusive and radiative heat transport in the powder, diffusive and convective heat transport in the melt pool, gravity effects, etc. In this study, a two-dimensional lattice Boltzmann model is formulated to investigate melting of a uniformly packed powder bed under the irradiation of laser beam during the selective laser melting process. In the model, phase change of individual powder particle is considered mesoscopically. The results give an insight into the details of heat transfer and melting in the powder bed and formation of the mushy zone. These mesoscopic results can be useful to set parameters of the powder bed in additive manufacturing processes. The model developed can be applied to any powder bed based additive manufacturing process.

Keywords Additive manufacturing · Selective laser melting · Heat transfer Mushy zone · Lattice Boltzmann method

Nomenclature

c	Lattice speed
c_s	Speed of sound
E	Energy density, i.e. the energy per unit volume

P. Gupta · A. Yadav · A. Kumar (✉) · N. Sinha
Department of Mechanical Engineering, Indian Institute of Technology Kanpur, Kanpur 208016, India
e-mail: arvindkr@iitk.ac.in

P. Gupta
e-mail: priyag@iitk.ac.in

A. Yadav
e-mail: anshuly@iitk.ac.in

N. Sinha
e-mail: nsinha@iitk.ac.in

L. J. Kumar et al. (eds.), *3D Printing and Additive Manufacturing Technologies*,
https://doi.org/10.1007/978-981-13-0305-0_8

$\mathbf{e_i}$	Microscopic velocity in *i*-direction
$f_i(\mathbf{x}, \mathbf{e_i}, t)$	Discrete probability distribution function
$f_i(\mathbf{x}, t)$	Particle distribution function in *i*-direction
$h_i(\mathbf{x}, t)$	E energy distribution function in *i*-direction
P_{laser}	Power of laser beam
q	Heat flux (W m^{-2})
t	Time (s)
w_0	Radial distance (m)
x	Distance (m)

Greek Symbols

η	Absorptivity
Ω	Particles collision operator in *i*-direction
ρ	Particle density
τ	Dimensionless relaxation times for the temperature field

1 Introduction

Nowadays, additive manufacturing (AM) technologies are being extensively used in many industries. AM offers a range of advantages compared to other existing manufacturing techniques. These include shorter time to market, greater production rate, versatility, highly accurate part, use of inexpensive materials, its capability to produce more functionality in the parts with novel design and improved engineered features. Selective laser melting (SLM) is one of the AM processes, which makes it possible to create fully functional parts directly from metals, ceramics and plastics without using any additional processing steps after the laser melting operation. SLM begins with a fully defined computer-aided design (CAD) model of the part to be made followed by its division into small cross-sections by software. The laser scans over the surface of powder layer deposited on the substrate previously. This is repeated layer by layer, hence making final required component.

Williams and Deckard [1] suggested a two-dimensional model to study process parameters in selective laser sintering (SLS) of materials mainly polymers based on finite difference method. Konrad et al. [2], and Xiao and Zhang [3] proposed a more realistic model in which the powder bed was divided into distinct zones having their individual effective thermal conductivity. Li and Gu [4] developed a three-dimensional model for studying thermal behaviour during SLM of commercially pure titanium powder. They also investigated effects of scan speed and laser power on thermal behaviour of powder. Xiao and Zhang [5] investigated the effects of various factors like existing sintered layers, porosity and initial temperature coupled with the laser intensity and scanning velocity. Their results clearly revealed that the liquid pool moves a little towards the negative direction of scanning and because of the higher scanning velocity, the shape of the liquid pool becomes shallower.

Kovaleva et al. [6] proposed a discrete grid model of heat transfer in granular porous medium to describe the processes of SLM. Moser et al. [7] proposed a model for simulating the melting phenomena of powder bed at particle level. For this, they developed a discrete element model (DEM) to generate the random packing structures of spherical particles. Antony et al. [8] investigated laser melting of stainless steel grade 316L powder on top of AISI 316L substrate using a pulsed Nd:YAG laser numerically and experimentally. They investigated a moving heat source problem and obtained transient temperature solutions. Körner et al. [9] investigated successive assembling in layers with a two-dimensional lattice Boltzmann model, which considering individual powder particles. Loh et al. [10] presented a numerical simulation and effective modelling on the Selective SLM process with aluminium alloy 6061.

The lattice Boltzmann method (LBM) has been developed as a new effective and alternative approach of computational fluid dynamics (CFD) and it has achieved significant success in simulating heat transfer and fluid flows [7–10]. In the LBM approach, kinetic equations for the particle distribution function are solved. Bhatnagar et al. [11] presented the Bhatnagar-Gross-Krook (BGK) model which is a kinetic theory approach for collision processes in gases. Palmer and Rector [12] proposed an algorithm for incorporating the effects of temperature into lattice Boltzmann simulations. He and Luo [13] presented a special discretised form of the Boltzmann equation. They discussed various approximations for the discretization of the Boltzmann equation in both time and phase space. Attar and Körner [14] investigated a multi-distribution function model, which was applied to simulate hydrodynamic flow and the coupled thermal diffusion–convection problem. Yan et al. [15] presented a parallel and efficient algorithm for simulating single-phase and multiphase fluid flows.

The aim of this work is to gain a better understanding of the laser beam and powder bed based AM processes with the help of a numerical model at a mesoscopic level. An LBM-based numerical model is used to investigate the melting of individual powder particles uniformly packed in the powder bed under the irradiation of a Gaussian laser beam during SLM process. The model developed can be applied to any powder bed based AM process whether it involves full melting or partial melting of powder particles.

2 Numerical Modelling

The LBM method was developed from Ludwig Boltzmann's kinetic theory of gases16]. The fundamental idea is that fluid can be considered as consisting of fine and randomly scattered particles. When these scattered particles collide with each other, they exchange energy and momentum. Particle streaming and billiard-like particle collision is used to solve the exchange of momentum and energy transport phenomena. The original idea of gas dynamics can be simplified with the use of LBM by decreasing the number of particles and restraining them to the nodes of a lattice. This Boltzmann transport equation [17] can be used to model this process, which is given by

$$\frac{\partial f_i(\mathbf{x},t)}{\partial t} + \mathbf{u}_i \cdot \nabla f_i(\mathbf{x},t) = \Omega_i \quad (1)$$

where particle distribution function in i-direction is given by $f_i(\mathbf{x}, t)$, particle velocity in i-direction is given by $\mathbf{u}$, and the particles collision operator in i-direction is given by $\boldsymbol{\Omega}$.

There are many LBM models available to model the process with the help of LBM in both two- and three-dimensional. For a two-dimensional modelling, there are nine possible directions in which a particle is restricted to stream, together with the one staying at rest. These velocities are called the microscopic velocities are denoted by $\mathbf{e_i}$, where $i = 1, \ldots, 9$. This model is a two-dimensional model, generally known as the D2Q9 model and involves nine velocity vectors. The lattice node of D2Q9 model with 9 velocities defined by $\mathbf{e_i}$ as follows:

$$\mathbf{e_i} = \begin{cases} (0,0) & i=1 \\ (\pm c,0)(0,\pm c) & i=2,\ldots,5 \\ (\pm c,\pm c) & i=6,\ldots,9 \end{cases}$$

where $c = \frac{\Delta x}{\Delta t}$ is the lattice speed. We associate a discrete probability distribution function $f_i(\mathbf{x}, \mathbf{e_i}, t)$, for each particle on the lattice. It can be simply written as $f_i(\mathbf{x}, t)$, $i = 1, \ldots, 9$, which explain the probability of streaming in one specific direction. Figure 1 clearly shows the systematic view of lattice node with nine velocity vectors.

Particle density (ρ) at macroscopic level density can be defined as a summation of microscopic particle distribution function

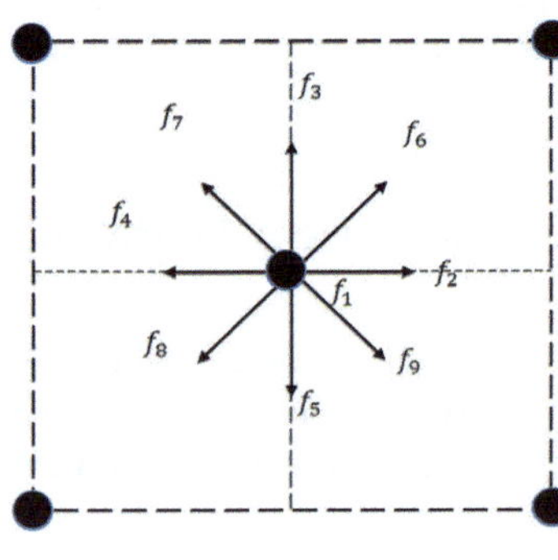

Fig. 1 Typical lattice nodes and velocity sets for D2Q9 model

$$\rho(\mathbf{x}, t) = \sum_{i=1}^{9} f_i(\mathbf{x}, t) \tag{2}$$

where $f_i(\mathbf{x}, t)$ is the particle distribution function in i-direction.

The energy density, i.e. the energy per unit volume (E) is given as

$$E(\mathbf{x}, t) = \sum_{i=1}^{9} h_i(\mathbf{x}, t) \tag{3}$$

where $h_i(\mathbf{x}, t)$ represents energy distribution function in i-direction.

The streaming and collision processes are the important steps in LBM, are given by

$$\underbrace{f_i(\mathbf{x} + \mathbf{e_i}, t + \Delta t) - f_i(\mathbf{x}, t)}_{\text{streaming}} = \underbrace{\frac{\Delta t}{\tau}(f_i^{eq}(\mathbf{x}, t) - f_i(\mathbf{x}, t)) + F_i}_{\text{collision}} \tag{4}$$

$$\underbrace{h_i(\mathbf{x} + \mathbf{e_i}, t + \Delta t) - h_i(\mathbf{x}, t)}_{\text{streaming}} = \underbrace{\frac{\Delta t}{\tau_h}(h_i^{eq}(\mathbf{x}, t) - h_i(\mathbf{x}, t)) + \Phi_i}_{\text{collision}} \tag{5}$$

$$f_i^{eq}(\mathbf{x}, t) = w_i \rho \left[1 + \frac{(\mathbf{e_i} \cdot \mathbf{u})}{c_s^2} + \frac{(\mathbf{e_i} \cdot \mathbf{u})^2}{2c_s^4} - \frac{(\mathbf{u})^2}{2c_s^2}\right] \tag{6}$$

$$h_i^{eq}(\mathbf{x}, t) = w_i E \left[1 + \frac{(\mathbf{e_i} \cdot \mathbf{u})}{c_s^2} + \frac{(\mathbf{e_i} \cdot \mathbf{u})^2}{2c_s^4} - \frac{(\mathbf{u})^2}{2c_s^2}\right] \tag{7}$$

where $f_i^{eq}(\mathbf{x}, t)$ and $h_i^{eq}(\mathbf{x}, t)$ are the equilibrium distributions functions. The speed of sound (c_s) is given by $c_s^2 = c^2/3$, where τ and τ_h are dimensionless relaxation times for the temperature field.

At the time of modelling of the process, streaming and collision are computed individually. These two terms become very important in boundary lattice nodes. The streaming step for the interior nodes is illustrated graphically in Fig. 2.

In the collision term of (4), the equilibrium distribution is given by f_i^{eq}, and the relaxation time towards local equilibrium is defined by τ. For simulating single-phase flows, it suffices to use BGK collision. Whose equilibrium distribution f_i^{eq} is given by [18]

$$f_i^{eq}(\mathbf{x}, t) = \omega_i \rho + \rho s_i(\mathbf{u}(\mathbf{x}, t)) \tag{8}$$

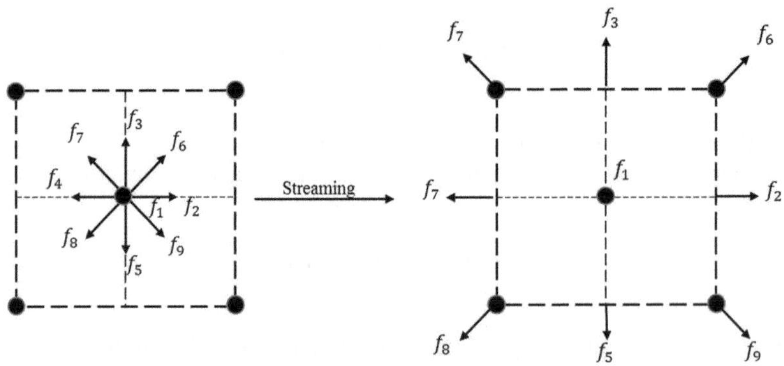

Fig. 2 Streaming process of a lattice node in D2Q9 model

where $\boldsymbol{s_i}$ $(\mathbf{u})$ and ω_i are defined as

$$s_i(\mathbf{u}) = \omega_i \left[3\frac{\mathbf{e_i} \cdot \mathbf{u}}{c} + \frac{9}{2}\frac{(\mathbf{e_i} \cdot \mathbf{u})^2}{c^2} - \frac{3}{2}\frac{\mathbf{u.u}}{c^2} \right]$$

$$\omega_i = \begin{cases} 4/9 & i = 0 \\ 1/9 & i = 1,2,3,4 \\ 1/36 & i = 5,6,7,8 \end{cases}$$

It may be noted that numerical issues can arise as $\tau \rightarrow 0.5$. In order to satisfy the imposed macroscopic boundary conditions during the streaming and collision step, the boundary nodes require special attention on the distribution functions. The algorithm is also summarised in the form of flowchart as shown in Fig. 3.

Fig. 3 Algorithm for implementation of LBM for the present model

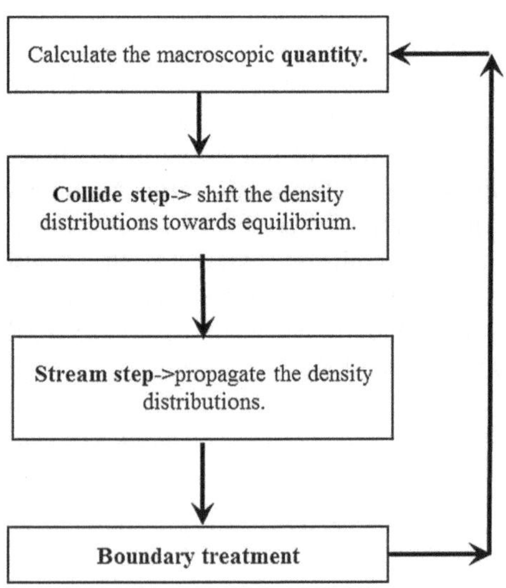

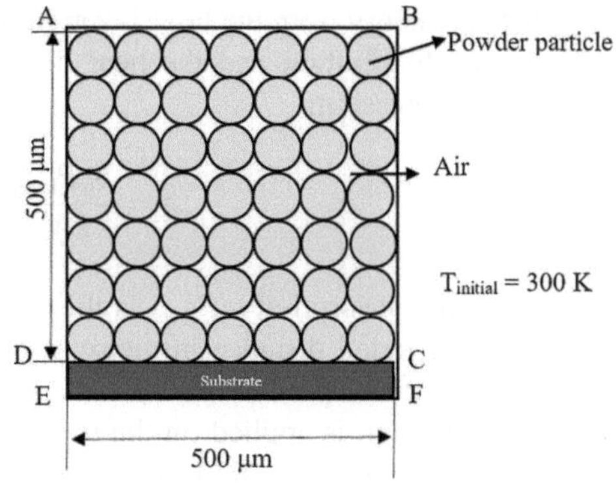

Fig. 4 Schematic of computational domain

2.1 Computational Domain

Figure 4 shows the schematic diagram of the computational domain considered. The present model considers all particles having same shape and size, and the powder bed having an ideally uniform distribution of the particles. The diameter of powder particle is 50 μm. The material of the powder particle is Ti-6Al-4V. A solid substrate has been considered for holding the powder particle. The substrate material is also considered same as the powder particle material and has a thickness of 15 μm. The thermo-physical properties of Ti-6Al-4V are given in Table 1. Air is considered in the gaps between the solid particles. The laser beam with Gaussian distribution of heat was applied to the top surface of the powder layer for 150 ms. In this model, it is assumed that the particle in powder bed are tightly packed such that after melting they retain their place.

Table 1 Thermo-physical property data [16]

Property of Ti-6Al-4V	Values
Liquidus temperature (K)	1923
Solidus temperature (K)	1876
Solid specific heat (J kg^{-1} K^{-1})	483.04 + 0.215T T $\leq$ 1268 412.7 + 0.1801T 1268 < T $\leq$ 1923
Liquid specific heat (J kg^{-1} K^{-1})	831
Thermal conductivity (W m^{-1} K^{-1})	1.25 + 0.015T T $\leq$ 1268 3.15 + 0.012T 1268 < T $\leq$ 1923 −12.75 + 0.024T T > 1923
Density (kg m^{-3})	4512
Latent heat of fusion (J K^{-1})	286,000
Thermal expansion coefficient (K^{-1})	1.1 × 10^5
Convective heat coefficient (W m^{-2} K^{-1})	10
Laser absorption coefficient	0.36

The heat energy from the laser beam in the SLM process can be approximated by a Gaussian distribution and the heat flux (q) on the powder bed is given by the following expression:

$$q = \frac{2\eta P_{\text{laser}}}{\pi w_o^2} e^{(-2x^2/w_o^2)} \quad (9)$$

where η is the absorptivity, P_{laser} is the power of laser beam, x is distance in metre and w_0 is the radial distance in metre at which energy density equals to e^{-2} at the centre of the laser spot.

The laser beam is applied on the top layer. In addition, at the top of the powder bed, convection and radiation losses have been considered. The initial temperature of the substrate and powder particle is taken to be 300 K. The sides of the domain are considered symmetric. In order to simplify the process some physical aspect such as evaporation, Marangoni effect and re-solidification have been neglected.

3 Results and Discussion

During the heat transfer in powder bed based AM processes; the heat energy coming from the laser is absorbed in the powder bed and heats the layer of powder particle. When the temperature of the metal become more than the solidus temperature, the particles starts to melt. LBM yields a numerical approach to model and simulate beam and powder bed layered AM processes at the mesoscopic level. In this work, a numerical tool based on LBM is used that allows prediction of local powder melting. The LBM is relatively easy to implement. The model composed of partial differential equations for heat transfer. For that, following two sets of lattice Boltzmann methodology have been defined: one for the density and the other for the temperature.

The fabricated parts quality may be affected by many factors such as powder bed characteristics, materials properties and process parameters. These parameters have significant influence on thermal field which is developed in the powder bed. A homogenous thermal field leads to better mechanical properties, microstructure, dimensional accuracy and surface finish. In this work, temperature-dependent properties of the material were considered. In simulations, a laser power of 20 W with beam spot radius of 0.4 mm was applied for a period of 150 ms on the top surface of the powder bed. The model is implemented using COMSOL Multiphysics™ software.

To obtain a better understanding of the SLM process, the results are presented for various time steps showing temperature plots in Fig. 5. At t = 60 ms melting of powder particles starts. As shown in Fig. 5a, at this time, melting front generates at the top of the powder layer and the maximum temperature reached is 1943 K. There are two black contours: the upper one is for liquidus front above which the

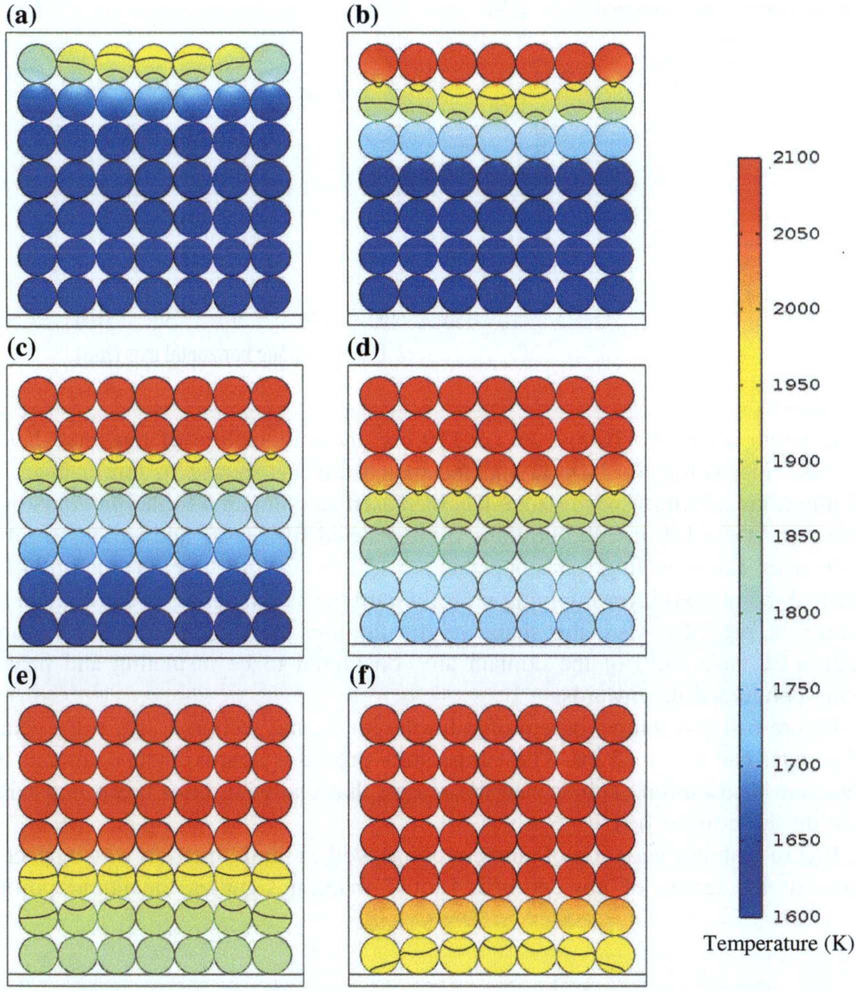

Fig. 5 Temperature maps in powder bed at time. **a** $t = 60$ ms, **b** $t = 80$ ms, **c** $t = 95$ ms, **d** $t = 110$ ms, **e** $t = 130$ ms, and **f** $t = 150$ ms. The black contours indicate liquidus (upper) and solidus (lower) front

temperature is above liquidus temperature (1923 K) and other one for the solidus front above which the temperature is above the solidus temperature (1876 K) of Ti-6Al-4V. At $t = 80$ ms, a deep melt front develops and moves to the second layer. At $t = 95$ ms, temperature of the top two layers is above liquidus temperature and the melting front had moved to third layer.

The gap between the solidus and the liquidus temperature line is known as mushy zone width. If this range is large, then defects for example porosity and shrinkage may generate in the product produced by the SLM process. This difference between liquidus and solidus shows a reasonable increment in the mushy

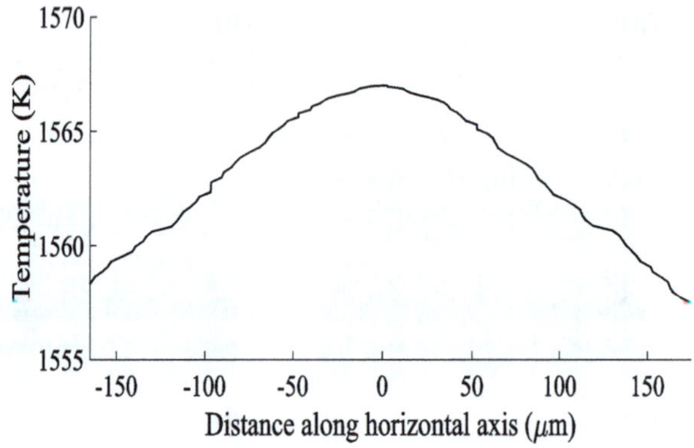

Fig. 6 Temperature variation at $t = 75$ ms along a horizontal section at the centre of powder bed

zone width after $t = 110$ ms. At $t = 110$ ms, the temperature of top three layers reached approximately 2100 K and the melt front has moved to fourth layer. In addition, the width of mushy zone has increased as compared to the previous time instants. At $t = 130$ ms, the liquidus front reached fifth layer and solidus front at sixth layer. Laser heating was stopped at $t = 150$ ms. At this time, the melting front reached the bottom layer implying no solid particles is remaining in the domain (as shown in Fig. 5f). Also, the shape of the liquidus front is converging towards bottom because sides of the domain are considered to be insulating and heat is being conducted downwards.

Figure 6 shows temperature profile for the horizontal section drawn at the centre of powder bed at $t = 75$ ms. The temperature profile is approximately same as the Gaussian distribution. This is due the reason that Gaussian laser beam is used to heat up the powder bed.

Figure 7 shows variation of temperature as well as liquid fraction with respect to time for the centre of powder bed. Liquid fraction signifies the liquid portion

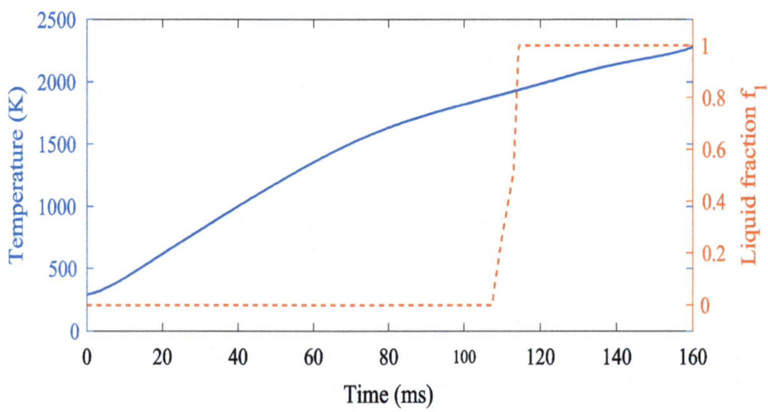

Fig. 7 Temperature and liquid fraction evolution at the centre of powder bed

present the control volume as compared to the solid portion. It varies between 0 and 1. The value of liquid fraction as 0 signifies that no liquid is present, while 1 means that control volume contains only liquid. When the range of liquid fraction is between 0 and 1, the region is known as mushy zone. This region should be thin otherwise defect will arise.

The results shown in Figs. 5, 6 and 7 provide a physical insight into the understanding of SLM process at mesoscopic level. The numerical simulation describes the influence of material properties and processing parameters, which cannot be simply understood from experiments.

4 Conclusions

In this study, simulation of powder bed based layered AM processes at mesoscopic scale is done with the help of LBM. LBM demonstrates a numerical approach to simulate local melting of powder bed. To understand the mechanism and parameters influencing SLM, mesoscopic simulation can prove to be an important key. Considering the temperature dependent thermo-physical parameters and the Gaussian distribution of the heat flux, a two-dimensional model is developed to investigate thermal behaviours during SLM of Ti-6Al-4V powder layers. The model based on LBM has been successfully applied to a powder bed containing seven layers of powder with seven particles in each layer.

The results show the heating and melting of powder particles at mesoscopic scale. The mushy zone is captured, which progresses downward with time. The symmetry in the progress of melting is well captured. The melting front gradually moves towards bottom layer because of heat accumulation. The model developed can be applied to any powder bed based AM process whether it involves full melting or partial melting of powder particles.

References

1. J.D. Williams, C.R. Deckard, Advances in modelling the effects of selected parameters on the SLS process. Rapid Prototyp. J. **4**, 90–100 (1998)
2. C. Konrad, Y. Zhang, B. Xiao, Analysis of melting and resolidification in a two-component metal powder bed subjected to temporal Gaussian heat flux. Int. J. Heat Mass Transf. **48**, 3932–3944 (2005)
3. B. Xiao, Y. Zhang, Analysis of melting of alloy powder bed with constant heat flux. Int. J. Heat Mass Transf. **50**, 2161–2169 (2007)
4. Y. Li, D. Gu, Thermal behaviour during selective laser melting of commercially pure titanium powder: Numerical simulation and experimental study. Addit. Manuf. **1**, 99–109 (2014)
5. B. Xiao, Y. Zhang, Laser sintering of metal powders on top of sintered layers under multiple-line laser scanning. J. Phys. D Appl. Phys. **40**, 6725–6734 (2007)
6. I. Kovaleva, O. Kovalev, I. Smurov, Model of heat and mass transfer in random packing layer of powder particles in selective laser melting. Phys. Procedia **56**, 400–410 (2014)

7. D. Moser, M. Cullinan, J. Murthy, Particle-scale melt modeling of the selective laser melting Process. In: Proceedings of the 27th Annual International Solid Freeform Fabrication Symposium—An Additive Manufacturing Conference, 8–10 Aug 2016
8. K. Antony, N. Arivazhagan, K. Senthilkumaran, Numerical and experimental investigations on laser melting of stainless steel 316L metal powders. J. Manuf. Process. **16**, 345–355 (2014)
9. C. Körner, A. Bauereiß, E. Attar, Fundamental consolidation mechanisms during selective beam melting of powders. Modelling Simul. Mater. Sci. Eng. 21, 085011 (2013)
10. L.E. Loh, C.K. Chua, W.Y. Yeong, J. Song, M. Mapar, S.L. Sing, Z.H. Liu, D.Q. Zhang, Numerical investigation and an effective modelling on the Selective Laser Melting (SLM) process with aluminium alloy 6061. Int. J. Heat Mass Transf. **80**, 288–300 (2015)
11. P. Bhatnagar, E.P. Gross, M.K. Krook, A model for collision processes in gases. I. Small amplitude processes in charged and neutral one-component systems. Phys. Rev. **94**, 511–525 (1954)
12. B.J Palmer, D.R. Rector, Lattice Boltzmann algorithm for simulating thermal flow in compressible fluids. J. Comput. Phys. **161**, 1–20 (1999)
13. X. He, L.S. Luo, Theory of the lattice Boltzmann method: from the Boltzmann equation to the lattice Boltzmann equation. Phys. Rev. E **56**, 6811–6817 (1997)
14. E. Attar, C. Körner, Lattice Boltzmann model for thermal free surface flows with liquid–solid phase transition. Int. J. Heat Fluid Flow **32**, 156–163 (2011)
15. Y.Y. Yan, Y.Q. Zu, B. Dong, LBM, a useful tool for mesoscale modelling of single-phase and multiphase flow. Appl. Therm. Eng. **31**, 649–655 (2011)
16. Z. Fan, F. Liou, *Numerical Modeling of the Additive Manufacturing (AM) Processes of Titanium Alloy, Titanium Alloys—Towards Achieving Enhanced Properties for Diversified Applications*. Dr. A. K. M. Nurul Amin Ed., InTech (2012)
17. Y.B. Bao, J. Meskas, Lattice Boltzmann method for fluid simulations, in *Courant Institute of Mathematical Sciences*, Apr 2011. www.cims.nyu.edu/~billbao/report930.pdf
18. Z. Guo, B. Shi, N. Wang, Lattice BGK model for incompressible Navier-Stokes equation. J. Comput. Phys. **165**(1), 288–306 (2000)

Effect of Process Parameters on Mechanical Properties of Solidified PLA Parts Fabricated by 3D Printing Process

Jagdish Khatwani and Vineet Srivastava

Abstract In rapid prototyping (RP), 3D printing is growing fast due to its ability to build different complex geometrical shapes and structures in least possible time. The mechanical behavior of 3D printed parts depends on the interaction of different process parameters and the raw material properties. In this work, the effect of process parameters, namely, nozzle diameter, layer thickness, and part bed temperature, has been studied on mechanical properties like tensile strength and flexural strength in 3D printing process. Material used in the study is solidified polylactic acid (PLA). It was observed that tensile strength and flexural strength increased with increase in part bed temperature. It was further observed that tensile strength decreased with increase with layer thickness whereas flexural strength increased. With respect to nozzle diameter, it was observed that tensile strength increased while flexural strength initially decreased and then increased with increase in nozzle diameter. SEM analysis has been done to evaluate the mechanism of failure of the parts.

Keywords 3D printing · Nozzle diameter · Layer thickness · Part bed temperature Tensile strength · Flexural strength

1 Introduction

Rapid prototyping (RP) technology provides important tools to fabricate geometrically complex functional parts. Compatibility of these RP technologies with computer-aided design (CAD) makes part fabrication faster. Least possible cycle time of design and fabrication is achieved through RP. On the basis of limits on the

---

J. Khatwani · V. Srivastava (✉)
Department of Mechanical Engineering, Thapar Institute of Engineering & Technology, Patiala 147004, Punjab, India
e-mail: vineet.srivastava@thapar.edu

J. Khatwani
e-mail: jkhatwani33@gmail.com

L. J. Kumar et al. (eds.), *3D Printing and Additive Manufacturing Technologies*,
https://doi.org/10.1007/978-981-13-0305-0_9

type and properties of the material that can be fabricated, many commercial RP systems are available nowadays in the market [1].

Fused Filament modelling (FFM) is one of the most popular 3D printing techniques based on Fused Deposition Modelling (FDM) process. FFM prints a three-dimensional object by depositing a continuous and constant flow of melted work material, which is carefully arranged into layer by layer, starting from the bottom up.

In past, there have been many attempts made to analyze the effect of different process parameters on the mechanical properties of the parts produced through different RP systems. A study made by Ahn et al. [2] showed that air gap and raster orientation affect tensile strength greatly. In FDM parts made with a negative air gap or overlap between roads, the typical tensile strength ranged between 65 and 72% and compressive strength ranged from 80 to 90% of the strength of injection moulded ABS respectively. Karalekas and Antoniou [3] showed that nonwoven fiber mats could improve the mechanical properties of specimens built using the stereolithography and vacuum casting techniques. This study showed that desirable increases in the elastic modulus and the mechanical strength are feasible providing that the reinforcement can be consolidated to the polymer material used. Ang et al. [4] investigated the mechanical properties and porosity relationships in fused deposition modelling. Air gap, Raster width, Build orientation, build lay down pattern, build layer were taken as process parameters. Compressive modulus of the part was only affected by two main parameters—air gap and raster width. Results shows that an increasing air gap caused a mean decrease in the compressive modulus and increasing raster width increase the compressive modulus. Jain et al. [5] used SLS machine and polyamide powder (PA 2200) as material and layer thickness, hatch spacing and scan speed as processing parameters. Parts were fabricated for different ranges of delay time and testing to see the variability in tensile strength. Study showed that if delay time is above or below an optimum value of delay time range, it will result in lower tensile strength. Bagisk et al. [6] pointed out that tensile and compression strengths are influenced by the orientation and the structure of the ABS manufactured parts in FDM. Percoco et al. [7] concluded that compressive strength is increased by 2–4%, when FDM prototypes treated with a solution of 90% dimethylketone and 10% water with raster angel and raster width is taken as process parameters. Croccolo et al. [8] concluded that greater contour will increase stiffness and strength but it will also increase the brittle behavior of the part fabricated by ABS. Theoretical model was developed between responses and parameters successfully.

Tymrak et al. [9] investigated the mechanical properties of ABS and PLA components made using various desktop open-source RepRap three-dimensional printers using build orientation and layer thickness as process parameters. The results showed that the average tensile strength and average elastic modulus of RepRap printed parts for PLA is higher than ABS parts. Afrose et al. [10] investigated the fatigue properties of FDM processed PLA standard tensile parts. These were based on ASTM D638 standard and were cyclically tested at 80, 70, 60, and 50% nominal values of the ultimate tensile stress. Study showed that parts in

X build orientation exhibit higher tensile stress, compared to those built in *Y* and 45° orientations. But under tensile cyclic loading condition, the parts in 45° build orientation show higher fatigue life than the parts in *X* and *Y* build orientations for the same percentage of applied static loads. Roberson et al. [11] investigated the effect of stress concentrator fabrication on 3D printed parts using ABS, PC, PC-ABS, and Ultem 9085 as material. Izod impact testing showed that samples printed on the *XY* plane at 45° were shown to have the greatest resistance to impact and samples printed vertically in the *ZXY* orientation had the lowest resistance to impact. Only samples printed from Ultem 9085 experienced a significant difference in impact strength when comparing stress concentrator fabrication method of specimens by printing and milling to develop stress concentrator.

From literature survey, it has been observed that maximum work regarding part strength is concerned on FDM, SL and SLS technologies. In 3D printing, one of the most widely used material is PolyLactic Acid (PLA), which has gained wide popularity due to its light weight and ready availability. PLA is being used for form and fit analysis and as casings for small equipments. It has also been observed that the analysis of strength of parts fabricated through Solidified PLA has not been done before. Therefore an attempt is being made to correlate the tensile strength and flexural strength of solidified PLA with controllable machine parameters.

2 Planning of Experiments

The strength of PLA parts is governed by a large number of interactive variables. In this study three controllable variables have been considered, namely, layer thickness, nozzle diameter, and part bed temperature to carry out experimental work. The experiments were conducted keeping these process parameters at various levels. The parts were fabricated on Protocentre 999 3D printer. On the basis of past literature reviews and capabilities of 3D printing machine, the range for each of the process parameter was selected. The range for nozzle diameter was selected from 0.3 to 0.5 mm, layer thickness 0.1 to 0.3 mm, and the part bed temperature from 51 to 53 °C respectively. The range of process parameters have been given in Table 1. The physical properties of PLA have been provided in Table 2. In this study, single variable experiments have been carried out.

Table 1 Range of process parameters

Parameters	Range
Layer thickness (mm)	0.1, 0.2, 0.3
Nozzle diameter (mm)	0.3, 0.4, 0.5
Part Bed temperature (°C)	51, 53, 55
Raster orientation (°)	45°/–45°
Extruder head speed	100 mm/sec
Temperature of extruder (°C)	220

Table 2 Physical properties of solidified PLA [12]

Parameters	Value
Grade	4043D
Density (g/cm^3)	1.24
Glass Transition Temperature (°C)	60
Melting point (°C)	160

Fig. 1 Specimen fabricated for tensile test (type 1)

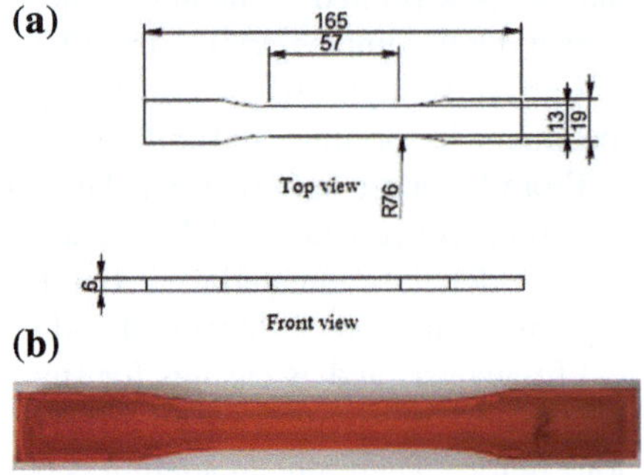

Figure 1 shows the specimen dimensions and specimen fabricated for tensile test. Tensile strength at break is determined on UTE 100 instrument according to ASTM D638 testing standards. Capacity of UTE 100 is 100 KN with minimum resolution of 0.05 KN. Crosshead speed at which testing was done is 5 mm/min. Flexural strength at yield is determined as per ASTM D790-10 shown in Fig. 2, in which three-point bending test is used for flexural strength determination. The specimen is supported by two supports and loaded in the middle by force until the test specimen fractures. Three-point bending test was performed on UTN 20, having capacity of 2000 kgf with minimum resolution of 4 kgf. The crosshead speed during testing was set to 2 mm/min.

Fig. 2 Specimen fabricated for flexural test

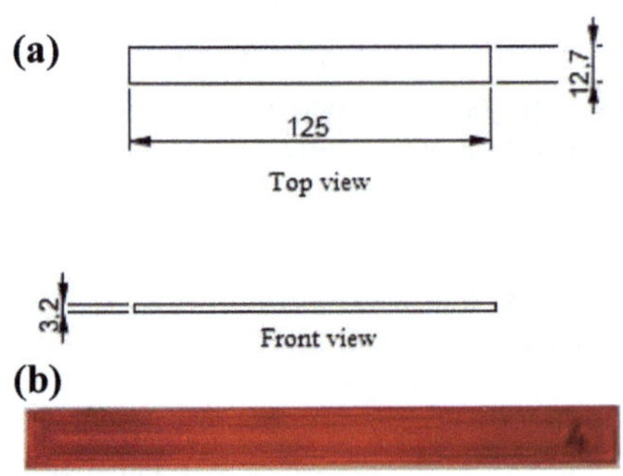

3 Results and Discussions

3.1 Analysis for Tensile Strength

From Fig. 3, it can be noted that tensile strength increases as layer thickness decreases. The strong inter layer bonding is one of the important factor for the strength. As the layer thickness decreases, more number of layers will be required [13]. More the number of layers required, higher would be time required to fabricate the part, which will result in more heat dissipation between layers and improve the bond formation between rasters.

During manufacturing of the parts, there is formation of voids between the adjacent layers, which is due to very high scanning speeds. Layer thickness is one of the important factor on which the volume of this voids depends. Figure 4 shows the scanning electron microscope (SEM) images for the 0.1 and 0.2 mm layer thickness. It is visible from Fig. 4b that voids dimensions are lower in 0.1 mm layer thickness as compared to the 0.2 mm layer thickness (Fig. 4a). Larger the void dimension, poor the inter layer bonding between layers will take place. Due to lower layer thickness, total volume of the voids taking place between the layers is minimum as compared to the higher layer thickness. Small air gap helps to provide strong bond between rasters and thus, improves strength.

Figure 5 shows that tensile strength increases as part bed temperature increases. Increase in part bed temperature provides strong inter-layer bonding. As the part bed temperature increases, there is increase in heat dissipation from one layer to another, which leads to post heating of layers which are already bonded. Due to this post heating of layers, greater diffusion of one layer to the adjacent layer occurs and hence improves the strength. On additional experimentation, when we approached the glass transition temperature, it was observed that the tensile strength started to decrease. The effective range of temperature where tensile strength increased was from 51 to 58 °C.

Fig. 3 Variation of tensile strength with layer thickness

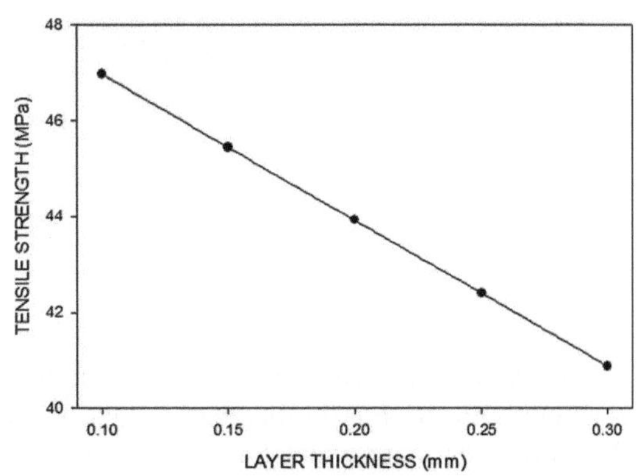

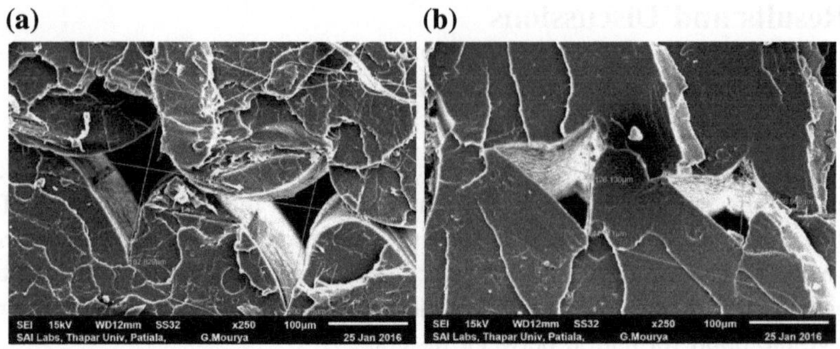

Fig. 4 SEM image of tensile failure of part, **a** void dimensions for 0.2 mm layer thickness, **b** void dimensions for 0.1 mm layer thickness

Fig. 5 Variation of tensile strength with part bed temperature

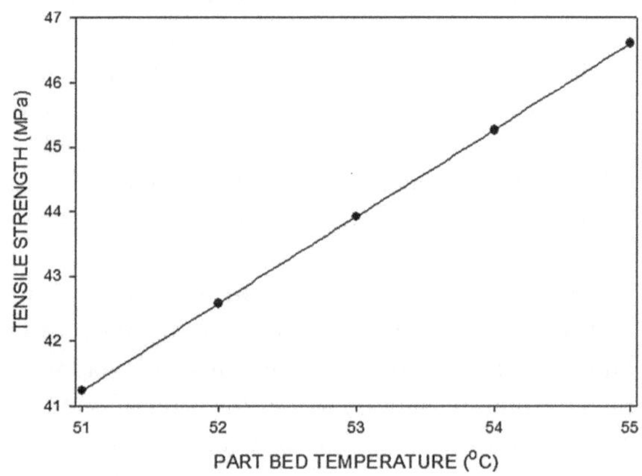

Fig. 6 Variation of tensile strength with nozzle diameter

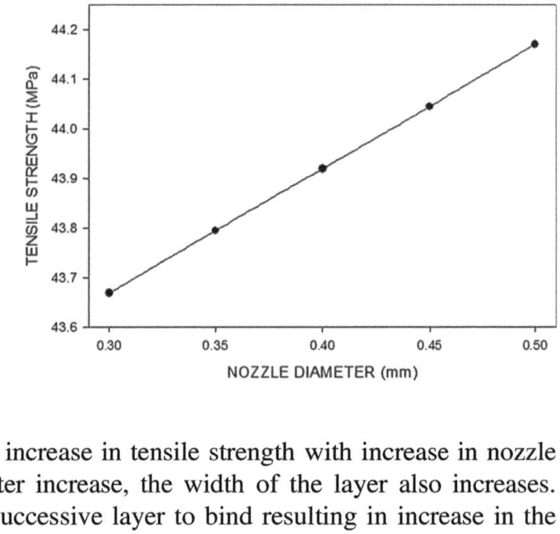

Figure 6 shows that there in increase in tensile strength with increase in nozzle diameter. As the nozzle diameter increase, the width of the layer also increases. This provides more width for successive layer to bind resulting in increase in the

Fig. 7 SEM image of tensile specimen showing the rupture of fibres

tensile strength. But the increase in strength is very less hence it is not a significant factor for tensile strength.

Figure 7 shows the SEM image of fractured surface. It can be clearly observed that failure has been caused because of rupturing due to pulling of fibres and rasters. It can be further observed that the material breakage of raster occurs in a plane approximately normal to a tensile stress.

3.2 Analysis for Flexural Strength

From Fig. 8, it can be observed that flexural strength of the part increases as the layer thickness increases. In flexural testing of the parts, the direction of force is normal to the direction of layers. Here raster orientation is 45°/−45° in all the parts which are fabricated. In case of flexural testing of the part, strength increases with the increase of raster angle and layer thickness. In flexural testing the top surface layers of the part will observe compressive stress and lower surface layers will observe tensile stress. The layer bonding of tensile stress side layer is slightly strong

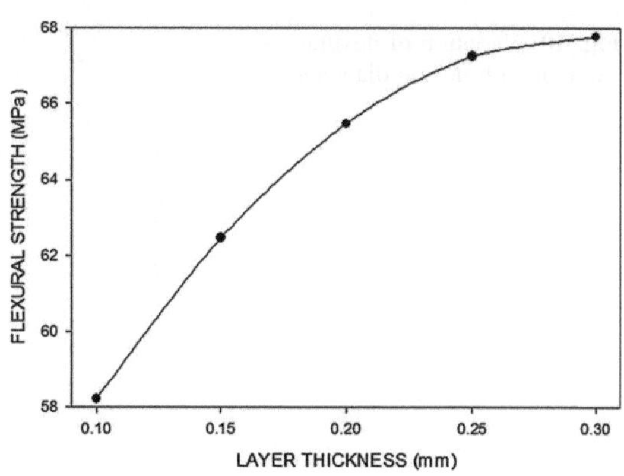

Fig. 8 Variation of flexural Strength with layer thickness

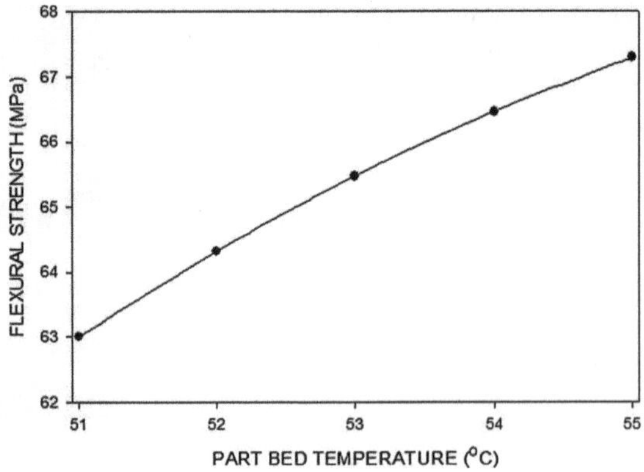

Fig. 9 Variation of flexural strength with part bed temperature

because of high temperature gradient from bottom to top layers. Due to increase in layer thickness with higher raster angle, reduction in the distortion of the layers because elevated raster angles generate smaller rasters which are difficult to bend [14].

Figure 9 shows the consequence of part bed temperature on the flexural strength of the part. It can be observed that flexural strength of the part increases with increase in part bed temperature. The reason for increase in strength is same as in case of tensile strength, which is the increase in heat dissipation from one layer to another. This leads to post heating of layers which are already bonded and results in improved strength.

Figure 10 shows the outcome of nozzle diameter on the flexural strength of the part. It shows that flexural strength decreases with increase in nozzle diameter firstly and then increases with increase in nozzle diameter. This may be because as the nozzle diameter initially increased, there were large voids observed between the successive layering, which despite more surface contact area, results in inferior strength. At larger nozzle diameter, these voids decreased in dimensions hence the increase in flexural strength.

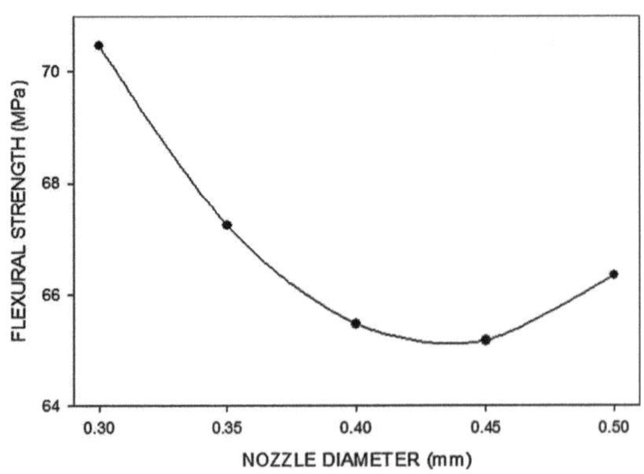

Fig. 10 Variation of flexural strength with nozzle diameter

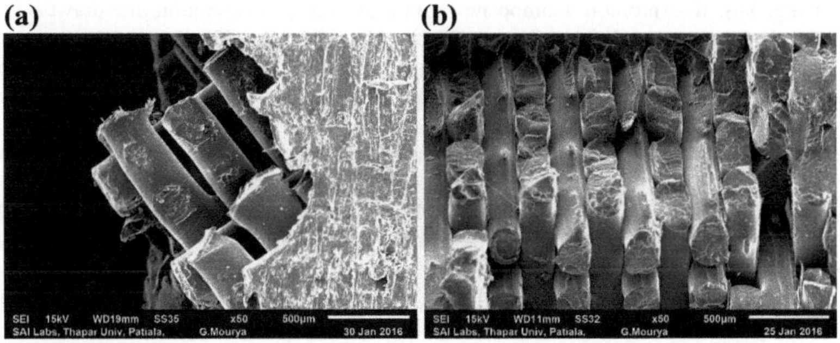

Fig. 11 SEM image of flexural specimen **a** side view, **b** top view

Fracture behavior of flexural specimen reveals that failure starts at the bottom side of the part, that is, the tensile side of the specimen which is shown in Fig. 11. However the pieces are seized together by non fragmented fibres of top side and the crack propagation in the direction of load is nearly straight. Compression side that is the top side of the specimen breaks by bending of the fibres. It can be further observed that the failure is caused because of tearing and breaking of rasters and material.

4 Conclusions

The relationship for layer thickness, part bed temperature, and nozzle diameter with respect to tensile strength and flexural strength for PLA has been analyzed in this paper. It has been observed that as the layer thickness increases, tensile strength decreased where as flexural strength increased. As the part bed temperature increased, tensile strength and flexural strength also increased. With respect to nozzle diameter, tensile strength increased while flexural strength initially decreased and then increased with increase in nozzle diameter. The failure in tensile testing has been caused because of rupturing due to pulling of fibres. Fracture behavior of flexural specimen reveals that failure starts at the bottom side of the part of the specimen.

References

1. D.T. Pham, S.S. Dimov, *Rapid Manufacturing. The Technologies and Applications of Rapid. Prototyping and Rapid Tooling* (Springer, London, 2001)
2. S.H. Ahn, M. Montero, D. Odell, S. Roundy, P.K. Wright, Anisotropic material properties of fused deposition modeling ABS. Rapid Prototyp. J. **8**(4), 248–257 (2002)

3. D. Karalekas, K. Antoniou, Composite rapid prototyping: overcoming the drawback of poor mechanical properties. J. Mat. Process. Tech. **153–154**, 526–530 (2004)
4. K.C. Ang, K.F. Leong, C.K. Chua, Investigation of the mechanical properties and porosity relationships in fused deposition modelling-fabricated porous structures. Rapid Prototyp. J. **12**(2), 100–105 (2006)
5. P.K. Jain, P.M. Pandey, P.V.M. Rao, Effect of delay time on part strength in selective laser sintering. Int. J. Adv. Manuf. Tech. **43**, 117–126 (2009)
6. A. Bagsik, V. Schoeppner, E. Klemp, FDM part quality manufactured with Ultem* 9085, in *Proceedings 14th International Scientific Conference on Polymeric Materials 2010*, Halle (Saale) (2010)
7. G. Percoco, F. Lavecchia, L.M. Galantucci, Compressive properties of FDM rapid prototypes treated with a low cost chemical finishing. Res. J. Appl. Sci. Engg. Tech. **4**(19), 3838–3842 (2012)
8. D. Croccolo, M. De Agostinis, G. Olmi, Experimental characterization and analytical modelling of the mechanical behaviour of fused deposition processed parts made of ABS-M30. Comput. Mater. Sci. **79**, 506–518 (2013)
9. B.M. Tymrak, M. Kreiger, J.M. Pearce, Mechanical properties of components fabricated with open-source 3-D printers under realistic environmental conditions. Mater. Des. **58**, 242–246 (2014)
10. F. Afrose, S.H. Masood, P. Iovenitti, M. Nikzad, I. Sbarski, Effects of part build orientations on fatigue behavior of FDM-processed PLA material. Prog. Addit. Manuf. https://doi.org/10.1007/s40964-015-0002-3
11. D.A. Roberson, A.R.T. Perez, C.M. Shemelya, A. Rivera, E. MacDonald, R.B. Wicker, Comparison of stress concentrator fabrication for 3D printed polymeric izod impact test specimens. Addit. Manuf. **7**, 1–11 (2015)
12. http://www.natureworksllc.com, TechnicalDataSheet _ 4043D _ 3D -monofilament_pdf
13. A.K. Sood, R.K. Ohdar, S.S. Mahapatra, Parametric appraisal of mechanical property of fused deposition modelling processed parts. Mater. Des. **31**, 287–295 (2010)
14. H. Persson, K. Adan, "Modeling and experimental studies of PC/ABS at large deformations," Master's Thesis, Division of Solid Mechanics, Lund University, Lund, Sweden, (2004)

Metal Powder Based Additive Manufacturing Technologies—Business Forecast

Vadlamannati Sriram, Vipin Shukla and Soumitra Biswas

Abstract Additive manufacturing (AM), a layer-wise manufacturing technique using CAD model and raw material as input poses new challenges to conventional manufacturing. Without any efforts required for material removal and tooling, the product build time and material consumption can be controlled and estimated effectively with high accuracy and quality. Further, AM has the necessary capability of manufacturing products as per their final dimensional and functional needs in one operation thereby cutting down the subsequent finishing and assembling steps. AM thus reduces total time for product manufacturing simultaneously eliminating capital and operational costs for further steps, inventory and material handling. However, process and material challenges are the techno-economic barriers for adapting AM in industries for commercialization. The paper has specifically chosen metal powders for analysis as their properties and process capabilities are better than non-metals. The data collected on the present status of the AM technologies globally (related to the sale of 3D printers both in numbers and revenues) based on the various reports are analysed to forecast market of 3D printers for the next 10 years. Final conclusions are made by predicting the values using regression technique. The paper analyses the global patent trends during last 15 years in various countries so as to assess the futuristic AM technologies.

Keywords Additive manufacturing · Metal powder · Regression
Patent analysis

V. Sriram · V. Shukla (✉) · S. Biswas
Technology Information Forecasting Assessment Council (TIFAC),
New Delhi, India
e-mail: shukla.vipin@gmail.com

V. Sriram
e-mail: sriram.srikiran@yahoo.com

S. Biswas
e-mail: soumitra2000@gmail.com

L. J. Kumar et al. (eds.), *3D Printing and Additive Manufacturing Technologies*,
https://doi.org/10.1007/978-981-13-0305-0_10

1 Introduction

The American Society for Testing and Materials (ASTM International) defines additive manufacturing as—"Process of joining materials to make object from 3D model data, usually layer upon layer, as opposed to subtractive manufacturing methodologies" [1].

Additive manufacturing (AM), a layer -wise manufacturing technique using CAD model and raw material as input, involves integration of a computer with a machine which builds the required model by performing series of operations. AM has faster rate for product development and permits dynamic changes in design and processing [2]. It also adds features such as design freedom, effective estimation of both materials and time required for finishing the parts. Recycling and eco-friendliness are the other reasons for AM to be commercially attractive benefitting the business growth and also for reducing the pollution. Figure 1 explains the generic AM process phases.

AM has a history of more than two decades, starting from the concept of Rapid Prototyping in succession to the first stereo-lithography technology which was patented by Charles W. Hull in 1984 and developed by the first rapid prototyping 3D systems in 1986 [3]. With the advancement in NC and CNC machines more AM technologies have emerged with larger application areas. While there are several ways to classify AM technologies, the easiest way could be to classify by type of raw materials used or by the type of process, e.g. laser technology, electron beam and thermal energy. Table 1 broadly presents the AM classification.

The seven AM processes in Table 1 have the capability of processing metals and non-metals in solid, liquid and semi-solid forms. The energy source for AM is either light or heat vis-a-vis friction (mechanical energy), which is the demerit for traditional manufacturing techniques affecting accuracy, quality and cost due to metal on metal machining. AM ensures high accuracy and quality with all the necessary dimensional and functional requirements in one manufacturing stage.

Fig. 1 AM process phases

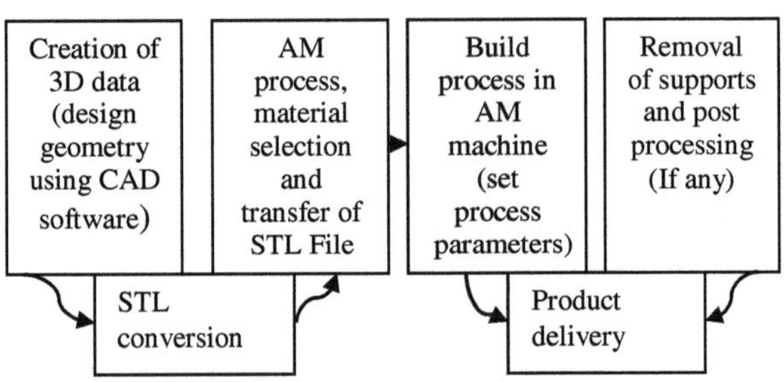

Table 1 AM classification [1, 14]

AM classification	Details
Material in liquid form	**UV curable resins/plastic, ceramic suspension, wax**
Associated AM binding technology	Photo polymerization and cooling—SLA, MJM
Material feed mechanism	Liquid resin in a vat (SLA), liquid polymer in jet (MJM)
Layering mechanism	Laser scanning, light projection, Ink-jet printing
Applications	Proto types, casting patterns
Material in powder form	**Thermoplastics, wax, metal powders, ceramic powders**
Associated AM binding technology	Partial melting (SLS), full melting (SLM, EBM, LMD, 3DP)
Material feed mechanism	Powder in bed, injection through nozzle
Layering mechanism	Laser scanning, electron beam scanning, binder printing
Applications	Prototypes, casting patterns, tooling, functional parts, repairing
Material in filament/paste	**Thermo plastics, wax, ceramic slurries**
Associated AM binding technology	Solidification—FDM, FFF, Robo casting
Material feed mechanism	Filament melted in nozzle (FDM), slurry in nozzle
Layering mechanism	Extrusion and deposition
Applications	Prototypes, casting patterns, Functional parts
Material in solid	**Paper, Plastic, Metal**
Associated AM binding technology	Bind sheets—LOM
Material feed mechanism	Rollers
Layering mechanism	Feed sheets are bonded by adhesive and cut using laser
Applications	Prototypes, casting models

AM process terminologies: *SLA* Stereo lithography, *MJM* Multi jet modelling, *SLS* Selective laser sintering, *SLM* Selective laser melting, *EBM* Electron beam melting, *LMD* Laser metal deposition, *3DP* 3D printing, *FDM* Fused deposition modelling, *FFF* Fused filament fabrication, *LOM* Laminated object manufacturing

2 AM Challenges—Techno-economic Barriers

Many research organizations of various countries prepared technology road maps [3] for additive manufacturing wherein they mentioned the barriers to additive manufacturing for its commercialization and ready replacement of conventional manufacturing techniques. Table 2 summarizes the technology challenges.

Economics for AM is difficult to ascertain for the wide ranging machine prices (from US$1000 for personal 3D printer to over US$73,000 for industrial 3D printer) [4]. Compared to conventional methods, additive manufacturing has the potential of reducing direct and indirect costs such as capital expenditure, machine footprint, manpower, material inventory and handling, product finishing, assembling, etc. While the initial investment would be high for AM system, its long-term

Table 2 AM challenges [6, 14, 15, 19, 20]

Technology challenges	Description
Process modelling	Software, process algorithms for process control to have consistency
Process control	Sensors, closed loop systems for controlling process parameters e.g. controlling laser power, controlling molten pool size, maintaining uniform part bed temperature
Process certification/ qualification	Non-availability of certified process parameters for a particular part and no certified specific testing procedure
Design	No specialized integrated software
Control of material characteristics	Powder particle size and distribution, microstructure
Material handling	Effect on mechanical and thermal properties due to phase transformations, environment and post processing
Material certification/ qualification	Non-availability of certified material characteristics and properties for a particular process
Standards	Non-availability of standards for both process and material, technology classification etc.

usage could be economically attractive. Another deterrent for large-scale adoption of AM technology has been the high cost and less availability of suppliers for metal powders.

3 Metal Powder for AM

The first rapid prototyping stereo-lithography technology used UV curable resins as the raw materials and thereafter research has much contributed in the processing of plastics. Metal AM was introduced during 1990–2000. Presently for additive manufacturing, 29 metallic materials are available; these include stainless steel, tool steels, Ti and its alloys, Inconel, Co–Cr alloy, Al alloys, etc. [5]. Process parameters and powder reuse are important concerns for achieving the required density and mechanical properties of the product. Densities and mechanical properties of parts made by AM are comparatively equivalent to parts manufactured by conventional techniques [6]. Metal powders of Ti Alloy and Co–Cr alloy find great applications in the bio-medical sector such as dental implants, prosthetics, etc. Table 3 gives the information about the AM metal powders collected from industrial material data sheets.

Metals have high melting points, mechanically stable on loading conditions. They are bio-compatible when compared to non-metals, have conductivity for heat, light and electricity and magnetic properties.

For precious metals like gold, silver and platinum for jewellery industries AM could be the ideal processing technique for design freedom, optimal material consumption and light-weight jewellery.

Table 3 AM Metals [16]

Metal	Presently used	Application area	Advantage
Ti alloys	Ti–6Al -4 V, Cp Ti	Medical, dental, aerospace, defence	High bio-compatible properties
Co alloys	Co–29Cr–6Mo, Co Base	Dental, medical prosthetics, investment casting, gas turbines	High strength, corrosion and wear resistant, with high hardness
Stainless steels and tool steels (Excellent hardness)	316L, PH-17, H13*	Oil and gas industry, tools*	Good mechanical properties and corrosion resistance
Inconel	625,718	Non magnetic, corrosion resistant	Turbine engine components, chemical process equipment and fuel exhaust systems
Al alloys	Al–Si–10 mg, Al–Si2	Aerospace, automotive	Low weight, good thermal and mechanical properties
Cu alloys	Cu–Sn10/ Bronze	Cooling channels and tool inserts	Good thermal conductivity

*Indicates—Tool Steel material and its application

3.1 Metal AM Capabilities

AM helps in attaining stability under high temperatures and phase transformation due to dynamic control of input energy and process environment. Metal additive manufacturing has greater advantages than using plastics and other materials as the metals have higher laser absorption power and stability under higher temperatures as compared to plastics (Table 4).

Table 4 Metal AM capability comparison

Additive manufacturing	Traditional manufacturing
Properties of a part can be varied from section to section	Not possible
No further operations required for making mechanical fits	Further operations or machining required to achieve best fits; not possible in one step
Functional assemblies	Not possible
High parts accuracy; not depending on machine setting like centring of the chuck, work piece setting	Not sure, will depend on the setting of tool and work piece
No metal to metal surface contact, no damage and loss	Contact necessary; damage to tool and work piece
Good mobility of AM systems for repair and/or production	Difficulty in mobility

4 AM Forecast

Forecast is an essential tool, which helps industries plan their resources and shape their business strategies. For researchers and government organizations, forecasting provides an idea on the present status and future scenarios of technology and business for investment decisions. It also helps researchers to understand futuristic technology developments and actions required therein.

The data were collected from reports on AM revenues and 3D printer units sold [4, 7, 8] for a complete understanding of the important factors, which helped in correlating the quantities. The factors considered are mainly year-wise revenues generated for AM products, equipment and AM material, AM service revenue, units of personal and industrial AM sold, etc. The objective has been to forecast the number of industrial metal AM units to be deployed for the next 10 years. The multiple regression technique was selected as the variable count has been more than one. This technique provides better prediction form multiple predictors, examines sophisticated research hypothesis with broader set of data available. Using multiple regressions we can also obtain the unique and individual contribution of independent variables on dependent variable.

The reason for choosing AM Service revenue as an independent factor compared to others as this has recorded good growth. Another factor like AM Metal material revenue has been the best independent variable in correlation for the Metal AM forecast.

The revenue estimates were calculated using compounded annual growth rate (CAGR) formulae. All the available data and estimated data are shown in Table 7. Table 5 shows the industrial commitments and opinions on 3D printing for the next 10 years. Figure 2 explains the AM market share sector wise. AM metal revenue (X_1) is the amount of metallic material sold purposely for additive manufacturing from the suppliers. AM service revenue (X_2) covers rapid prototyping services, customised production and maintenance of 3D printing. Based on X_1, X_2 as independent variables, metal 3D printer sales (Y) can be the resultant output. Hence multiple regressions on X_1 and X_2 were applied to forecast Y and trend is plotted as shown in Fig. 3 and Table 6.

4.1 *Revenue Estimates for 2016–25 Using CAGR*

Before applying multiple regressions on the independent variables to forecast metal 3D printers for the next 10 years, complete data set of independent variables X_1 and X_2 during 2009–2025 were collected. CAGR was calculated for 2009–2015 by the following formula:

Table 5 AM market remarks [17]

Company name	Imp Remarks
Siemens	Predicts that 3D printing will be 50% cheaper and up to 400% faster in the next five years. Siemens is also predicting the global market for 3D printing will reach €7.7B ($8.3B) by 2023
Boeing	Already installed over 200 different types of flying production parts on 16 different aircraft
GE aviation	Designed an entirely new nozzle that integrates 18 separate parts into one. The new nozzle is 25% lighter and 5 times more durable
3D systems	Sales into design and manufacturing increased 27% from 2013 to 2014, growing to $609.8 M in sales. Sales into healthcare increased 80%, from $71.7 M in 2013 to $129.3 M in 2014. The consumer segment of 3D systems business grew by 26% in the last year, from $34.8 M in 2013 to $43.8 M in 2014
Gartner	Prototyping (24.5%), product development (16.1%) and innovation (11.1%) are three most common reasons for companies to pursue printing
PWC	PwC estimates 67% of manufacturers are already using 3D printing. Of these, 28.9% are exploring how 3D printing can be optimally integrated into their production processes. 24.6% are using 3D printing for prototyping

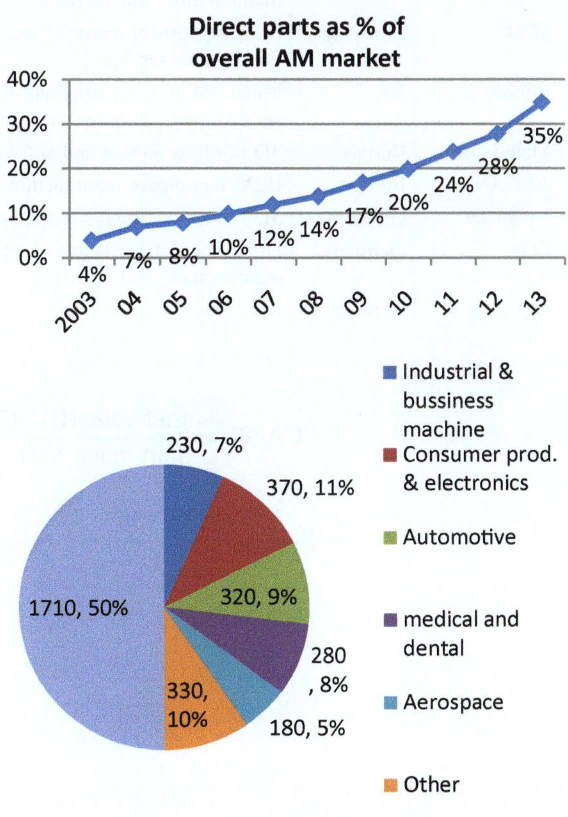

Fig. 2 *Source* Wohler's associates report 2013 [13], sector wise AM market: Roland Berger report 2013 [7]

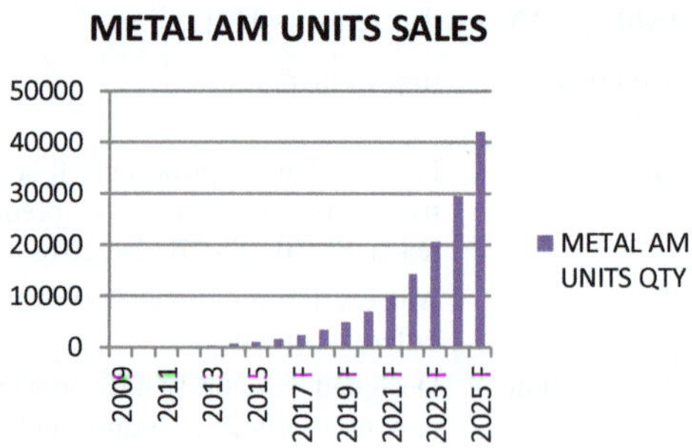

Fig. 3 Metal AM units sales and forecast *F*

Table 6 Global AM revenues [18]

Company	Country	Business domain	Revenue in million $
Stratasys	USA	Polyjet abd FDM—two technologies being pursued for equipment and product manufacture	750
3D systems	USA	SLA, SLS, Direct metal 3D printing—equipments manufacturer, materials and services	650
Exone	USA	Binder jetting (*S* print, *M* Print)—equipment manufacturer and services	43.9
SLM	Germany	SLM equipment manufacturer and Rapid prototyping service	36
Renishaw	UK	Metal 3D printer equipment manufacturer and service, metal powder supplier	494
Materialise	Europe	3D printing service and software	81
ARCAM	Europe	EBM equipment manufacturer	39
Voxel Jet	Germany	3D printer equipment manufacturer and services	18
EOS	Germany	Direct metal laser sintering equipment manufacturer and service	49
Alpha form	Germany	Services	30

$$\mathrm{CAGR} = \frac{\mathrm{End\ value(EV)}}{\mathrm{Start\ value(SV)}}^{\frac{1}{\mathrm{Time}(T)}} - 1$$

$$X_1(\mathrm{CAGR}) = \frac{156^{\frac{1}{7}}}{12} - 1 = 0.44(44\%)$$

$$X_2(\text{CAGR}) = \frac{3824.35^{\frac{1}{7}}}{918.44} - 1 = 0.23(23\%)$$

The following formula was used to estimate X_1 and X_2 for 2016–2025

$$\text{Estimate Value} = \text{SV} + (1 + \text{CAGR})^T$$

$$X_{1,2016}(\text{Estimate Value}) = 12 + (1 + 0.44)^8 = 221.86$$

$$X_{2,2016}(\text{Estimate Value}) = 918.44 + (1 + 0.23)^8 = 4811.62$$

4.2 Multiple Regression

$Y_{cal} = A + BX_1 + CX_2$ is the multiple regression Eq. (1) [9].

X_1 and X_2	Independent variables
A	'*y*' Intercept
B	The change in *Y* for each increment in X_1
C	The change in *Y* for each increment in X_2
N	Total no of observations, 7

Using the equation the positive or negative relation between the Metal 3D printer sales to revenues was observed as generated by conducting regression for $R^2 = 0.995$ and Multiple regression coefficient (R) = 0.998. The value of R being close to +1 infers that the two independent variables have strong positive/direct relationship with *Y*. Further, coefficients *A*, *B* and *C* were determined using respective formulas and found $A = -13.866$, $B = 6.954$, $C = 0.033$.

Substituting these coefficients in (1) equation becomes

$$Y_{cal} = -13.866 + (6.954)X_1 + (0.033)X_2$$

Calculating *Y* for 2016,

$$Y = -13.866 + (6.954)\,(221.86) + (0.033)\,(4811.62) = 1688\,\text{Units}$$

Similar exercise repeated for the years 2017–2025 and the sales forecast (*Y*) obtained is shown in Table 7. Figure 3 shows Metal 3Dprinter units.

Table 7 Forecast table

	CAGR 44%	CAGR 23%	
Years/revenues in US$M	Metal supplies (X_1)	AM services (X_2)	Metal AM Units (Y)
2009	12.00	918.44	115
2010	13.50	1232.72	125
2011	18.00	1737.44	160
2012	24.90	2187.67	190
2013	32.60	2638.11	345
2014	104.00	3156.37	790
2015	156.00	3824.35	1212
2016	221.86	4811.62	**1688**
2017	319.48	5918.30	**2403**
2018	460.05	7279.51	**3426**
2019	662.47	8953.79	**4888**
2020	953.96	11013.16	**6983**
2021	1373.71	13546.19	**9986**
2022	1978.14	16661.82	**14292**
2023	2848.52	20494.03	**20471**
2024	4101.86	25207.66	**29342**
2025	5906.68	31005.42	**42084**

5 AM Patents

Study of patents helps understand contributions by the researchers in a specific field. Patent studies also give directions to technology updates and assess existing technology gaps. The key words such as additive manufacturing, 3D printing, selective laser melting, electron beam melting, direct metal deposition, metal powders for additive manufacturing were used to search abstracts and full texts in the USPTO, WIPO database [10, 11]. Tables 8 and 9 represent the collected data related to additive manufacturing and 3D printing patent publications year wise to country wise.

B33Y is the new patent group introduced in international patent classification in 2015 for mainly specifying the patents under additive manufacturing [11]. Table 9 shows subgroups of B33Y and patents filed therein.

Keyword query syntax given in patent database for searching patent publication.

ALLTXT:(EN_ALLTXT:((((3D Near Printing) OR (Additive Near Manufat*) OR ((Selective OR Electron Beam OR Direct metal) AND (“laser melting”)) OR (“binder jetting” Near2 model*) OR (“Three dimension” Near Printing))))) AND DP:[01.01.2000 TO 01.01.2016]

Table 8 Regional metal AM patent publications trend [10]

Year	USA	PCT	EP	CD	CN	KR	OTH
2000	62	19	17	13	1	1	7
2001	98	57	23	23	–	–	9
2002	177	32	31	13	1	1	7
2003	161	46	33	30	–	–	7
2004	210	59	42	16	3	1	5
2005	209	57	42	21	7	1	25
2006	217	67	43	25	7	4	17
2007	198	83	52	19	5	4	18
2008	276	127	43	31	9	2	14
2009	245	87	54	31	14	7	4
2010	295	125	48	48	14	6	8
2011	356	160	60	37	6	7	8
2012	430	203	60	19	14	10	18
2013	592	313	66	19	53	14	27
2014	1005	696	116	15	425	12	15
2015	1781	1242	184	9	666	26	28
Sum	6312	3373	914	369	1225	96	217

Table 9 AM patents group wise (2000–2016) [10, 11]

IPC	Patents
B29C: Covers technologies related to joining and shaping of plastics	2034
B29C 33/000: Moulds or cores	
B22F: Manufacturing of articles from metallic powders	696
G06F: Digital data processing related technologies	1178
B41J: Typewriters; selective printing mechanisms, e.g. ink-jet printers, thermal printers	838
B33Y: Additive manufacturing, i.e. manufacturing of three-dimensional [3-d] objects by additive deposition, additive agglomeration or additive layering, e.g. by 3-d printing, stereo lithography or selective laser sintering [USPTO]	
B33Y10/00: Processes of additive manufacturing	22
B33Y30/00: Apparatus for additive manufacturing; details thereof or accessories therefore	35
B33Y40/00: Auxiliary operations or equipment manufacturer, e.g. for material handling	3
B33Y50/00: Data acquisition or data processing for additive manufacturing	4
B33Y70/00: Materials specially adapted for additive manufacturing	7
B33Y80/00: Products made by additive manufacturing	3
B33Y99/00: Subject matter not provided for in other groups of this subclass	–

Total of 12,506 patents were published from 2000 to 2016 and main countries have been USA, PCT (Patent Cooperation Treaty), EP (Europe), JP (Japan), CN (China), KR (Korea), CD (Canada), OTH (Others including Russia, Germany, etc.).

6 Conclusions

Metal additive manufacturing would enjoy very high potential application in next 10 years with the sale of total units around 42,000 as forecast in this article. The AM services revenue of approx US$ 31,000 million interprets that most of the service industries would move towards this technology. While for the first 10 years (2009–2020) over 20,000 units of metal additive manufacturing units would be sold, its sale would hit another 20,000 units in the next 5-year period itself as observed in Table 7. The reasons for this drastic increase will be the selling price and intellectual property availability of the technology. Already many public funded research organizations and industries are working on reducing the system costs. According to the Gartner forecast, average selling price for metal AM system has fallen by 23% over the last 10 years and would reduce more in the coming years. As many patents are getting expired by 2015 (patents of MIT and Stratasys, 3D systems on SLA, SLS already expired in 2014), new comers would have better opportunities to enter the technology area.

Figure 4 shows the progress of USA in R&D activities in the field of additive manufacturing. USA has an organization named "America makes" [12] for dedicated work on R&D aspects for AM and generating their technology roadmaps. US initiatives should be benchmarked by other countries for making forays into AM technology.

The number of patent publications as given in Table 9 infers that most of the work is currently focused on plastics as raw material (B29C) having 2034 patents. Only 696 patents were under B22F category, which signifies the usage of metal powder for additive manufacturing. Hence, the metal additive manufacturing technology is still at the nascent stage thus offering early entry opportunity for researchers.

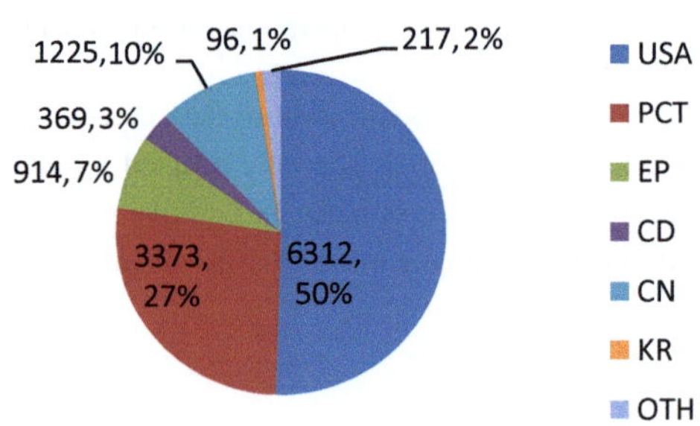

Fig. 4 Country-wise metal AM patent publication

The other groups G06F, B41J, which deal with the data processing such as STL file conversion from CAD to STL and STL file errors when transferred to machine, etc. and B41J is about processing mechanism and parameters for a particular process. Research on to these groups helps overcome the existing technical challenges for both the process and material thus reducing the barriers for commercialization.

While all the above predictions and analysis promise bright future for the additive manufacturing, commercialization of the products and machines are at early stage due to the technology gaps as explained in Table 2.

Lack of standards and certification procedure of the products and equipment for metal AM are other crucial factors for their adoption by many of the industries and product users.

References

1. ASTM F42 Committee on Additive manufacturing technologies. Available: http://www.astm.org/COMMITTEE/F42.htm
2. K. Jain Prashant, K. Senthilkumaran, M. Pandey, P.V.M. Rao, Advances in materials for powder based rapid prototyping, in *Proceedings of International Conference on Recent Advances in Materials and Processing*, PSG-tech Coimbatore, India, 15–16 Dec 2006
3. I. Gibson, D. Rosen, B. Stucker, *Additive Manufacturing Technologies*, 2nd edn. (Springer, New York) ISBN: 978-3-662-46089-4 (ebook)
4. C. mewalla, C. wood, 3D printing who's who and who's impacted, CM research report on d SYNC. Issue No. 60, 9 Apr 2013
5. M. Qian, Metal powder for additive manufacturing. JOM **67**(3), 2015, https://doi.org/10.1007/s11837-015-1321-z (The Minerals, Metals & Materials Society)
6. J.-P. Kruth, M. Badrossamay, E. Yasa, J. Deckers, L. Thijis, J. Van Humbeck, Part and material properties in selective laser melting of metals, in *Proceedings of the 16th International Symposium on Electro machining* (*ISEMXVI*), Shanghai, China, 19–23 Apr 2010
7. H. Bikas, P. Stavropoulos, G. Chryssolouris, Additive manufacturing methods and modelling approaches: a critical review, Int. J. Adv. Manuf. Technol. (2015), https://doi.org/10.1007/s00170-015-7576-2 (Springer)
8. R. Berger report on Additive manufacturing: a game changer for the manufacturing industry? Munich November 2013
9. J. Higgins (ed.) *The Radical Statistician: A Beginners Guide to Unleashing the Power of Applied Statistics in The Real World*, 5th edn. (2006)
10. WIPO patent scope data base result: https://patentscope.wipo.int/search/en/result.jsf
11. USPTO patent database: http://appft.uspto.gov/netahtml/PTO/search-adv.html
12. America makes established (2012). Available: https://americamakes.us/about/overview
13. Image of Wohler's associates report 2013. Available: https://blog.adafruit.com/wp-content/uploads/2013/05/WohlersAssociates.jpg
14. Additive Manufacturing: Pursuing the Promise—DOE's Advanced Manufacturing Office (AMO). 672. Available: https://www1.eere.energy.gov/manufacturing/pdfs/additive_manufacturing.pdf
15. K. Jurrens, Measurement Science Roadmap for Metal-Based Additive Manufacturing, National Institute of Standard and Technology (2013). Available: http://events.energetics.com/NIST-AdditiveMfgWorkshop/index.html

16. DMLS material specifications for AM metals. Available: https://www.stratasysdirect.com/materials/direct-metal-laser-sintering/
17. L. Columbus, 2015 roundup of 3D Printing market forecasts and estimates, Forbes/tech report, 31 Mar 2015
18. AM Company wise revenues Available: http://3dprintingindustry.com/2015/03/19/fiscal-2014-revenue-results-3d-printings-top-10-guns/
19. W.E. Frazier, Metal additive manufacturing: a review. J Mater. Eng. perform. JMEPG (2014) **23**(6), 1917–1928, 8 Apr 2014
20. K.P. Cooper, Layered manufacturing: challenges and opportunities, in *Material Research Society Symposium Proceedings*, vol. 758:23–34, Boston, Massachusetts, 3–5 2002

Design and Development of Drug Delivery System for Chronic Wound Using Additive Manufacturing

Mohan Pushparaj, Rajesh Ranganathan and Sivakumar Ganesan

Abstract The paper focuses on applying additive manufacturing in a healthcare sector. In healthcare wound healing remains a challenging clinical problem, for chronic wounds rather than in acute wound. It shows that chronic wound would take much time to heal. In general wound dressing product available in market for chronic wounds have irregular pores. The model of the wound dressing is categorized depending upon the type of wound; size of the wound. The product will be customized for person to person. Through additive manufacturing technology customization of the product can be provided. The aim of this paper is to design and develop a wound dressing model with pores for chronic wound using additive manufacturing technology. The outcome of the paper would result in customized product for chronic wound.

Keywords Additive manufacturing · Drug delivery · Acute and chronic wound · Wound healing

1 Introduction

Additive manufacturing is a layer-by-layer manufacturing of end-user product directly from CAD model [1].The main advantages of the additive manufacturing are efficient use of raw material and good geometrical accuracy [2]. In the additive manufacturing a design in form of CAD file convert into STL file can be directly manufactured as a finished product without need of cutting tool and fixtures [3]. As a result of Additive manufacturing shows the possibility of make parts with complexity and design innovation.

M. Pushparaj (✉) · R. Ranganathan · S. Ganesan
Department of Mechanical Engineering, Coimbatore Institute of Technology, Coimbatore 641014, India
e-mail: mohanpus@gmail.com

R. Ranganathan
e-mail: drrajeshranganathan@gmail.com

L. J. Kumar et al. (eds.), *3D Printing and Additive Manufacturing Technologies*,
https://doi.org/10.1007/978-981-13-0305-0_11

Through additive manufacturing customization can be supported, which produces the product in respond to the customer requirement, availability of customization, enabling low volume production to be achieved efficiently [4].

Additive manufacturing technology is successfully implemented in the sector like automotive, aerospace and health care [5]. Additive manufacturing technology applied in healthcare industry involves the direct conversion of 3D data into product for consumer [6]. By using this technology, customization of healthcare product for individual's needed is satisfied. Healthcare is a key sector finding its ground for supporting individuals physical and mental health, and makes comfort to the individuals around the world [7].

Among the health care practices the most predominant vital are the trauma care or wound care [8]. The wound can be classified into acute and chronic wound which is shown in Fig. 1. The wounds are categorized based on their healing time, acute wounds are the ones that undergoes step by step healing within forecasted time period [9]. Meanwhile, chronic wounds are the ones that does not follow the healing stages which resulting in a prolonged wound period [10]. Frykberg and Banks [11] through their research shown that the chronic wounds failed to heal within four weeks and subsequently there is no improvement, and requires further treatment to heal the wound.

The product development for these treatments play a major role in the treatment of chronic wounds and is being scrutinized. The need for the development of product for wound healing because of normal wound dressing having irregular pores and also provides a customized based on wound size and type of wound.

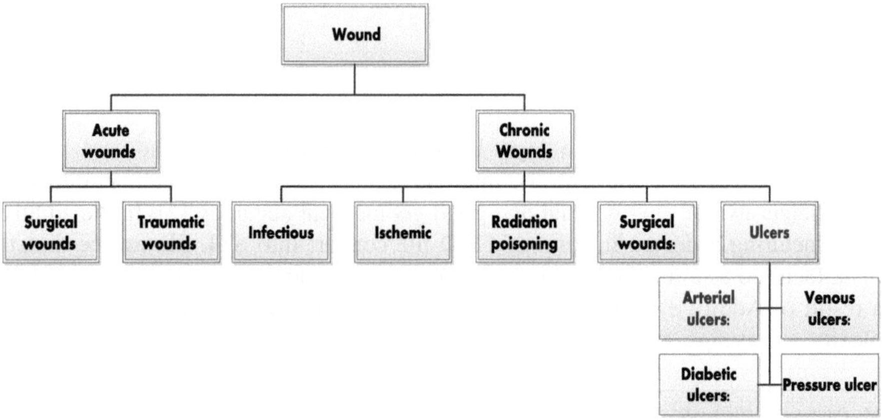

Fig. 1 Types of wounds

2 Literature Review

Additive manufacturing technique is applied in several fields it can be used for verification of a design of the object or to make functional test. Furthermore, investigators have studied about the fabrication of functional parts that can be custom fit and developed to incorporate various drug delivery technologies.

Chu et al. [12] conducted a study to fabricate 3D functional parts through Nano composite deposition system using bio compactable polymer resin as material and it is used for bio application. Similarly, Chu et al. [13] through their drug delivery system made it with a bio compatible material and biodegradable thermoplastic which is mixed with anti-cancer drug and bio-ceramic material. The drug delivery test shows that drug release can be controlled up to 50 days by bio-ceramic application. Additive manufacturing printing methods permit to scan and build a physical model of defective bones from patients and make the bone transplant by characters such as density, pore shape, size, and pore interconnectivity which are important parameters as it results with high mechanical strength and reduce inflammation caused by micro debris [14].

Giannatsis and Dedoussis [6] paper point out that additive manufacturing used in the drug delivery, applied in micron scale is expected to make possible fabrication of controllable drug-discharge. Experimental research shows that 3D porous chitosal scaffolds was fabricated with various pores architecture and mechanical properties of various anisotropic with 50 µm diameter [15]. A study conducted by Xu et al. [16] shows a series of micro porous silicon rubber membrane with different pore size, which are used for wound healing in wound dressing.

The experiment is to measure the gas flow of air born particle in open surgical wound then the particles enter the wound with a diameter of less than 5 µm, and it results in reducing the infection rate in open surgical wound [17]. The study confirms that by using pores to control the absorbency of the wound dressing with chemical composition CNF is a potential solution as it increases the wound healing ability [18].

From the literature review, it was observed that pores are more important for healing chronic wound which needs continuous drug delivery system based on level of severity of the wound.

3 Design and Manufacture of the Product

3.1 Modelling

The models of the drug delivery product are designed in the 3D modelling software 'SOLIDWORKS' as per the standards. In which dimensions of the pore size is 0.3 mm or 300 µm as shown in Fig. 3, the primary material used is Vero clear. Figure 2 shows the design phase considering the design specification, criteria and

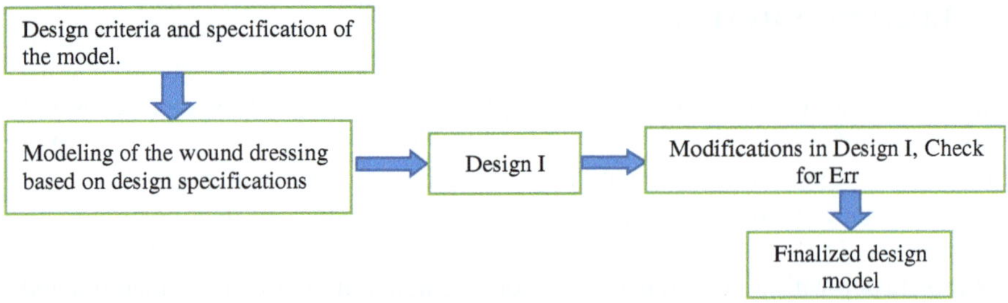

Fig. 2 Design phase of the product

dimensions of the product. Then the finalized model is selected based on the design specifications after checking for its dimensional, functional capabilities.

3.2 Manufacturing Method

Connex 3D printing is a polyjet technology which is like dot-matrix printing, as dot-matrix printing involves dropping of ink on the paper, in this Connex printer jet laying liquid polymer is spread in the build tray then cured with UV light.

By selecting the model material and support material, a part is build. Support material provides as a support with respective projections for manufacture of complex geometries. The support materials are removed easily with water jet. Connex printing machine has many benefits. It gives smooth surface, fine precision and highest accuracy of the model. This machine has material options with maximum of 120 materials. The accuracy of the machine is up to 200 µm for full size models and it depends up on geometry, build parameters and part orientations. The printer has an accuracy of build model with horizontal build layer thickness of 16 µm. This printer gives the ability to print the model with representative aesthetic such as transparency, rigid and flexible. Figure 3 show the designed product developed using the 3D printing AM technology.

3.3 Manufacture of the End Product

Table 1 shows the estimation time of the product to be manufactured and also describes the printing methods for the wound dressing model with pore size of 0.3 mm. There are three types of printing mode, namely high quality, high speed and digital material. By choosing high quality mode the quality of the product are functionally good. For printing the model with 0.3 mm pore size, material consumption for high quality build mode was 8 g for Vero clear digital material and support consumption as 11 g. The estimated time taken for printing the model is 0.28 min.

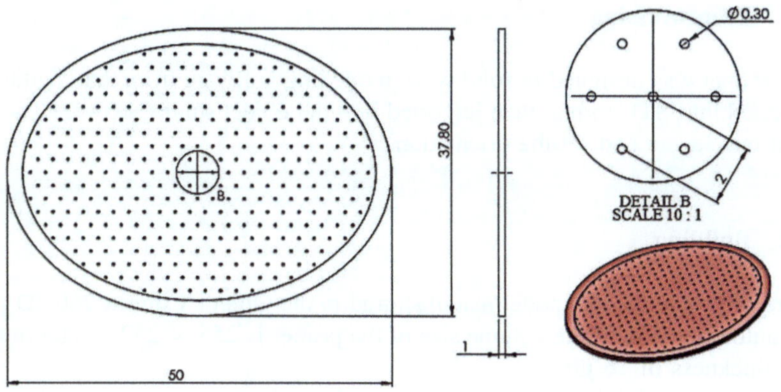

Fig. 3 Dimensions of drug delivery model

Table 1 Details prototypes to be manufactured

S. No.	Wound dressing model with pores size (mm)	Printing mode	RGD 810 Vero clear composite (g)	Support material full cure 705 resin support (g)	Time taken in minutes
1	0.3	High quality	8	11	00.28
2	0.3	High speed	7	10	00.17
3	0.3	Digital material	7	11	00.33

For manufacturing of the product it has undergo three stages (1) Preprocessing, (2) Building, (3) Post processing. The design model is then converted to STL file and checked for errors in the Objet Connex software. The sequence of operation for manufacturing of the prototype is shown in Fig. 4.

Fig. 4 Stages of manufacturing

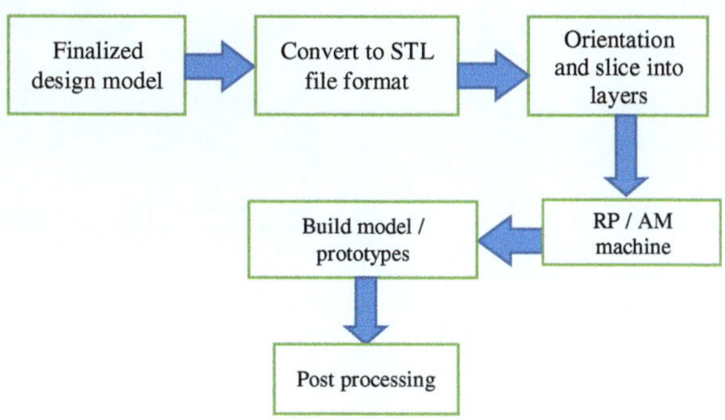

3.3.1 Preprocessing

The product was designed in solid work modelling software from that digital file is converted into STL format then imported into the object studio software The file is sliced into layers and set the orientation.

3.3.2 Building

Figure 5 represents, the model manufactured in the Objet's Connex 260 3D printer with auto orientation. The volume size of the printer is 255 × 252 × 200 mm with layer thickness of 16 μm.

3.3.3 Post Process

These support material were removed in POWERBLAST High-pressure Water Cleaner, with 20–120 bar. Figure 6 finally shows the drug delivery model using additive manufacturing.

Fig. 5 Auto orientation of product

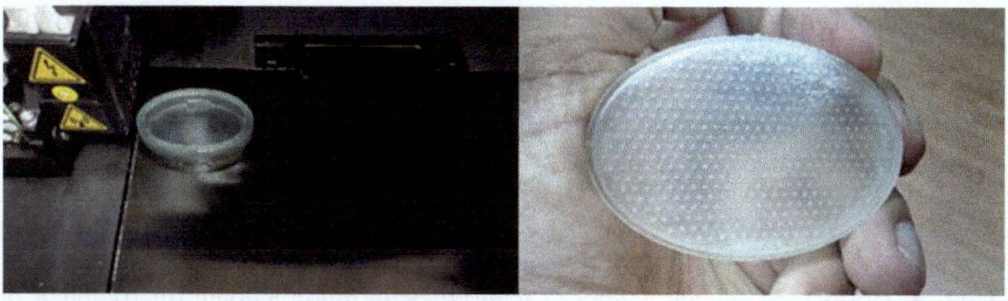

Fig. 6 The wound dressing model

4 Conclusion

This paper identifies the possible difficulties faced in chronic wounds and through additive manufacturing a solution is provided. In order to overcome the difficulties a separate wound dressing model with pores are designed and manufactured. In order to develop the prototype, design criteria's are met and included in the model. The model is then manufactured through additive manufacturing technology, which uses polyjet technology. For this kind of application objet connex 260 3D printing machine is selected. The prototype of wound dressing with 0.3 mm pores are manufactured in 3D printing machine to reduce the wound healing time and provides a customized based on wound size and type of wound.

References

1. S.H. Huang et al., Additive manufacturing and its societal impact: a literature review. Int. J. Adv. Manuf. Technol. **67**(5–8), 1191–1203 (2013)
2. A.R. Mashadi, B. Esmaeilian, S. Behdad, Impact of additive manufacturing adoption on future of supply chains, in *International Manufacturing Science and Engineering Conference (MSEC)* (2015), pp. 1–10
3. V. Petrovic et al., Additive layered manufacturing: sectors of industrial application shown through case studies. Int. J. Prod. Res. **49**(4), 1061–1079 (2011)
4. P. Reeves, C. Tuck, R. Hague, *Additive Manufacturing for Mass Customisation* (Springer, London, Dordrecht, Heidelberg, New York, 2011)
5. I. Campbell, D. Bourell, I. Gibson, Additive manufacturing: rapid prototyping comes of age. Rapid Prototyping J. **18** (2012)
6. J. Giannatsis, V. Dedoussis, Additive fabrication technologies applied to medicine and health care: a review. Int. J. Adv. Manuf. Technol. **40**(1–2), 116–127 (2009)
7. World Health Organization, Anniversary of smallpox eradication. Geneva, 18 June 2010 (2010) No Title
8. V.J. Jones, The use of gauze: will it ever change? Int. Wound J. **3**(2), 79–86 (2006)
9. M.C. Robson, D.L. Steed, M.G. Franz, Wound healing: biologic features and approaches to maximize healing trajectories. Curr. Probl. Surg. **38**(2), A1–A140 (2001)
10. T. Velnar, T. Bailey, V. Smrkolj, The wound healing process: an overview of the cellular and molecular mechanisms. J. Int. Med. Res. **37**(5), 1528–1542 (2009)

11. R.G. Frykberg, J. Banks, Challenges in the treatment of chronic wounds. Adv. Wound Care **4** (9), 560–582 (2015)
12. W.-S. Chu et al., Fabrication of a biodegradable drug delivery system with controlled release made of PLGA/5-FU/hydroxyapatite. Rapid Prototyping J. **14**(5), 293–299 (2008)
13. W.S. Chu et al., Fabrication of bio-composite drug delivery system using rapid prototyping technology. Key Eng. Mater. **342–343**(January), 497–500 (2007)
14. K.V. Wong, A. Hernandez, A review of additive manufacturing. ISRN Mech. Eng. **2012**, 1–10 (2012)
15. S. Jana, A. Cooper, M. Zhang, Chitosan scaffolds with unidirectional microtubular pores for large skeletal myotube generation. Adv. Healthc. Mater. **2**(4), 557–561 (2013)
16. R. Xu et al., A silicone rubber membrane with a specific pore size enhances cell proliferation and wound regeneration via the Wnt/β-catenin pathway. J. Tissue Eng. Regenerative Med. (2017)
17. P. Kokhanenko et al., Carbon dioxide insufflation deflects airborne particles from an open surgical wound model. J. Hosp. Infect. **95**(1), 112–117 (2017). Available at: http://linkinghub.elsevier.com/retrieve/pii/S0195670116305321
18. A.A. Jack et al., The interaction of wood nanocellulose dressings and the wound pathogen *P. aeruginosa*. Carbohyd. Polym. **157**, 1955–1962 (2017)

Design and Development of Orthosis for Clubfoot Deformity

Chandrasekeran Vivek and Rajesh Ranganathan

Abstract This paper deals in identifying a complex geometry problem related to health care sector. Paper focuses on identifying a problem in health care which is identified in the area of deformities. Non-surgical treatment for congenital talipes equino varus (clubfoot) deformity known as Ponseti method involves many complications thereby leading to the recurrence of the deformity. In this regard, design and development of customized orthosis that can be used as an alternate solution for Ponseti method for treating clubfoot deformity through AM technology is carried out in this work.

Keywords Clubfoot · Additive manufacturing · Customized orthosis

1 Introduction

Recent advancements in additive manufacturing (AM) show that it is capable of producing complex geometries with high degree of customization [1]. Application areas of AM technology includes mechanical, aerospace, automobile, electronics, garments, health care, and biomedical engineering [2]. Demand for customized healthcare products are increasing day by day [3]. The potential requirements for customized products in health care includes; prosthetic, orthosis, which can be more accurately produced through AM technology [4].

Deformity is considered as abnormal growth in the shape of the body due to genetic and hormone disorder leading to structural changes which can be either congenital or acquired deformity. Agrawal et al. [5], inferred that congenital talipes equinovarus (CTEV or Clubfoot) is a complex muscular skeletal congenital foot

C. Vivek (✉) · R. Ranganathan
Department of Mechanical Engineering, Coimbatore Institute of Technology, Coimbatore 641014, India
e-mail: vivekkct@gmail.com

R. Ranganathan
e-mail: drrajeshranganathan@gmail.com

L. J. Kumar et al. (eds.), *3D Printing and Additive Manufacturing Technologies*,
https://doi.org/10.1007/978-981-13-0305-0_12

deformity affecting both the feet which immobilizes a person. Clubfoot deformity is complex three axis foot deformity present at birth where the forefoot appears to be rotated inward position. Treatment methods for clubfoot involves surgical and non-surgical methods, the most commonly used is the non-surgical treatment methods. Xue-Cheng et al. [6], conducted a study in that, customized foot orthosis for clubfoot deformity is designed, and identified is that a significant reduction in the pressure at heel region and at lateral foot.

Siapkara et al. [7] suggested that, in order to correct clubfoot deformity the usage of non-surgical techniques methods like casting, taping, stretching, soft bandages and splinting are widely considered. Non-surgical treatment involves, Thomas wrench equipment, French function method and Ponseti method are the techniques related to manipulation of the foot, varied casting procedure and bracing of the foot according to the severity of the deformity. Drawbacks of these non-surgical treatments are weight of the cast, casts (long plasters), cast pressure, infection, cast problems, and bracing complications [8, 9]. In this paper, a customized orthosis is identified for a suitable health care problem and through AM technology design and development of the orthosis is carried out.

2 Literature Review

Need for custom parts is very high in health care as it needs to cater the individual requirements, in the increased population. Medical industries seek advanced technology to meet customized products and efficient products at a cost-effective way. AM can be used as an effective tool to create personalized products according to individual requirements, thereby reducing material wastages and delivering products quickly to the market.

AM technologies are applied in the medical/surgical domain for building models that provide visual and tactile information in the applications like operation planning, surgery rehearsals, training and prosthesis design. Classification of health care is grouped into major five categories, such as hospital, pharmaceuticals, medical equipments, and medical insurance. Customization of the products can be applied to medical equipment category, which includes orthopedic, orthosis and implants. This paper is limited to medical equipment category as it involves in manufacture of the customized orthosis. Orthopedic problems in children are common, which can be congenital, developmental or acquired. The following section focuses on congenital deformity as it occurs right from the birth and treatment methods for correction of the deformity includes non-surgical methods.

2.1 Congenital Malformations

Congenital defects or malformation or deformity are termed as defects that are present at the time of birth and mostly related to physiologic functions and metabolic disorders. According to World Health Organization, causes of congenital malformations are due to genetic, environmental and those of unknown causes. Brent [10] suggested that malformations are classified based on the severity and categorized as deformations, disruptions and dysplasia which fits into the category of multifactorial diseases. The cause of congenital malformations can be divided into five categories as single gene defects, chromosomal abnormalities, multifactorial disorders, teratogenic factors and those of unknown causes. Congenital defects affects the muscular—skeletal system, altering the position of the body structures and can be reversed only with orthopedic maneuvers [11]. The method of treatment of congenital deformity involves surgical treatment and non-surgical treatment and most commonly used is the non-surgical treatment.

According to Siapkara et al. [7], correcting tibial deficiency is usually carried out by surgical corrections that affect the lower limb. Narayanan et al. [12], pointed out that the Ponseti method of correcting clubfoot deformity involves maintaining the foot at the correct position after initial casting treatment procedure, this is by usage of braces. Jeevan et al. [13], indicated that the shape of the foot is rotated inside and looks like club shaped, after the initial treatment of varied serial casting methods, an Ankle Foot Orthosis (AFO) is to be used after the final cast procedure is completed.

Janicki et al. [12], suggested that wearing foot abduction orthosis during maintenance phase of the treatment leads to less recurrence of foot when compared with ankle foot orthosis. Hayek et al. [14], inferred that need for new brace design could prevent the foot from slipping and dislodgement of the foot from the brace, thereby delivering more comfort to the patients. In a study conducted by Desai et al. [15], revealed that biomechanical effects on tissues is the prime factor for considering the different types of braces to achieve the full functionality of the braces.

Clubfoot is characterized ankle and foot equinus, hind foot varus and fore foot adduction along the three axis of the foot so that clubfoot is termed as complex congenital deformity [16]. Severity of the deformity is classified based on Dimeglio and Fig. 1 gives the range of motion of the foot characterized by clubfoot.

Meena et al. [18], pointed out that the clubfoot deformity treatment methods involve varied serial casting procedure for correcting the foot. Cast is produced after each manipulation of the foot is achieved. By using point cloud data of the deformed foot Jain et al. [19], designed and developed an Ankle foot orthosis for clubfoot deformity. According to Jain et al. [19], for preventing the foot recurring back to original position orthotic devises are used which enhances mobility. In a study conducted by Khas et al. [20], silicone rubber clubfoot model was fabricated and tested for mechanical properties such as Young's modulus, Poisson's ratio and with rubber model developed a devise to measure the deformity.

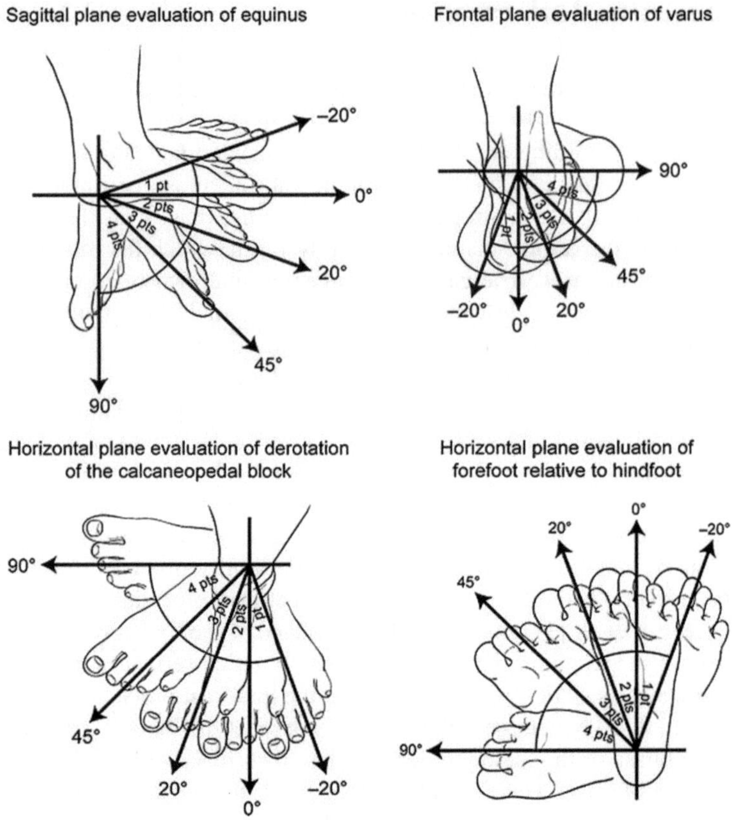

Fig. 1 Classification of clubfoot based on Dimeglio [17]

Congenital Talipes Equinovarus (CTEV) is a complex three axis foot deformity affecting both the foot and non-surgical treatment methods are advised for treating the deformity which involves manipulating the foot, cast procedures and bracing protocols. Zhao et al. [21] identified that the braces will be more effective for reducing or preventing the foot returning back to original position.

Clubfoot is a complex three axis musculoskeletal abnormality where the affected foot appears to have been rotated internally at the ankle [22]. Foot is rotated along coronal, sagittal and transverse axis of plane (with adduction–abduction, inversion–eversion and dorsiflexion–plantarflexion range of motions) with severity ranging from mild to severe [23].

2.2 Club Foot Treatment

The non-surgical method of treatment for club foot are the French function and Ponseti method. French function method or physiotherapy method involves gentle manipulation, stretching, taping, and splinting to maintain the correction of the foot. McGroggan and Dunlop [24], stated that splint is used till the child reaches 2 to 3 years old in order to prevent relapse. Dobbs and Gurnett [25] indicated that child has to undergo physical therapy for two months and needs frequent follow up which provides discomfort.

McGroggan and Dunlop [24] pointed out that Ponseti method as shown in Fig. 2 is gentle sequential manipulation of the foot performed to achieve a plantigrade, functional foot, and gradual casting of the foot for over 4–6 weeks to obtain full correction of the foot. Ponseti method undergoes two stages of treatment procedures, treatment phase—undergoes casting procedure and maintenance phase—undergoes bracing protocols.

Wallander [26] indicated that cavus, adductus, varus, and equinus must be treated in order and the process is repeated weekly for 6 weeks until the foot is achieved a full correction. Clubfoot affects three bones namely calcaneus, talus, and navicular. Treatment phase indicates that aligning the forefoot with midfoot and hindfoot by rotating along three planes every week aligning the foot thereby

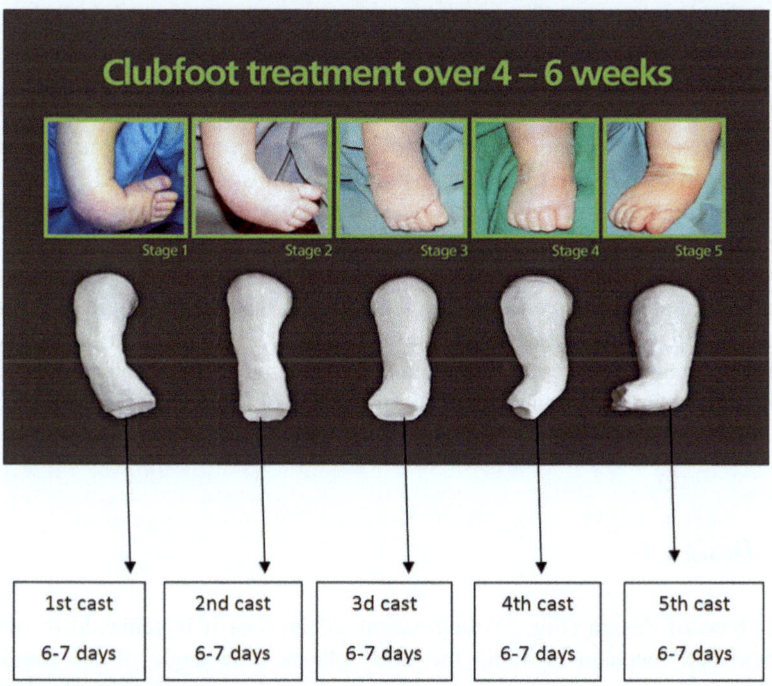

Fig. 2 Ponseti treatment

achieving adduction, abduction, inversion, eversion, dorsiflexion range of motions. Dobbs and Gurnett [25] summarized that cavus should be corrected first by supinating the first metatarsal in a single cast. At the end of final cast a small surgical procedure called tenotomy is carried out, allowing the ankle to be positioned.

Dobbs and Gurnett [25] and Staheli et al. [16] inferred that after the last cast and tenotomy procedure, 70° of abduction 15° of dorsiflexion of the foot is achieved while the final cast is retained for three weeks before removing. Once the final cast is removed, maintenance phase is initiated by wearing a foot abduction orthosis in order to maintain the foot at the corrected position.

The complication observed in Ponseti method is casting errors, slippage of cast, pressure sores, and weight of the cast. On other hand, bracing complications include improper fittings of brace, weight of the brace and difficult to wear and due to this the foot recurrence happens due to bracing complications [8].

From the literatures it was identified that Ponseti treatment of clubfoot deformity is identified to have casting and bracing compliances which leads to the relapse of the corrected clubfoot, thereby it was suggested that there is a need for an alternative solution for the treatment of clubfoot deformity in order to overcome the above difficulties. A customized orthosis through additive manufacturing technology can provide an alternative solution for the treatment of clubfoot deformity which can overcome the difficulties and reduce the relapse of the corrected foot.

3 Design of the Customized Orthosis for Clubfoot

A customized orthosis can be provided by incorporating additive manufacturing technology to correct the foot. The customized orthosis is designed in such a way that the manipulation of the foot is done weekly and to hold and rotate the foot along particular planes (coronal, sagittal, and transverse plane) for correcting [27].

Important criteria that are needed to design the new customized orthosis are (i) to rotate the foot along the planes, (ii) lock the foot in that desired angle, (iii) maintain adduction–abduction, inversion–eversion, extension—dorsiflexion angle, (iv) feasibility to change the external rotation angle, (v) suitable for unilateral and bilateral clubfoot, (vi) eliminating or minimizing the usage of fasteners, (vii) comfortness to wear, easy to operate and for removal, (viii) less weight and compact in size.

3.1 *Design I*

In this type of design (Fig. 3) for rotation of the foot it is achieved by utilizing spherical ball mechanism along the three planes. The angle of the rotations is provided on the top and side face of the socket mechanism. The drawbacks of this model are (i) the abduction angle was achieved at 45° for 8 position and it is based

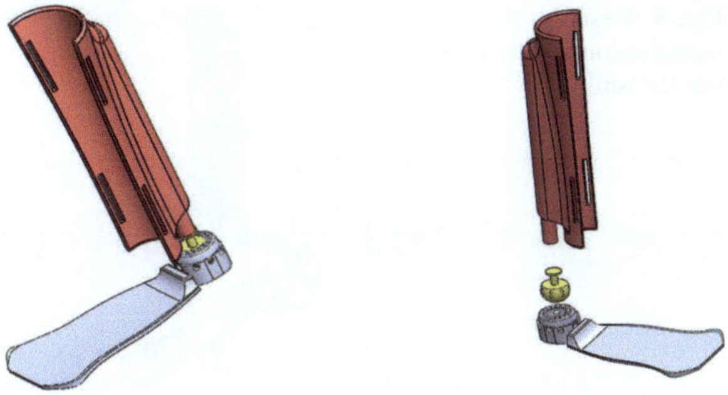

Fig. 3 Preliminary design—spherical ball mechanism

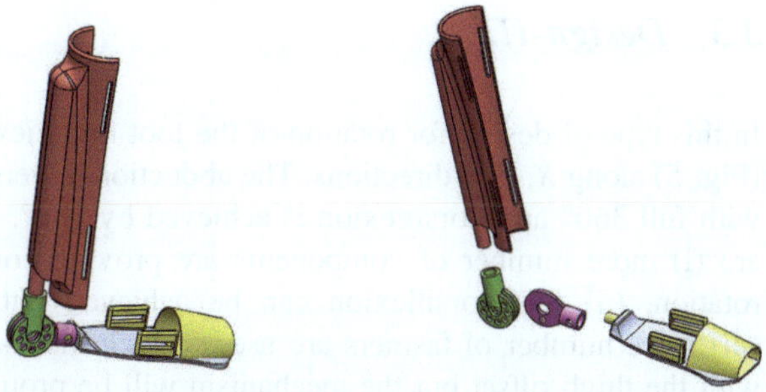

Fig. 4 Design modification-I—hinged mechanism

on the number of holes provided on the spherical ball, (ii) inversion–eversion, dorsiflexion angle cannot be achieved since the ball is half spherical, (iii) foot is not aligned with the thigh offset and (iv) fastners are used to lock the mechanism, (v) difficulty in connecting spherical ball with thigh portion.

3.2 *Design-II*

In this type of design for rotation of the foot is achieved by hinged mechanism (Fig. 4) along *X*, *Y*, *Z* directions. The abduction, inversion-eversion can be achieved with full 360° and dorsiflexion is achieved up to 51° in 7 steps. The drawbacks of this model are (i) number of holoe are to be provided on the hinged mechanism (ii) foot is not aligned with the thigh offset and (iii) fastners are used to lock the mechanism, (iv) hinged mechanism will be protruding outside which will be discomfort for the children, (v) chances of foot slipping from the devise.

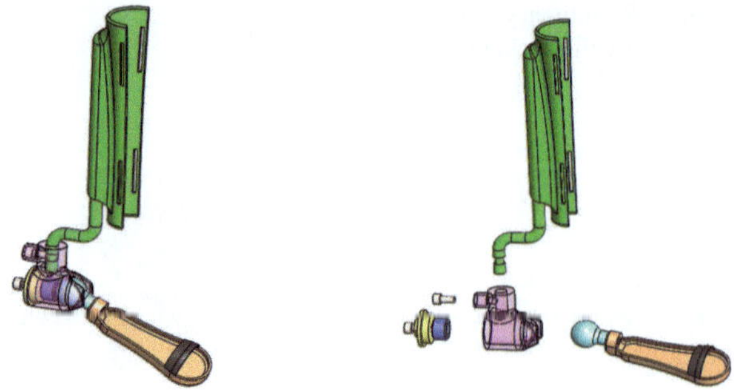

Fig. 5 Design modification-II—ball socket mechanism

3.3 *Design-III*

In this type of design for rotation of the foot is achieved by ball socket mechanism (Fig. 5) along *X*, *Y*, *Z* directions. The abduction, inversion-eversion can be achieved with full 360° and dorsiflexion is achieved by 180°. The drawbacks of this model are (i) more number of components are provided on the ball socket to hold the rotation, (ii) full dorsiflexion can be achieved but limited to +90° and −90°, (iii) more number of fastners are used to lock the mechanism, (iv) foot is aligned with the thigh offset but the mechanism will be protruding outside, (v) chances of shaft to break along the *X* and *Y* directions, (vi) manufacturing of full spherical is difficult. From the design iteration it is found out that Design I, II and III requires more number of fasteners to hold the rotation of the foot after manipulating to specified angle. For manufacturing of the prototype Design I are selected.

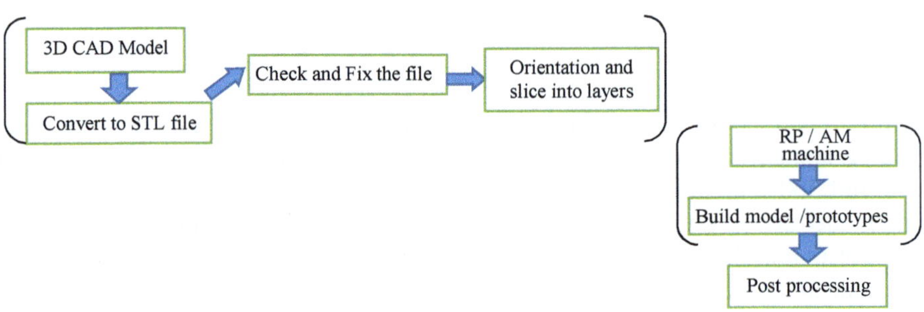

Fig. 6 Block diagram for manufacturing sequences

4 Manufacturing of Customized Orthosis

The manufacturing methods for developing the customized orthosis were identified based on the material, geometrical complexity and machine availability. In order to develop a geometrical complexity type product, the conventional manufacturing method are not suitable for this particular applications, thereby additive manufacturing technology are selected. AM technology is used for fabricating complex 3D parts from numerical data by adding one layer over the other. Prototypes developed using AM technology can be used for design and form-fit verifications, direct tooling, and patterns for casting applications [28].

For developing the prototype, Design I is selected based on the design criteria and specification. For producing the prototype of the Design I, Stratasys uPrint SE machine is selected which works on the Fused Deposition Modeling technique which uses ABS plastics as the base material and support material in the form of wires.

4.1 Machine Specifications

uPrint SE from Stratasys has a build volume size of (8 × 8 × 6 in.) with a layer thickness of 0.254 mm. It is loaded by model material as ABSplus in ivory and support as low grade ABS which is breakaway water soluble support structures.

4.2 Manufacturing of the Prototypes

A sequence of operations as illustrated in Fig. 6 that are needed for manufacturing of the prototype which includes 3 stages alike, (1) preprocessing, (2) manufacturing, and (3) post processing.

For every RP/AM machine a standard preprocessing software are available which converts the given CAD design model into machine language STL file format (Standard tessellation language) and it is checked for errors and then proper orientation of the model are carried out. Figure 7 describes the preprocessing of the design I model is carried out which uses model material as solid fill and support material as smart fill, the build volume of the model is reduced to 50%. Figure 8 illustrates the orientation of the model in such away that model and support material are not wasted much

Figures 9 and 10 illustrates the time required to build the prototype of the Design I which takes 2 h and 47 min with model material consumption as 0.99 in^3 and support material consumption as 0.81 $in.^3$ for this possible best orientation and the prototype are printed in the machine. Figure 11a shows the prototype of the machine build from uPrint SE and the prototype is then ready for the post

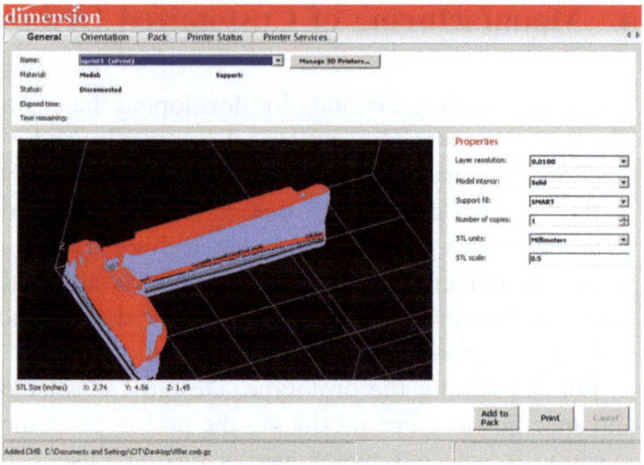

Fig. 7 Preprocessing of the design model

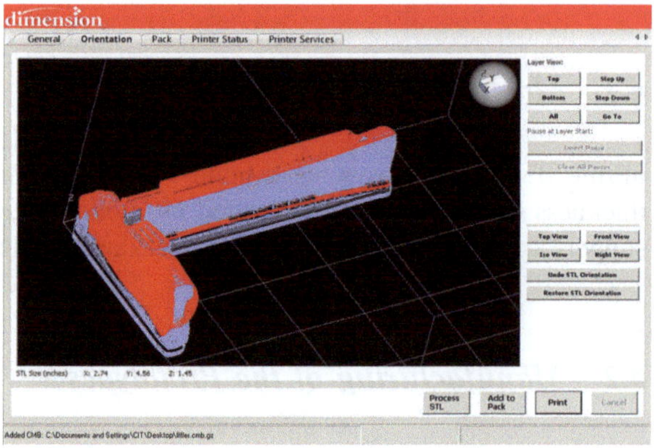

Fig. 8 Orientation of the design model

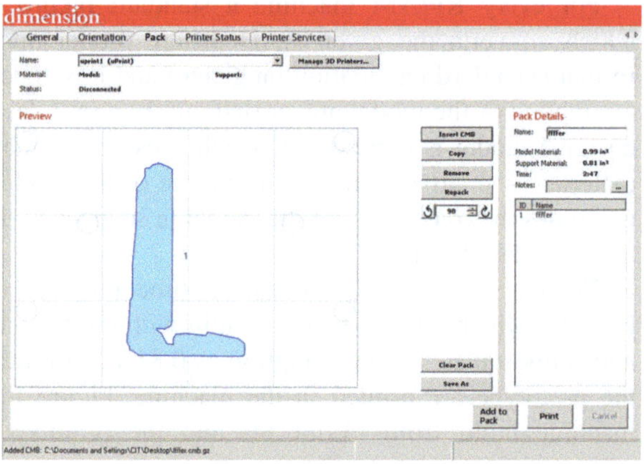

Fig. 9 Printing of the design model

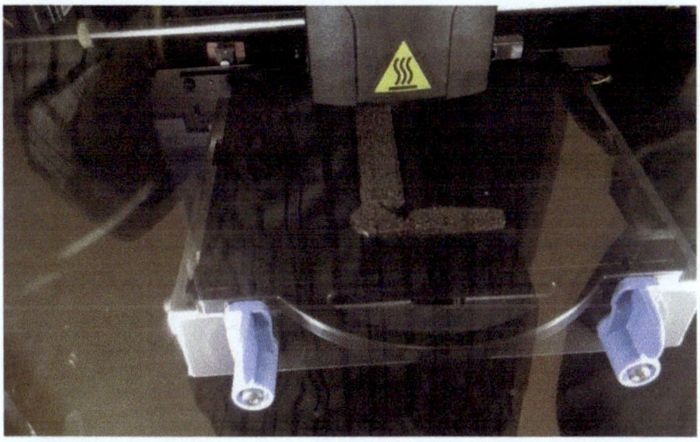

Fig. 10 Manufacturing of the model

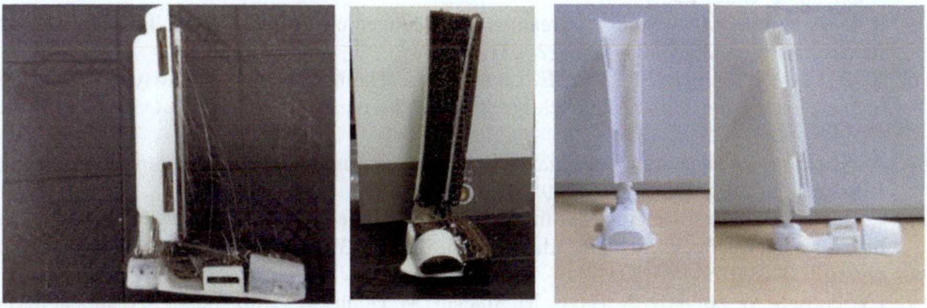

Fig. 11 a, b Prototype of the model

processing stage where the support material are removed in the water soluble tank heated at 70 °C with the water diluted with Sodium hydroxide granules and then immersed in the tank for 2 h for support removal. The final prototype after support removed in Fig. 11b.

5 Conclusion

Deformities related to foot were studied and found that the literature study was concentrated in congenital talipes equino varus (Clubfoot) deformity. Ponseti non-surgical method, which employ POPs and braces, is used predominantly for clubfoot treatment and major drawback of this method is relapsing of the foot. With the intention of avoiding the above-mentioned drawback of Ponseti method, a customized orthosis solutions are to be provided. A model of the customized orthosis is designed based on the constraints and drawbacks of the Ponseti method. The specified design criteria's are met and are incorporated in the model. A model of the customized orthosis for Design I were fabricated using additive manufacturing technology.

References

1. E.A. Nasr, A.K. Kamrani, *Computer—Based Design and Manufacturing* (Springer, New York, USA, 2007)
2. C.K. Chua, K.F. Leong, *Rapid Prototyping: Principles and Applications in Manufacturing* (World Scientific, 2000)
3. E. Minvielle, M. Waelli, C. Sicotte, J.R. Kimberly, Managing customization in health care: a framework derived from the services sector literature. Health Policy **117**, 216–227 (2014)
4. M.C. Goiato, M.R. Santos, A. Moreno, Prototyping for surgical and prosthetic treatment. J. Craniofac. Surg. **22**(3) (2011)
5. R.A. Agrawal, M.S. Suresh, Treatment of congenital clubfoot with Ponseti method. Indian J. Orthop. **39**(4), 244–247 (2005)
6. X.-C. Liu, C. Tassone, R. Rizza, E. Linford, J.G. Thometz, R. Lyon, Newly designed foot orthosis for children with residual clubfoot after Ponseti casting. J. Prosthet. Orthot. **26**, 38–42 (2014)
7. A. Siapkara, R. Duncan, Congenital talipes equinovarus: a review of current management. J. Bone Joint Surg. Br. **89-B**(8), 995–1000 (2007)
8. C. Radler, The Ponseti method for the treatment of congenital club foot: review of the current literature and treatment recommendations. Int. Orthop. **37**(9), 1747–1753 (2013). https://doi.org/10.1007/s00264-013-2031-1
9. I.V. Ponseti, M. Zhivkov, N. Davis, M. Sinclair, M. Dobbs, J.A. Morcuende, Treatment of the complex idiopathic clubfoot. Clin. Orthop. Relat. Res. (Lippincott Williams & Wilkins) (2006)
10. R.L. Brent, *Environmental Causes of Human Congenital Malformations: The Pediatrician's Role in Dealing with These Complex Clinical Problems Caused by a Multiplicity of Environmental and Genetic Factors*, vol. 113, no. 4. https://doi.org/10.1542/peds.113.4.s1.957
11. A. Manea, P. Timisoara, Methods to diagnose congenital malformations in newborns. Prenat. Diagn. **XII**(47), 10–15 (2009)
12. J.A. Janicki, J.G. Wright, S. Weir, U.G. Narayanan, A comparison of ankle foot orthoses with foot abduction orthoses to prevent recurrence following correction of idiopathic clubfoot by the Ponseti method. J Bone Joint Surg. [Br.] **93-B**, 700–704 (2011)
13. R.R. Jeevan, E. Vijayaragavan, A. Kiruba, 3Dimensional modeling of an ankle foot orthosis for clubfoot deformity. Int. J. Biomed. Res. **2**(3), 171–180 (2011)
14. Y. Hemo, E. Segev, A. Yavor, D. Ovadia, S. Wientroub, S. Hayek, The influence of brace type on the success rate of the ponseti treatment protocol for idiopathic clubfoot. Journal of Children's Orthopaedics, **5**(2), 115–119 (2011). https://doi.org/10.1007/s11832-010-0321-3
15. L. Desai, F. Oprescu, A. Dimeo, J.A. Morcuende, Bracing in the treatment of children with clubfoot: past, present, and future. The Iowa orthopaedic journal. **30**(April), 15–23 (2010)
16. L.T. Staheli, M. Benson, J. Cheng, J. Burns, C. Conrad, A. Crawford, S. Demuth, P. D, Practice of pediatric orthoapedics. Growth (Lakeland), 1–448
17. D.M. Eastwood, The clubfoot: congenital talipes equinovarus, in *Children's Orthopaedics and Fractures*, ed. by M. Benson, J. Fixsen, M. Macnicol, K. Parsch (Springer, London, 2009)
18. S. Meena, P. Sharma, S.K. Gangary, L.K. Lohia, Congenital clubfoot. J. Orthop. Allied Sci. **2**(2)

19. M.L. Jain, S. Govind Dhande, N.S. Vyas, Virtual modeling of an ankle foot orthosis for correction of foot abnormality. Robot. Comput. Integr. Manuf. Elsevier **27**(2), 257–260 (2011)
20. Kanwaljit S. Khas, Pulak M. Pandey, Alok R. Roy, Rapid manufacturing of a clubfoot model imitating soft tissues and bones. J. Virtual Phys. Prototyping **8**(3), 2013 (2013)
21. D. Zhao, J. Liu, L. Zhao, Z. Wu, Relapse of clubfoot after treatment with the Ponseti method and the function of the foot abduction orthosis. CiOS Clin. Orthop. Surg. **6**(3), 245–252 (2014)
22. W. Chen, F. Pu, Y. Yang, J. Yao, L. Wang, H. Liu, Y. Fan, Correcting congenital talipes equinovarus in children using three different corrective methods: a consort study. Medicine **94** (28), e1004 (2015). https://doi.org/10.1097/MD.0000000000001004
23. A.M. Wainwright, T. Auld, M.K. Benson, T.N. Theologis, The classification of congenital talipes equinovarus. J. Bone Joint Surg. Br. **84**(7), 1020–1024 (2002)
24. J. McGroggan, G. Dunlop, A literature review of non-surgical intervention for the treatment of idiopathic talipes equinovarus: a podiatric perspective. Foot Ankle Online J. (2010). https://doi.org/10.3827/faoj.2010.0310.0002
25. M.B. Dobbs, C.A. Gurnett, Update on clubfoot: etiology and treatment. Clin. Orthop. Relat. Res. **467**(5), 1146–1153 (2009). https://doi.org/10.1007/s11999-009-0734-9
26. H.M. Wallander Congenital clubfoot. Aspects on epidemiology, residual deformity and patient reported outcome. Acta Orthop. Suppl. **81**(339), 1–25 (2010). https://doi.org/10.3109/17453671003619045
27. K.S. Khas, P.M. Pandey, A.R. Ray, Design and development of a device to measure the deformities of clubfoot. Proc. Inst. Mech. Eng. [H] **229**(3), 194–204 (2015). https://doi.org/10.1177/0954411915574758
28. N. Guo, M.C. Leu, Additive manufacturing: technology, applications and research needs. Front. Mech. Eng. **8**(3), 215–243 (2013). https://doi.org/10.1007/s11465-013-0248-8

Optimization of Selective Laser Sintering Process Parameters on Surface Quality

M. Akilesh, P. R. Elango, A. Achith Devanand, R. Soundararajan and P. Ashoka Varthanan

Abstract Selective laser sintering (SLS) empowers the fast, flexible, cost-efficient, and easy manufacture of prototypes for various application of required shape and size by using powder based material. The physical prototype is important for design confirmation and operational examination by creating the prototype unswervingly from CAD data. In SLS procedure optimization of construction parameters of good responses, will also help to save time and material. In this work, optimal SLS process parameters, by varying the laser power, bed temperature and layer thickness on surface quality of Length, Depth and Surface roughness for the designed part by using Polyamide and also evaluate the part quality by using Coordinate measuring machine (CMM). The experimentations were carried out rendering to the Taguchi parametric strategy L9 at various combinations of process parameters and arithmetical optimization method ANOVA was used to decide the optimal levels and to find the percentage of contribution of the process parameters. The results show that the Laser power is the most important factor followed by the Bed Temperature and Layer thickness for maximizing the Length and Depth, Minimizing Surface roughness of the SLS processed Polyamide. This optimized process capability paves the way for the society.

Keywords SLS process parameter Optimization

1 Introduction

Rapid Prototyping (RP) technique is one among the manufacturing activities that leads in reduction of product development time [1]. Some of the successfully commercialized solid freeform fabrication processes include Stereolithography

M. Akilesh (✉) · P. R. Elango · A. A. Devanand · R. Soundararajan · P. A. Varthanan
Department of Mechanical Engineering, Sri Krishna College of Engineering & Technology, Coimbatore, India
e-mail: akileshakil07@gmail.com

L. J. Kumar et al. (eds.), *3D Printing and Additive Manufacturing Technologies*,
https://doi.org/10.1007/978-981-13-0305-0_13

(SL), Fused Deposition Modeling (FDM), Selective Laser Sintering (SLS), Laminated Object Manufacturing (LOM), Three Dimensional Printing (3D-P) and Laser Engineered Net Shaping (LENS) [2]. One of the important techniques is stereo-lithography [3]. Prototype making is an important step in product development involving all stages of manufacture [4]. The equipment gives engineers and designers an opportunity to put their ideas and concepts in the prototype and check for their correctness in a short period of time and with minimal price [5]. Almost all RP process are based on a layered manufacturing approach in which objects are fabricated as a sequence of horizontal cross-sections, each one being formed separately from the appropriate raw materials and merged to preceding layers until the object is finalized [6]. The RP technology also needs several supporting technologies such as solid modeling, database management, and electronic networking for data exchange [7]. Rapid Prototyping shelters a range of technologies that create physical 3D objects from 3D data sources output by solid modeling CAD systems, 3D laser scanners, and CT/MRI scanners [8]. The technology is presently controlled by "additive" processes established by companies such as 3D systems, Stratasys, EOS, ZCorpand Object Geometries [9]. These build models by linking a range of materials including liquid plastics, resins, and sheet paper/metal—layer by layer using thin, horizontal cross-sections of the computer model to drive the RP equipment [10]. It is recommended that one or more parts may be selected for evaluating a particular system or benchmarking it against other systems or routes, followed by an economic comparison of different materials, geometry, quality, operation, and the various costs [11]. The best surface finish obtained is produced of SGC machines, and followed by SLA, LOM, FDM, and SLS. The RP machines are continuously being upgraded to improve the surface finish without sacrificing build speed [12]. The idea is to optimize the process parameters of stereolithography machine for the newly developed proto model by using solid works. Taguchi method is one of the best optimizing technique for optimizing process parameters [13]. Part quality of the prototype in the RP technique mainly depends upon the parameters such as hatch cure depth, layer thickness, orientation, laser power, temperature, hatch spacing [14]. Thus an effort was made to study and optimization of process parameters leading the system which are linked with developed part characteristics of dimensional accuracy, surface roughness, and microhardness by using the Taguchi method [15]. Based on the above facts, a statistical methodology based on the Taguchi and ANOVA techniques was implemented in this present work to govern the degree of importance of each process parameter on the Length, Depth and Surface roughness of SLS processed nylon 12 [16].

2 Experimental Methods and Methodology

2.1 Processing of Specimen

The major experimental apparatus adopted includes, Sinterstation 2500 + Selective Laser Sintering Machine where it use the Nylon—Duraform PA:12 (Polyamide) to build geometrical shape of the work piece. The drawing of part is given below as Fig. 1.

The experimental building material espoused is Nylon—DuraForm PA:12 (Polyamide) as powder. The experimental building models designed based on L9 OA. The STL format is generated by PRO-E Creo 2.0 and sends to the Sinterstation 2500 + Selective Laser Sintering Machine. The various preprocessing stages such as STL verification, Laser power, Bed Temperature and Layer Thickness, building interior structure form, supporting method and building deposition path are to be followed by the layer slicing process to produce the building path according to the shape of part. Building quality characteristics or features include the Larger-the-better for length and depth also nominal-the-better for surface roughness for Dimensional accuracy [17–20]. One of the SLS machines offered by Sinterstation 2500 + can build parts of size 381 × 330 × 457 mm. On the other hand, the desktop RP machines are designed to build small size or scaled down models within an envelope of 200 × 200 × 200 mm. For larger parts, an alternative may be to fabricate and glue several segments together, instead of installing a larger machine. In this investigation, the Nylon—DuraForm PA:12 (polyamide) was used as the base material. Due to high elongation and high abrasion resistance it commercial is widely used in electronic industries and automobile industries. The Nylon—DuraForm PA:12 of size 50 × 30 × 2 mm is used in this experiment.

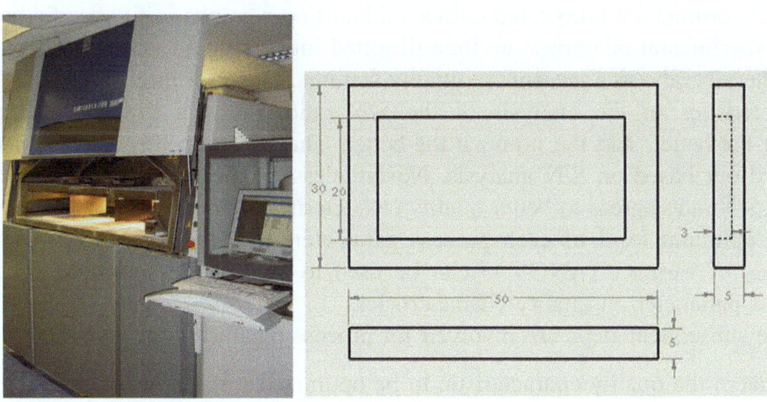

Fig. 1 Experimental setup with CAD model

2.2 Selection of SLS Process Parameters and Their Levels

In the present examination, three process parameters, i.e., Laser power, Bed Temperature, and Layer thickness were considered. The chemical formula for Nylon—DuraForm PA:12 is $(C_{12}H_{23}NO)_n$ used in this investigation. Among various SLS process parameters like hatch cure depth, layer thickness, orientation, laser power, bed temperature, hatch spacing, Pilot experiments were carried out using nylon material for a deserved shape and size to determine the influenced process parameters and working range of the SLS process one factor at a time approach. From the outcomes, the range of the process parameters such as the Laser power was designated as 12.5–17.5 W, the Bed Temperature was selected as 128–138 °C and the Layer Thickness was selected as 0.12–0.18 mm.

2.3 Taguchi Quality Engineering

Taguchi methods established by Dr. Genichi Taguchi refer to the quality engineering techniques that personified both statistical process control (SPC) and new quality related management procedures. Taguchi examined the engineering problems with a statistical approach. He suggested that the engineering optimization of a process should be carried out in three step approaches namely the system design, the parameter design and the tolerance design. Taguchi method customs orthogonal arrays from the design of experiments theory to study a large number of variables with a small number of experiments. The orthogonal arrays diminish the number of experimental configurations to be studied. The resulting decision from the examination will be true within the range of the experimental region well-defined by the control factors. Orthogonal arrays are not exclusive to Taguchi, but were in use, much earlier. However, Taguchi has altered their use by providing tabulated sets of standard orthogonal arrays and corresponding linear graphs to fit specific projects. The experimental outcomes are then distorted into a signal-to-noise (S/N) ratio. It uses the S/N ratio as a measure of quality features. There are three classes of quality characteristics in the analysis of the S/N ratio, i.e., the lower-the-better, the higher-the-better, and the nominal-the-better. The S/N ratio for each parameter is worked out based on S/N analysis. Nevertheless, of the quality characteristics, a larger S/N ratio agrees to better quality characteristics. Therefore, highest S/N ratio is the optimum level of each process parameter. Besides, a statistical technique analysis of variance (ANOVA) can be used to control the significance of each process parameter on quality characteristics.

The subsequent steps are involved for process parameters optimization

1. Govern the quality characteristic to be optimized.
2. Recognize the noise factors and test conditions.
3. Recognize the control factors and their alternativelevels.

4. Design the matrix experiment and describe the dataanalysis procedure.
5. Conduct the matrix experiment.
6. Analyze the data and regulate the optimum levels for control factors.
7. Forecast the performance at these levels.

2.4 Selection of Orthogonal Array (OA)

Experimentations have been carried out using Taguchi's L9 Orthogonal Array (OA) experimental design which comprises of nine combinations of laser power, bed temperature, and layer thickness. It contemplates three process parameters (without interaction) to be varied in three discrete levels. As per Taguchi experimental design idea, a set of three levels allocated to each process parameter has two degrees of freedom (DOF). This gives a total of 6 DOF for three process parameters designated in this work. Thus we have a total of 8 DOF for the factors for the present tests. The nearest three level orthogonal array available, sustaining the criterion of selecting the OA is L9 having 8 DOF.

2.5 Process Parameter and Responses

Fabricated model (nine samples) taken for checking the dimensional accuracy like Length, Depth and Surface roughness, which are to be done Coordinate Measuring Machine as shown in Fig. 2 and average value was noted in the given Table 1.

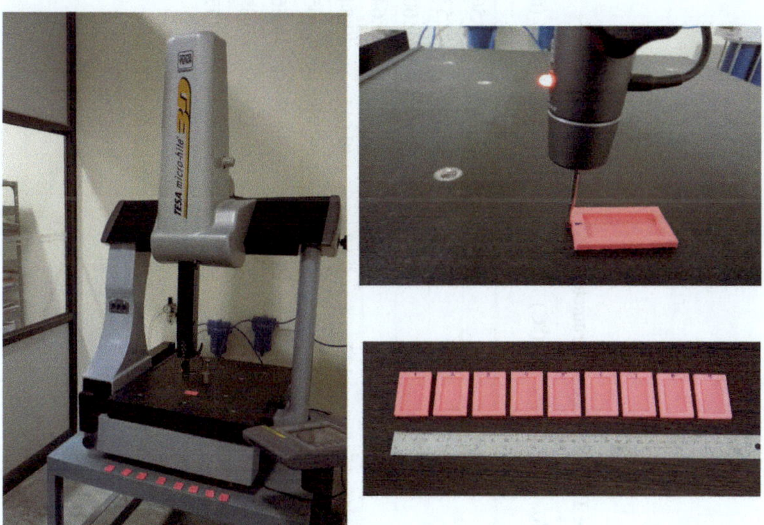

Fig. 2 Co-ordinate measuring machine with prepared specimen

Table 1 SLS process parameters and responses

S. No.	LP (W)	BT (°C)	LT (mm)	L1 (mm)	L2 (mm)	L-Mean (mm)	D1 (mm)	D2 (mm)	D-Mean (mm)	Ra1 (μm)	Ra2 (μm)	Ra –Mean (μm)
1	12.5	128	0.12	49.54	49.564	49.552	3.0119	2.667	2.8394	1.46	1.45	1.455
2	12.5	133	0.15	49.588	49.534	49.561	3.002	3.011	2.982	1.57	1.57	1.570
3	12.5	138	0.18	49.562	49.603	49.5825	2.9987	2.7878	2.8932	1.79	1.77	1.780
4	15	128	0.15	49.603	49.643	49.623	3.048	3.3214	3.1847	1.64	1.66	1.650
5	15	133	0.18	49.616	49.545	49.5805	2.9998	3.112	3.0559	1.75	1.75	1.750
6	15	138	0.12	49.589	49.654	49.6215	3.0412	3.012	3.0266	1.71	1.72	1.715
7	17.5	128	0.18	49.534	49.567	49.5505	2.7887	3.2124	3.0005	1.52	1.51	1.515
8	17.5	133	0.12	49.521	49.589	49.556	2.6656	3.1234	2.8945	1.48	1.47	1.475
9	17.5	138	0.15	49.627	49.578	49.6025	3.0121	3.2122	3.1121	1.79	1.78	1.785

3 Results and Discussion

3.1 Taguchi Analysis of Length (L) on SLS Process

Finding the important process parameter with maximum signal and minimum noise ratio for machining the composites is a stimulating task. ANOVA technique is monitored to complete the stimulating task. ANOVA accounts that the *F*-value of this Length model is 21.69, 13.83, 6.32, 48.11 for laser power, bed temperature, layer thickness which suggests that, this model is substantial. There is only a 0.01% chance that this "Model *F*-Value" arises due to noise. Values of "$P > F$" less than 0.0500 postulates that model terms are significant. In this experimental case, all the input parameters and their collaboration such as laser power, bed temperature, layer thickness are significant model terms. The "Predicted *R*-Squared" of 97.7% is in judicious agreement with the "Adjacent *R*-Squared" of 90.7%."Adequate Precision" shows the signal-to-noise ratio. The results show that laser power is highly subsidized (Figs. 3, 4; Tables 2, 3, 4, 5, 6 and 7).

From this experiment we conclude that the length gradually increases from laser power 12.5–15 W and decreases when laser power is increased above 15 W. When seeing bed temperature, the length decreases gradually from 128 to 133 °C and increases when temperature increases until 138 °C. Layer thickness contributes only to a small extent. The length increases from thickness between 0.12 and 0.15 mm and decreases when thickness is decreased. This is same in S/N ratio analysis. The bed temperature contributes about 32.5% in change of length when compared to layer thickness and bed temperature and for S/N value the laser power

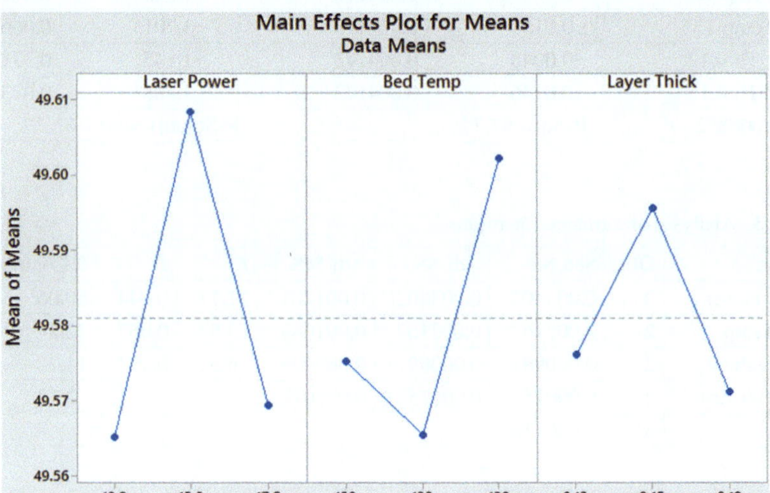

Fig. 3 Length main effects plot for means

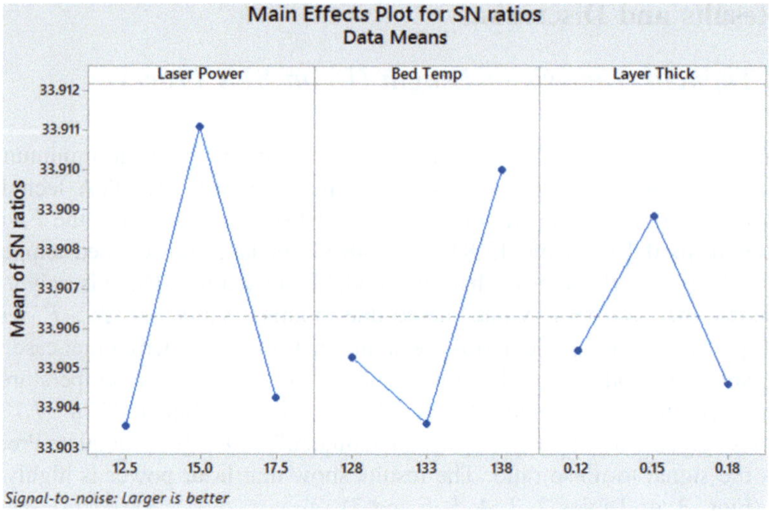

Fig. 4 Length main effects plot for SN ratio

Table 2 Estimated model coefficients for means

Term	Coef	SE Coef	*T*	*P*
Constant	49.5809	0.002951	16802.959	0.000
Laser Po 12.5	−0.0158	0.004173	−3.781	0.063
Laser Po 15.0	0.0274	0.004173	6.563	0.022
Bed Temp 128	−0.0058	0.004173	−1.385	0.300
Bed Temp 133	−0.0154	0.004173	−3.701	0.066
Layer Th 0.12	−0.0048	0.004173	−1.145	0.371
Layer Th 0.15	0.0146	0.004173	3.488	0.073
S = 0.008852	R-Sq = 97.7%		R-Sq(adj) = 90.7%	

Table 3 Analysis of variance for means

Source	DF	Seq SS	Adj SS	Adj MS	*F*	*P*	%Contribution
Laser Power	2	0.003402	0.003402	0.001701	21.71	0.044	50.06
Bed Temp	2	0.002167	0.002167	0.001083	13.83	0.067	32.2
Layer Thick	2	0.000991	0.000991	0.000495	6.32	0.137	14.7
Residual Error	2	0.000157	0.000157	0.000078			2.33
Total	8	0.006716					

donates 50%. The optimal process parameters value for essential length value in Nylon—DuraForm PA:12 material is 15 W Laser power, 138 °C Bed temperature and 0.15 mm layer thickness.

Table 4 Response table for means

Laser		Layer	
Level	Power	Bed Temp	Thick
1	49.57	49.58	49.58
2	49.61	49.57	49.60
3	49.57	49.60	49.57
Delta	0.04	0.04	0.02
Rank	1	2	3

Table 5 Estimated model coefficients for SN ratios

Term	Coef	SE Coef	*T*	*P*
Constant	33.9063	0.000517	65593.194	0.000
Laser Po 12.5	−0.0028	0.000731	−3.778	0.063
Laser Po 15.0	0.0048	0.000731	6.561	0.022
Bed Temp 128	−0.0010	0.000731	−1.383	0.301
Bed Temp 133	−0.0027	0.000731	−3.703	0.066
Layer Th 0.12	−0.0008	0.000731	−1.147	0.370
Layer Th 0.15	0.0026	0.000731	3.488	0.073
S = 0.001551	R-Sq = 97.7%		R-Sq(adj) = 90.7%	

Table 6 Analysis of variance for SN ratios

Source	DF	Seq SS	Adj SS	Adj MS	*F*	*P*	%Contribution
Laser Power	2	0.000104	0.000104	0.000052	21.69	0.044	50.48
Bed Temp	2	0.000067	0.000067	0.000033	13.83	0.067	32.52
Layer Thick	2	0.000030	0.000030	0.000015	6.32	0.137	14.56
Residual Error	2	0.000005	0.000005	0.000002			2.42
Total	8	0.000206					

Table 7 Response table for signal–to-noise ratios

Laser		Layer	
Level	Power	Bed Temp	Thick
1	33.90	33.91	33.91
2	33.91	33.90	33.91
3	33.90	33.91	33.90
Delta	0.01	0.01	0.00
Rank	1	2	3

3.2 Taguchi Analysis of Depth (D) on SLS Process

Finding the substantial process parameter with maximum signal and minimum noise ratio for machining the composites is an interesting task. In order to complete this interesting task ANOVA technique is followed. ANOVA reports that the *F*-value of this Length model is 35.28 1.03, 40.48 for laser power, bed temperature, layer thickness which infers that, this model is significant. There is only a 0.01% chance that this "Model *F*-Value" occurs due to noise. Values of "$P > F$" less than 0.0500 specifies that model terms are significant. In this experimental case, all the input parameters and their interface such as laser power, bed temperature, layer thickness are significant model terms. The "Predicted *R*-Squared" of 98.7% is in practical agreement with the "Adjacent *R*-Squared" of 94.9%."Adequate Precision" shows the signal-to-noise ratio. From the results it shows that the layer thickness has exceedingly contributed (Figs. 5, 6; Tables 8, 9, 10, 11, 12 and 13).

From this experimentation, we accomplish that the depth gradually increases from laser power 12.5–15 W and decreases when laser power is increased above 15 W. As far as bed temperature is concerned, the depth decreases gradually from 128 to 133 °C and increases when temperature increases until 138 °C. The layer thickness contributes only to a fewer percentage. The depth increases from thickness between 0.12 mm to 0.15 mm and decreases when thickness is decreased. This is same in S/N ratio analysis. The laser power contributes about 47% in change of depth. In S/N analysis the layer thickness contributes about 50%. The optimal process parameters value for required depth value in Nylon—DuraForm PA:12 material is 15 W Laser power, 138 °C Bed temperature and 0.15 mm layer thickness.

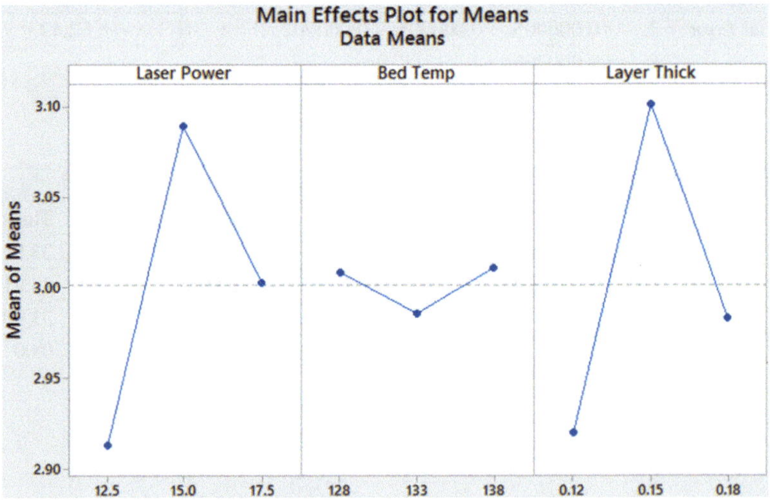

Fig. 5 Depth main effects plot for means

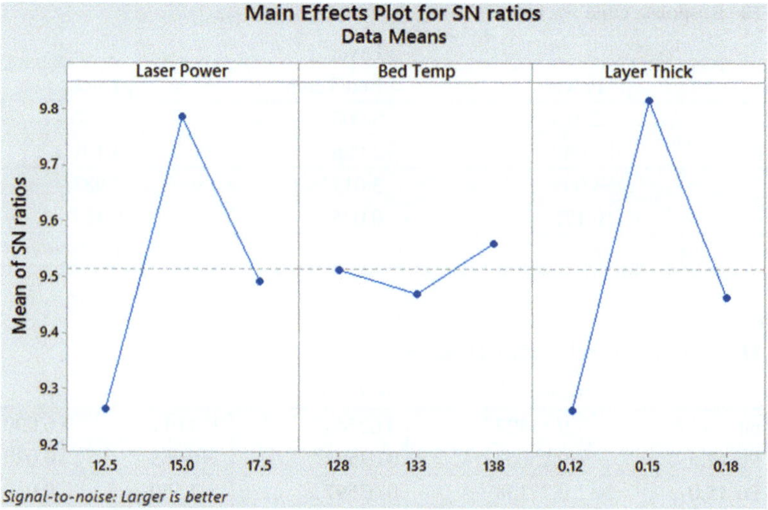

Fig. 6 Depth main effects plot for SN ratio

Table 8 Estimated model coefficients for means

Term	Coef	SE Coef	*T*	*P*
Constant	3.00151	0.005910	507.896	0.000
Laser Po 12.5	−0.08844	0.008358	−10.583	0.009
Laser Po 15.0	0.08756	0.008358	10.476	0.009
Bed Temp 128	0.00672	0.008358	0.804	0.506
Bed Temp 133	−0.01588	0.008358	−1.900	0.198
Layer Th 0.12	−0.08133	0.008358	−9.731	0.010
Layer Th 0.15	0.09961	0.008358	11.918	0.007
S = 0.01773	R-Sq = 99.4%		R-Sq(adj) = 97.5%	

Table 9 Analysis of variance for means

Source	DF	Seq SS	Adj SS	Adj MS	*F*	*P*	%Contribution
Laser Power	2	0.046468	0.046468	0.023234	73.92	0.013	47
Bed Temp	2	0.001143	0.001143	0.000572	1.82	0.355	1.15
Layer Thick	2	0.050609	0.050609	0.025304	80.50	0.012	51.1
Residual Error	2	0.000629	0.000629	0.000314			0.63
Total	8	0.098848					

Table 10 Response table for means

Laser		Layer	
Level	Power	Bed Temp	Thick
1	2.913	3.008	2.920
2	3.089	2.986	3.101
3	3.002	3.011	2.983
Delta	0.176	0.025	0.181
Rank	2	3	1

Table 11 Estimated model coefficients for SN ratios

Term	Coef	SE Coef	T	P
Constant	9.51327	0.02544	374.011	0.000
Laser Po 12.5	−0.25053	0.03597	−6.965	0.020
Laser Po 15.0	0.27158	0.03597	7.550	0.017
Bed Temp 128	−0.00228	0.03597	−0.063	0.955
Bed Temp 133	−0.04344	0.03597	−1.208	0.351
Layer Th 0.12	−0.25150	0.03597	−6.992	0.020
Layer Th 0.15	0.30216	0.03597	8.400	0.014
S = 0.07631	R-Sq = 98.7%		R-Sq(adj) = 94.9%	

Table 12 Analysis of variance for SN ratios

Source	DF	Seq SS	Adj SS	Adj MS	F	P	%Contribution
Laser Power	2	0.41090	0.41090	0.205450	35.28	0.028	45.36
Bed Temp	2	0.01195	0.01195	0.005973	1.03	0.494	1.31
Layer Thick	2	0.47137	0.47137	0.235684	40.48	0.024	52.03
Residual Error	2	0.01165	0.01165	0.005823			1.28
Total	8	0.90586					

Table 13 Response table for signal-to-noise ratios

Laser		Layer	
Level	Power	Bed Temp	Thick
1	9.263	9.511	9.262
2	9.785	9.470	9.815
3	9.492	9.559	9.463
Delta	0.522	0.089	0.554
Rank	2	3	1

3.3 *Taguchi Analysis of Surface Roughness (Ra) on SLS Process*

Finding the major process parameter with maximum signal and minimum noise ratio for machining the composites is an inspiring task. In order to complete this inspiring task ANOVA technique is followed. ANOVA reports that the *F*-value of this Length model is 4.88, 14.92, 6.49 for laser power, bed temperature, layer thickness which denotes that, this model is significant. There is only a 0.01% chance that this "Model *F*-Value" happens due to noise. Values of "$P > F$" less than 0.0500 indicates that model terms are significant. In this experimental case, all the input parameters and their interaction such as laser power, bed temperature, layer thickness are noteworthy model terms. The "Predicted *R*-Squared" of 96.3% is in reasonable agreement with the "Adjacent *R*-Squared" of 85.3%."Adequate Precision" shows the signal-to-noise ratio. From the outcomes it shows that bed temperature has highly donated (Figs. 7, 8; Tables 14, 15, 16, 17, 18 and 19).

From this experiment we conclude that the Ra gradually increases from laser power 12.5–15 W and decreases when laser power is increased above 15 W. When considering bed temperature, the Ra increases as the value of temperature increase from 128 to 138 °C. When seeing layer thickness, the Ra increases as the layer thickness increases. But in S/N ratio, the S/N value is lesser at 15 W and higher at 18 W. The S/N value decreases as temperature decreases. The S/N ratio decreases as the layer thickness decreases. The bed temperature contributes about 50% in both Ra value and S/N value. The required optimum process parameter value for Ra value in Nylon—DuraForm PA:12 material is 15 W laser power, 138 °C bed temperature, 0.18 mm layer thickness.

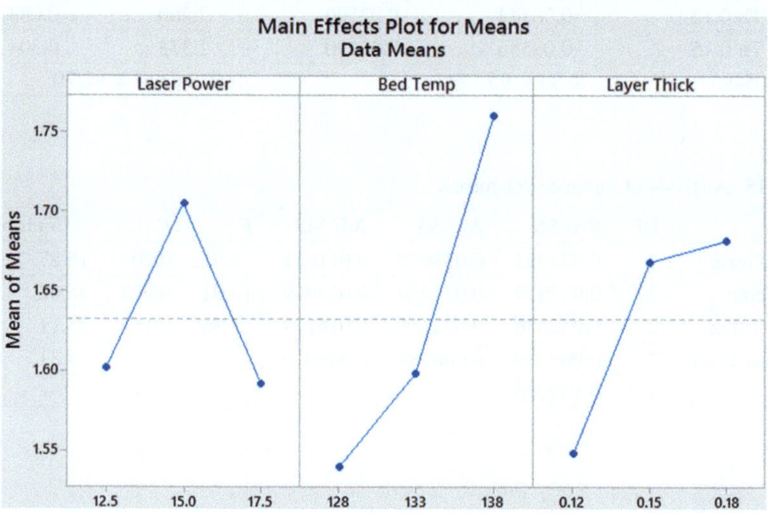

Fig. 7 Surface roughness main effects plot for means

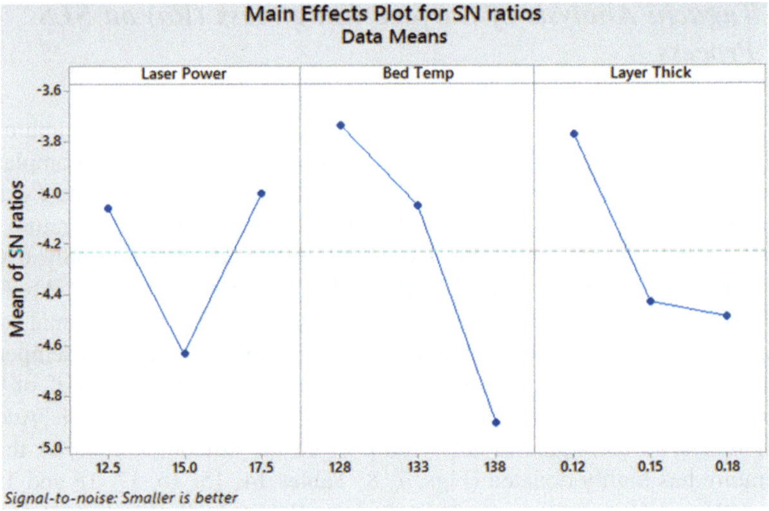

Fig. 8 Surface roughness main effects plot for SN ratio

Table 14 Estimated model coefficients for means

Term	Coef	SE Coef	*T*	*P*
Constant	1.63278	0.01832	89.143	0.000
Laser Po 12.5	−0.03111	0.02590	−1.201	0.353
Laser Po 15.0	0.07222	0.02590	2.788	0.108
Bed Temp 128	−0.09278	0.02590	−3.582	0.070
Bed Temp 133	−0.03444	0.02590	−1.330	0.315
Layer Th 0.12	−0.08444	0.02590	−3.260	0.083
Layer Th 0.15	0.03556	0.02590	1.373	0.304
S = 0.05495	R-Sq = 95.7%		R-Sq(adj) = 82.7%	

Table 15 Analysis of variance for means

Source	DF	Seq SS	Adj SS	Adj MS	*F*	*P*	%Contribution
Laser Power	2	0.023622	0.023622	0.011811	3.91	0.204	16.87
Bed Temp	2	0.077939	0.077939	0.038969	12.91	0.072	55.68
Layer Thick	2	0.032356	0.032356	0.016178	5.36	0.157	23.11
Residual Error	2	0.006039	0.006039	0.003019			4.31
Total	8	0.139956					

Table 16 Response table for means

Laser		Layer	
Level	Power	Bed Temp	Thick
1	1.602	1.540	1.548
2	1.705	1.598	1.668
3	1.592	1.760	1.682
Delta	0.113	0.220	0.133
Rank	3	1	2

Table 17 Estimated model coefficients for SN ratios

Term	Coef	SE Coef	*T*	*P*
Constant	−4.2330	0.09057	−46.736	0.000
Laser Po 12.5	0.1717	0.12809	1.340	0.312
Laser Po 15.0	−0.3990	0.12809	−3.115	0.089
Bed Temp 128	0.4945	0.12809	3.861	0.061
Bed Temp 133	0.1814	0.12809	1.416	0.292
Layer Th 0.12	0.4601	0.12809	3.592	0.070
Layer Th 0.15	−0.2006	0.12809	−1.566	0.258
$S = 0.2717$	R-Sq = 96.3%		R-Sq(adj) = 85.3%	

Table 18 Analysis of variance for SN ratios

Source	DF	Seq SS	Adj SS	Adj MS	*F*	*P*	%Contribution
Laser Power	2	0.7210	0.7210	0.36052	4.88	0.170	17.89
Bed Temp	2	2.2029	2.2029	1.10144	14.92	0.063	54.66
Layer Thick	2	0.9580	0.9580	0.47898	6.49	0.134	23.77
Residual Error	2	0.1477	0.1477	0.07383			3.66
Total	8	4.0295					

Table 19 Response table for signal-to-noise ratios

Laser		Layer	
Level	Power	Bed Temp	Thick
1	−4.061	−3.738	−3.773
2	−4.632	−4.052	−4.434
3	−4.006	−4.909	−4.493
Delta	0.626	1.170	0.720
Rank	3	1	2

4 Conclusion

The effect of SLS process variables such as Laser power, Bed temperature and Layer Thickness on the Length, Depth, Surface roughness of Polyamide has been calculated. The experimentations were carried out as per Taguchi parametric design concept and the results were analyed using ANOVA technique. The following conclusions have been drawn from the present investigation.

1. The SLS process parameters were optimized and the optimal values of the process parameters were found to be 15 W for Laser Power, 133 °C for Bed Temperature, and 0.15 mm for Layer thickness.
2. The ANOVA was carried out to inspect the influence of process parameters on the Length, The Laser power (50.06%) has the key influence on the length, shadowed by the Layer thickness (14.7%) and the bed temperature (32.2%). For Depth the layer thickness (51.1%) has the major influence followed by laser power (47%) and the bed temperature (1.15%). For surface roughness, Bed temperature (55.6%) has the major influence followed by layer thickness (23.11%) and the laser power (16.07%) of polyamide (Nylon—DuraForm PA:12).

References

1. D. Ahn, H. Kim, S. Lee, Fabrication direction optimization to minimize post-machining in layered manufacturing. Int. J. Mach. Tools Manuf. **47**, 593–606 (2007)
2. B.H. Lee, J. Abdullah, Z.A. Khan, Optimization of rapid prototyping parameters for production of flexible ABS object. J. Mater. Process. Technol. **169**, 54–61 (2005)
3. C.J. Tzeng, Y.H. Lin, Y.K. Yang, M.C. Jeng, Optimization of turning operations with multiple performance characteristics using the Taguchi method and Grey relational analysis. J. Mate Proc. Tech. **209**, 2753–2759 (2009)
4. N. Jawahar, U. Chandrasekar, K.N. Ramanathan, K. Chockalingam, Establishment of process model for part strength in stereolithography. J. Mater. Process. Technol. **208**, 348–365 (2008)
5. S. Li, R. Zhang, F. Lin, R. Wu, Q. Lu, Z. Xiong, X. Wang, Y. Yan, Rapid prototyping and manufacturing technology: principle, representative technics, applications, and development trends. Tsinghua Sci. Tech. **14**, 1–12 (2009)
6. L. Chen, Y. He, Y. Yang, S. Niu, H. Ren, The research status and development trend of additive manufacturing technology. Int. J. Adv. Manuf. Technol. 2016
7. C.C. Chang, Direct slicing and G-code contour for rapid prototyping machine of UV resin spray using PowerSOLUTION macro commands. Int. J. Adv. Manuf. Technol. **23**, 358–365 (2004)
8. A. Bowyer, The self-replicating 3D printer—manufacturing for the masses, in *Eighth national conference on rapid design, prototyping and manufacture* (2007)
9. D.N. Silva, M. Gerhardt de Oliveira, E. Meurer, M.I. Meurer, J.V. Lopes da Silva, A. Santa Barbara, Dimensional error in selective laser sintering and 3D-printing of models for cranio maxillary anatomy reconstruction. J. Craniomaxillofac Surg. **36,** 443e449 (2008)
10. A. Nizam, R.N. Gopal, L. Naing, A.B. Hakim, A.R. Samsudin, Dimensional accuracy of the skull models produced by rapid prototyping technology using stereolithography apparatus. Arch. Orofac. Sci. 1: 60e66 (2006)

11. A. Boschetto, L. Bottini, Accuracy prediction in fused deposition modeling. Int. J. Adv. Manuf. Technol. **73**(5–8), 913–928 (2014)
12. B.M. Tymrak, M. Kreiger, J.M. Pearce, Mechanical properties of components fabricated with open-source 3-D printers under realistic environmental conditions. Mater. Design. **58**, 242–246 (2014)
13. M. Lukić, J. Clarke, C. Tuck, W. Whittow, G. Wells, Printability of elastomer latex for additive manufacturing or 3D printing. J. Appl. Polym. Sci. **133**, 4 (2016)
14. K.L. Chalasani, B.N.Grogan, A. Bagchi, C.C. Jara-Almonte, A.A. Ogale, R.L. Dooley, An algorithm to slice 3D shapes for reconstruction in prototyping systems, in *Proceedings of the 1991 ASME Computers in Engineering Conference,* pp. 209–216 (1991)
15. G. Cardano, R. Giannoccaro, A.D. Ludovico, E.L. Bohez, S.L. Campanelli, Statistical analysis of the stereolithographic process to improve the accuracy. Comput. Aided Des. **39**, 80–86 (2007)
16. B.P. Conner, G.P. Manogharan, K.L. Meyers, An assessment of implementation of entry-level 3D printers from the perspective of small businesses. Rapid Prototyping J. **21** (5), 582–597 (2015)
17. H.C. Kim, S.H. Lee, Reduction of post-processing for stereolithography systems by fabrication-direction optimization. Comput. Aided Des. **37**, 711–725 (2005)
18. J.G. Zhou, D. Herscovici, C.C. Chen, Parametric process optimization to improve the accuracy of rapid prototyped stereolithography parts. Int. J. Mach. Tools Manuf **40**, 363–379 (2000)
19. P. Alexander, S. Allen, D. Dutta, Part orientation and build cost determination in layered manufacturing. Comput. Aided Des. **30**(5), 343–356 (1998)
20. W. Cao, Y. Myiamoto, Direct slicing from AutoCAD solid models for rapid prototyping. Int. J. Adv. Manuf. Technol. **21**, 739–742 (2003)

[illegible]

Reconstruction of Damaged Parts by Integration Reverse Engineering (RE) and Rapid Prototyping (RP)

Anuj V. Dongaonkar and Rajesh M. Metkar

Abstract This paper presents an integrated design cycle of reverse engineering (RE) and rapid prototyping (RP) for reconstruction of a damaged parts in industry and production systems by taking gear wheel in a rotating machinery as an example. In order to replace the damaged part or destructive part, it is required to change urgently with new part to avoid the financial and production loss to the industry. This can be possible with the help of 3D scanning and additive manufacturing. 3D scanning help to capture digital information of parts from space to CAD software. The physical to electronic translation is done by a layered digitizing method called Deconstruction and semi-automated software tools are used to create a CAD virtual solid model. Conventional CAD tools are used to manipulate the model of damaged part and reconstruct damaged area to form new one by CAD tools as necessary. A newly created 3D solid model is translated from virtual to physical model by using additive manufacturing process. Here FDM process is used. A designer can then use hand tools to form the model by carving, shaping, and joining additional material. In this paper, a broken and damaged gear wheel of rotating machinery have been taken. This damaged model is scanned, then gear is reconstructed in CATIA V5 software and exact model of gear wheel is formed. This model is then printed by using 3D printing machine (uPrint SE Plus). Hence, by integration of reverse engineering and RP techniques, the gear model is prepared which can be used in rotatory machinery or casted for more numbers if required.

Keywords Reverse engineering · Rapid prototyping · Damaged parts Computer-aided design

A. V. Dongaonkar (✉)
Mechanical Engineering Department, Government College of Engineering, Amaravti, Maharashtra, India
e-mail: anujvd@gmail.com

R. M. Metkar
Government College of Engineering, Mechanical Engineering Department, Amaravti, Maharashtra, India
e-mail: rajeshmetkar@yahoo.co.in

L. J. Kumar et al. (eds.), *3D Printing and Additive Manufacturing Technologies*,
https://doi.org/10.1007/978-981-13-0305-0_14

1 Introduction

Generally, differentiation of engineering products can be done as new parts, modified parts and renewal and spare parts such as damaged or broken one. Due to the performance requirement, greater complex shape and burden of completion activity, the new parts and replacement of damaged parts development is became difficult. For replacement of worn or damaged parts, import the part or develop the technology is an easier way in small period. Financial loss and more time in procurement and shipment is main problem. Technology development may make one becoming self-reliant for long period, but not useful for one or two items. If there are not any spare parts available, it is time-taking process [1].

That is undesirable in production routine. From many advanced technologies one of the technologies is reverse engineering in service operations in industry. So to optimize repair process time to reduce financial, and time loss in production, reverse engineering and rapid prototyping is one of best technologies.

The main aim of this work is to give a demonstration of reverse engineering of parts whose geometric, material, and manufacturing information is misplaced or not available (drawings) and to replace broken, damaged and warn out parts to reduce financial and production loss to the industry. The backbone is on use rapid prototyping technologies and CAD or CAE that can significantly improve the replacement of damaged or broken parts. The primary goal of the practical investigations is to create a suitable path for reverse engineering of replacement of damaged or broken parts that are needed in small quantity in no time period.

A real-life case study of a gear wheel of a rotating machinery was given here to authenticate the route and express the experience comes out by using these technologies.

2 Experimental Process for Replacement of Damaged Parts

Reverse engineering significantly transpired as the answer to give spares for replacing broken or damaged parts for which there is an unavailability of technical engineering drawings [2]. Imported part without data or lost or misplaced drawings can be a case. Reverse engineering of broken parts can be a less expensive solution for immediate replacement and to fabricate additional spares to keep the product over a longer period in comparison with importing. A solid model of the part is the mainstay for this replacement parts in reverse engineering. By the help of standard formats such as STEP, IGES, VDA and STL, exporting from or importing of model data can be done in CAD/CAM/CAE systems. The three most salient steps of data in reverse engineering activities connect to the CAD model generation, material identification, and rapid manufacturing [3].

Fig. 1 Process of reconstruction of the damaged part

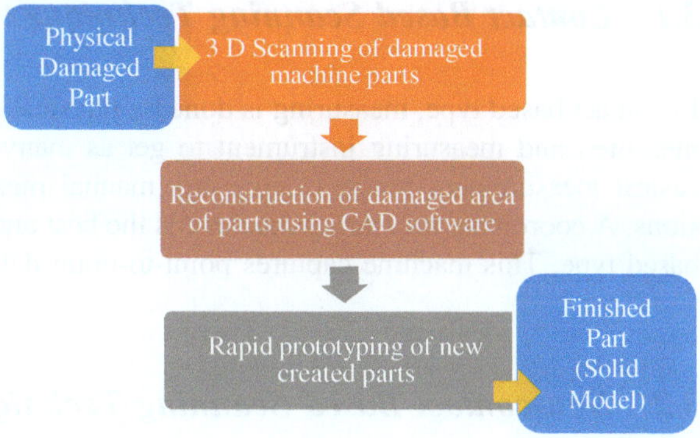

Basically, the experimental process of work of reconstruction of the damaged part consists of the three step procedure. The process can be

1. 3D scanning of damaged machine parts
2. Reconstruction of damaged area of parts using CAD software (Fig. 1).
3. Rapid prototyping of new created parts

3 Reverse Engineering (RE)

The definition of reverse engineering is as the operation of acquiring a geometric CAD model by scanning or digitizing given products or parts by obtaining 3D point cloud data [4]. The process of digitally capturing the physical entities of a component, referred to as Reverse engineering (RE) is defined as the operation of digitally capturing the geometry physically of a parts [5]. RE is the method of reproducing new geometry from an existing part by digitizing and editing an existing CAD data of part [6].

The scan for a Virtual Model (damaged part) is carried out using a 3D scanner. 3D scanner is a device that analyses a real world object or environment to on its shape and possibly its appearance [7]. Generally, these scanning processes differ in many aspects but the basic and digitalizing steps remain the same. It involve three main steps [8]:

- exploration (acquisition of points)
- manual or automatic alignment; and
- Fusion or processing of the geometrical points pack for conversion into solid state.

Usually, there are two ways of scanning, contact based and noncontact based.

3.1 Contact Based Scanning Technique

In contact based type, measuring is done by physical contact between surface being measured and measuring instrument to get as many dimensions as possible. The easiest measurement process comprises manual measurement of a part's dimensions. A coordinate measuring machine is the best and accurate approach in contact based type. This machine captures point-to-point data of measurement.

3.2 Non-contact Based Scanning Technique

The device capture the data of part without any contact between them in non-contact based type. It can be sub-divided into: Active and passive technique. The phenomenon of structured lighting and reflection is used in the active technology. The point data is captured using lasers, optics, X-rays and charge-coupled device (CCD) sensors in non-contact scanners Other types of infrared radiation are also used [9].

4 Formation of Modified Model

The result of all digitizing methods is point cloud data or cloud of points (COPs), which are random and unstructured collection of *x*, *y*, *z* coordinates [10]. A transfer of 3D point′s cloud data from 3D Scanner into 3D CAD software is next step for further modifications and reconstruction of scanned surface of part from scanner and formation new CAD model. '*.stl*' file—file format accepted as universally for data transfer between 3D CAD software [11].

In this step of work the point cloud data is to be transfer into editable form as it only interpret surface points, not the final surfaces. By using number of operations, final surface overlaying point's cloud with a mesh is done. After the number operations for modification in damaged model in CAD software, the final new modified 3D model is formed which is base for further operations. The CAD software which can be used further modification and reconstruction for a damaged part CATIA V5, SolidWorks, Geomagic Design X, Ansys SpaceClaim [12].

5 Rapid Prototyping (RP)

Rapid prototyping can be defined as an additive process based on the philosophy of layered technology in which cross-sectional layers are deposited one after other by plastic, wax or paper to create part [13]. The CAD model is sliced into the cross-sectional layers of parts and printed using one of RP technology available [14].

All techniques involves the same basic step process: (i) Conversion of the CAD model to STL format, (ii) slicing the part STL file into cross-sectional layers nearly 0.01–0.7 mm thin, (iii) construction of the layers of the model one after another as sliced, and (iv) post processing such as cleaning and curing [15].

A variety of rapid prototyping technologies have emerged. Fused Deposition Manufacturing (FDM) is process used in work as RP technique [16]. In FDM, through a fine nozzle a model material is extruded out as filament and deposited as initial position set by software. A thin cross-sectional slice of part is laid down with filament by nozzle on platform in movement *X–Y* plane. For the next layer deposition platform moved down relative to nozzle and next layer is deposited on previous one. Bonding between two slices takes place due hotness of extruded filament. Support structure is built by second nozzle with different material for part where needed. As layer by layer the part is completed, it is post processed and support material is removed [17].

A solid 3D printed model is formed which serve as masterpiece for casting or moulding process for further production new parts [18].

6 Case Study—Gear Wheel

Damage of machine parts is a serious concern in the production. It affects production flow and causes financial losses due machine failure or fault. A gear wheel in a rotating machinery was selected to examine the application of computer aided reverse engineering technologies for rapid development of one-off complex parts required for replacement process. This replacement part has not any technical data or manufacturing drawings available with it.

The study involved three steps as follows:

STEP 1—3D scanning of damaged machine parts
STEP 2—Reconstruction of damaged area of parts using CAD software
STEP 3—Rapid Prototyping of new created parts (Fig. 2).

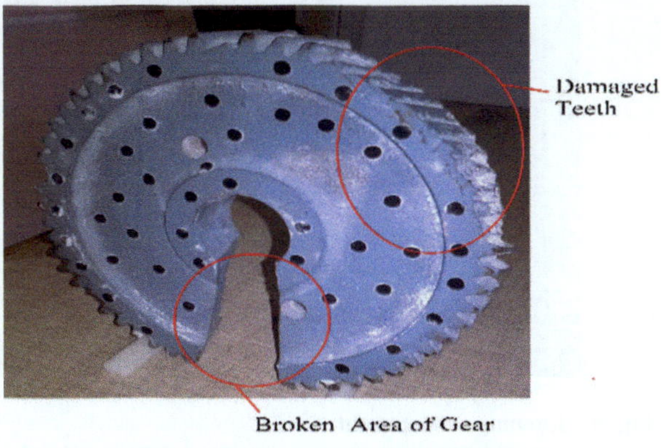

Fig. 2 Diagram of broken and damaged area of gear wheel

6.1 3D Scanning of Damaged Gear Wheel

A CAD model generation of damaged or broken gear wheel is the first step. For obtaining the geometry of damaged body part, Range Vision 3D Scanner is used as 3D scanning device. Due to the use of the structured light principle, new 3D scanner Smart provides highly accurate and detailed scans. Its accuracy is up to 0.1%, resolution—up to 0.12 mm. Smart is fully equipped, easy in use and fits perfectly for scanning of different-sized objects: from 4 cm to 1 m [19]. Smart can scan with markers that make alignment automatically and simplifies the scanning process. RangeVision Smart ScanCenter is software used for scanning process in scanner and the manual in processing the scan data in RangeVision ScanMerge software. When calibrating the scanner makes snapshots of the calibration plate in 11 different positions. Calibration field is pre-measured with high accuracy. Part is prepared by spraying developer spray which create a coat white colour and markers are stick to part and turning table as suitable as shown in Fig. 3.

The project of gear with scanned markers is created in the RangeVision Smart ScanCenter. After the first scan shot rotate the table slightly and Scan the object in the next position, if the scan during stitching is in obviously incorrect position, it is clear that such scan should be deleted—select it in the tree and press Delete. In this way by taking necessary scanned shots for complete scanning of part by rotating table is to be done and ScanMerge is opened for further processing of the cloud points of scanned part.

After the global registration editing of scanned part is done selection mode (rectangle, brush or lacco) and unwanted part deleted using proper selection mode as shown in Fig. 4. Function for filling gaps allows to automatically fill the gaps on unscanned area, remaining on the model. Use this feature only for closed models, scanned from all sides.

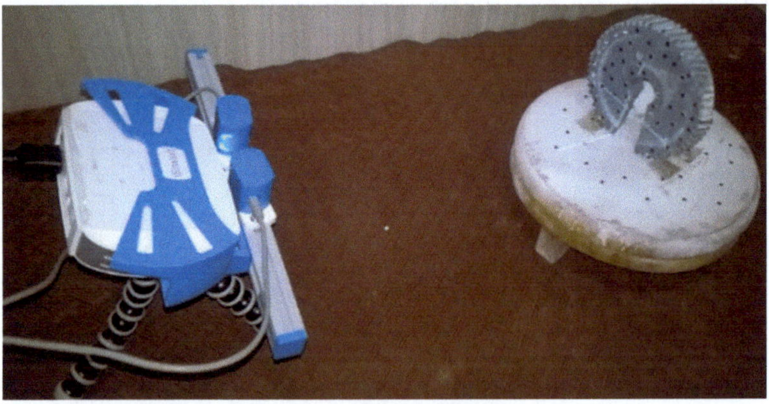

Fig. 3 Scanning arrangement

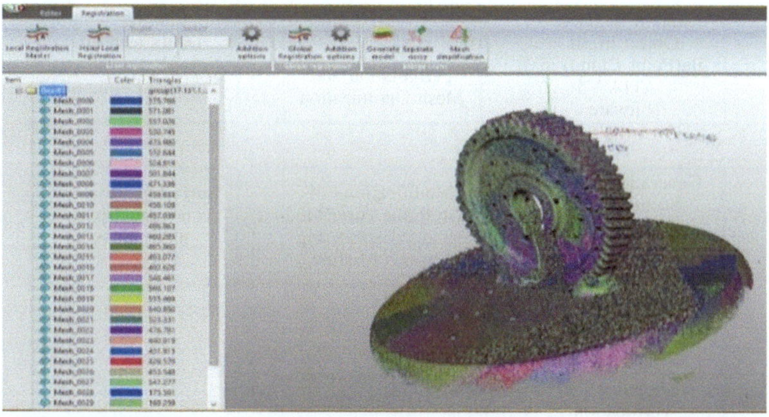

Fig. 4 After completion of scanning in RangeVision ScanMerge

Fig. 5 Final .stl file of gear wheel

After the end of the work on processing scan data save the results. The model can now be exported in one of the following formats: *stl, obj, ply*. Figure 5 shows STL model of new part is reconstructed (Fig. 6).

6.2 *Reconstruction of Broken and Damaged Area of Gear Wheel*

For further modifications on cloud points scanned data, CATIA V5 is used as editing CAD software. In the CATIA V5, there is specialized module for the .stl file

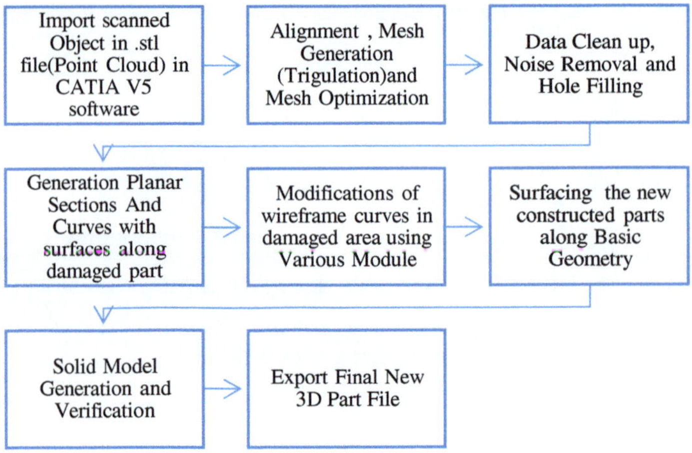

Fig. 6 Representation of reconstruction damaged area of parts workflow

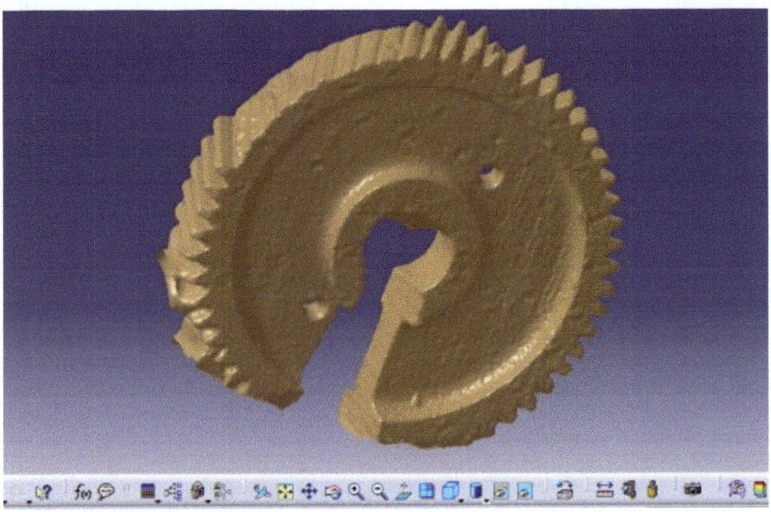

Fig. 7 Import .stl file CATIA V5

editing and modifications can be done easily. Mainly, the basic three module of CATIA software are used as (i) Digitized Shape Editor (ii) Generative Shape Design (iii) Quick Surface Reconstruction. The reconstruction process of part goes through the following steps to get the final product as solid 3D model as shown in Fig. 7.

The scanned gear wheel is imported in Digitized Shape Editor module of CATIA V5 for modification of broken area of gear.

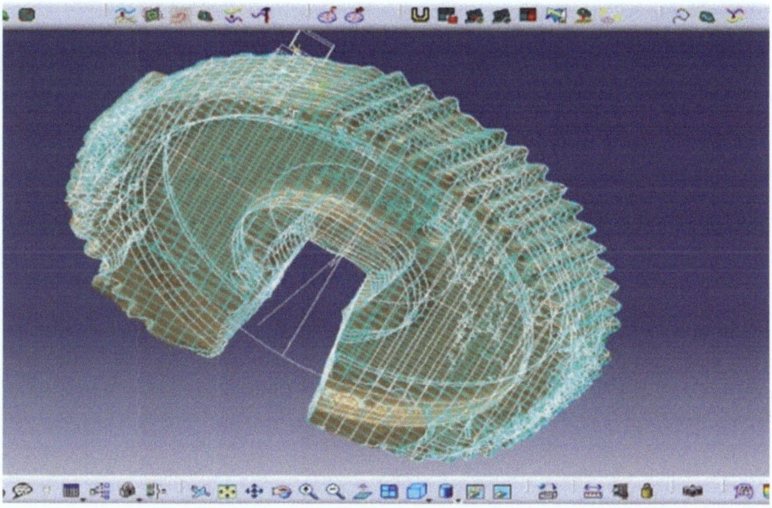

Fig. 8 Mesh generation, optimization with planar sections curves

The alignment of the gear wheel is according to the coordinate axes and *X*–*Y* plane by the 'Align by Best Fit' tool in the module. Scanned point cloud data is then used for mesh creation on scanned point by 'Mesh Creation' tool in the module by generating 3D mesh on points. As generated mesh contains large number triangles in formed mesh of part, it creates lot of trouble during the editing of wheel. The optimization of triangles is done using 'Mesh Optimization' tool in module. This optimization by to 90% causes reduction and simplified mesh file of gear wheel to be modified.

The generated meshed gear wheel contains corrupted triangles, duplicated triangles, non-manifold edges, inconsistent orientation, and non-manifold vertices and noise in the mesh as shown in Fig. 8. 'Mesh Cleaner' is the tool used for Data clean up and noise removal from mesh file of gear. The hole is filled using the 'Fill Holes' tool easily.

As cleaned mesh file of gear wheel is created, 'Planar Sections' tool created curves along the scanned mesh on planes along different axis at suitable distance as per given. The planar sections are created along *X*-axis, *Y*-axis and *Z*-axis also. With 'Curve From Scan' tool, create the curves from planar sections for modifications on this curves. This gives us complete structure of damaged and broken part as wireframe structure.

As the wire frame type structure is formed, Generative Shape Design module in CATIA is used for further modifications in the broken area and damaged teeth of gear wheel. So broken area is shown, the reconstruction of curve can be done, joining new area with suitable lines and circles, the unwanted part deleted. The gear teeth profile from curves which is generated on scanned gear wheel is newly created as shown in Fig. 9 and the profile projection of gear teeth is done to second limit of

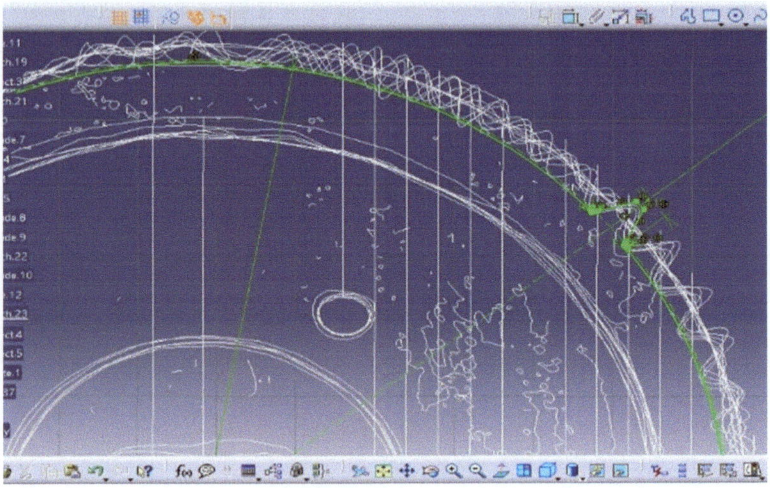

Fig. 9 Gear teeth profile creation along curves

Fig. 10 Wireframe structure of gear

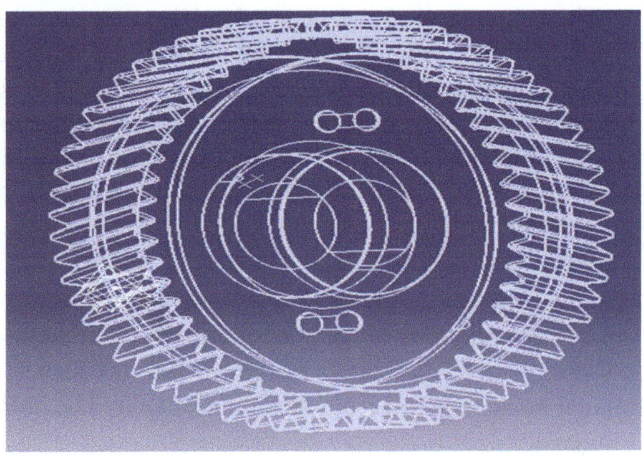

the gear by angle 7° and a complete single teeth is created of gear as wireframe structure.

As single teeth is created of gear wheel, the whole is structure is rotated around 360° and total gear teeth profile of wheel is created in wireframe structure. The complete reconstructed new wireframe model is generated from scanned gear wheel as shown in Fig. 10.

The process of surface generation along the wireframe structure is done and surface 3D model is generated in Generative Shape Design module. As Surface geometry of new modified gear wheel is created. It is converted to the solid 3D model as shown in Fig. 11 and part verification is done to avoid any mistake in the modifications and reconstruction of damaged and broken area.

Fig. 11 CAD model of gear wheel

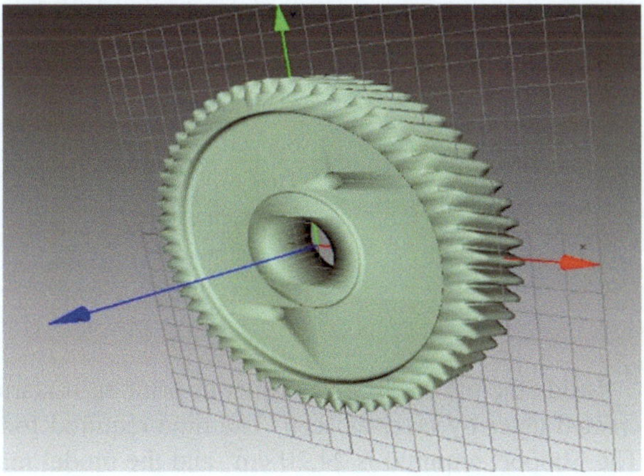

Fig. 12 STL model of gear

Solid 3D model of new gear wheel can be exported in various file format as per the requirement for further processes like STL, IGES, STEP, etc. file format. It is exported in STL file for rapid prototyping of new 3D model as shown in Fig. 12.

6.3 *Rapid Prototyping of New Created Parts*

'uPrint SE Plus' is used as RP machine which made by US based company 'Strasys' which uses FDM Technology to build in real ABSplus thermoplastic.

The .stl file of gear wheel model is imported to the CATLYST EX build-preparation software. CATLYST EX software slices and positions a 3D CAD file and generates a route to extrude ABS material and calculates any need of support material to the model (Fig. 13).

Fig. 13 3D printed model of gear

Table 1 Summary of properties gear wheel

Part	Gear wheel
Model material	188.86 cm^3
Support material	24.20 cm^3
Time	7 h 24 min
Layer thickness	0.3350 mm
Model interior	Solid
Support fill	Smart

Inside the uPrint SE plus printer as printing is started, 3D model of gear wheel is created layer by layer as seen. The time required for model of gear is 7 h 24 min. The Support required is 24.20 cm^3 and the model material required is 188.86 cm^3. As the process of printing of new model is completed, support material is to be broken away or dissolves it in solution cleaning solution for support (water and detergent), and the solid gear wheel model is ready to use for further process (Table 1).

7 Conclusions

On the basis of this work it is given a breakthrough to use of RE processes in product design, and services of machines and equipment and their repair work. As in the case study a solid 3D model of gear wheel is created. 3D model is plays an important role as backbone and importing it into the different file format such as STL, IGES, STEP which serves as platform for CAD/CAM and simulation. 3-D model of gear wheel can be used in different process like CNC Code Generation,

Rapid Prototyping, Tooling Die and Mould Design, Manufacturing Drawing (2D Drawing), Manufacturing Process Simulation or Optimization (Casting, Machining, Etc.). 3D printed model of gear wheel could be used as master pattern for casting process to make number new parts from mould generated. So, masterpiece generation is easier for complicated parts.

This study gives an integrated design framework reverse engineering and rapid prototyping for reconstruction of damaged parts in industry and production systems. This investigation gives a new path for reduction in time consumption in damaged product development.

References

1. M. Dubravcik, S. Kender, Application of reverse engineering techniques in mechanics system Services. Procedia Eng. **48**, 96–104 (2012)
2. T. Varady, R.R. Martin, J. Cox, Reverse engineering of geometric models-an introduction. Comput. Aided Des. **29**(4), 255–268 (1997)
3. M. Sokovic, J. Kopac, RE (reverse engineering) as necessary phase by rapid product development. J. Mater. Process. Technol. **175**, 398–403 (2006)
4. S.C. Kumawat, A.L. Gawali, Application of CAD in rapid prototyping technology. J. Eng. Res. Stud. E-ISSN 0976-7916
5. B. Gurumoorthy, Reverse engineering of solid models. J. India Inst. Sci. **76**, 93–107 (1996)
6. Y.T. Xie, W.L. Sun, J. Yang, Application of reverse engineering and rapid prototyping on new products development. New Technol. New Process **11**, 25–27 (2007)
7. K.A. Igle, *Reverse Engineering* (McGraw-Hill Inc., New York, 1994), pp. 1–35
8. The Easy3D website online. Available at http://www.easy3d.co.il/il/
9. A.P. Valergaa, M. Batistaa, R. Bienvenidoa, S.R. Fernández-Vidala, C. Wendtb, M. Marcosa, Reverse engineering based methodology for modelling cutting tools. Procedia Eng. **132**, 1144–1151 (2015)
10. D.K. Pal, L.S. Bhargava, B. Ravi, U. Chandrasekhar, Computer-aided reverse engineering for rapid replacement of parts. Defense Sci. J. **56**(2), 225–238 (2006)
11. S. Ashley, From CAD art to rapid metal tools. Mech. Eng. **119**, 82–87 (1997)
12. A. Kumar, P.K. Jain, P.M. Pathak, Identification of wear in gear teeth by reverse engineering approach. Int. J. Precis. Technol. **4**(1/2), 46 (2014)
13. X. Yan, P. Gu, A review of rapid prototyping technologies and systems. Comput. Aided Des. **28**(4), 307–318 (1996)
14. P.M. Pandey, Rapid prototyping technologies, applications and part deposition planning. Researchgate
15. C.K. Chua, K.F. Leong, *Rapid prototyping: Principles and application in manufacturing* (Wiley, Singapore, 1997), pp. 17–24
16. S.O. Onuh, Y.Y. Yusuf, Rapid prototyping technology: applications and benefits for rapid product development. J. Intell. Manuf. **10**, 301–311 (1999)
17. S. Upcraft, R. Fletcher, The rapid prototyping technologies. Assembly Autom. **23**(4), 318–330 (2003)
18. J.C. Ferreira, N.F. Alves, Integration of reverse engineering and rapid tooling in foundry technology. J. Mater. Process. Technol. **142**, 374–382 (2003)
19. The RangeVision website online. Available at http://rangevision.com/en/

The Impact of Additive Manufacturing on Indian GDP

J. Avinash, K. UdayKiran and K. Srujita

Abstract The 3D printing is popularly known as an Additive Manufacturing (AM) technology. In this process, a three-dimensional object is created by laying down successive layers of a material. In the recent years, this revolutionary method is one of the most eye-catching technological innovations. In addition, this technology has numerous potential applications in various industries. As a result, this process has a great technological and economical influence on nation's growth. This current paper presents rapid prototyping, rapid manufacturing and the latest technologies available to fabricate 3D components in particular to engineering components. In case of electronic Industries, products such as sensors and switches have been made using 3D printing technology Optimization of printing parameters during 3D printing makes a component with improved physical and mechanical properties. As compared to conventional manufacturing technologies such as melting, casting, rolling, forging, etc., AM technologies produces materials with low cost production. An attempt has also been made to study the influence of AM and its direct impact on Indian GDP. Moreover, this detailed analysis would benefit various Indian academic, research institutes, manufacturing industries, etc.

Keywords Additive manufacture · 3D printing · Rapid prototyping
Indian manufacturing · GDP

J. Avinash (✉) · K. UdayKiran
Department of Mechanical Engineering, CMR Institute of Technology,
Bangalore 560037, India
e-mail: avinash.j306@gmail.com

K. Srujita
Department of Electronics & Communication, PES Institute of Technology-BSC,
Bangalore 560100, India

L. J. Kumar et al. (eds.), *3D Printing and Additive Manufacturing Technologies*,
https://doi.org/10.1007/978-981-13-0305-0_15

1 Introduction

1.1 Additive Manufacturing

Additive Manufacturing (AM) is an advanced manufacturing method which builds 3D objects by adding material layer by layer to produce the object. Additive manufacturing processes utilize the data from a computer-aided design.

(CAD) file that is later converted to a stereo-lithography (STL) file. These materials could be metals, polymers, composites, or other powders to "print" a scale of functional components, layer by layer, including complex structures that cannot be manufactured by any other techniques. These new methods, while still progressing are forecasted to create an acute impact on manufacturing [1]. The segment wide repercussions of this ability have caught the creative impulses of investors. Improvements in additive manufacturing are essential to make AM a cost-effective and significant method of production.

1.2 3D Printers

The 3D printing is a method of additive manufacturing that performs rapid prototyping using a CAD (Computer-Aided Design). 3D Printing is becoming more and more affordable. The basic principles include materials cartridges, flexibility of output and translation of code into a visible pattern. The stereo-lithography was developed by Charles Hull in the year 1984. The 3D printing occurred in 2006 with the commencement of an open source operation named Reprap, which was aimed at developing a self-replicating 3D printer Versatility to manufacture a large scale of products, where 3D printing can be done near to the point of consumption, indirectly having a massive impact to the supply chains and business model, many operations can be eradicated like industrial distribution, warehousing and retail. An important feature of 3D printing is that a single equipment can produce different products and also create components that cannot be produced using conventional means.

Prototyping also gives designers the privilege to touch and test the product in the early design stage itself, thus avoids changes incurred in the later stage of the process thus saving time and money, significantly plus manufacturers can consequently minimize the development lifecycle [2]. The 3D printing permit precise control of the material being used; A designer can remodel the structure of the product for a phenomenal effect. These 3D printers can possibly work in dual ways: a material is initially melted and is ejected through a mini nozzle on to the build area, where the material hardens and rises the body up progressively (layer by layer) or a powdered material is made to settle on the bed. This process demands for post-production processes such as cleaning the surplus powder, baking to attain strength and hardness.

Furthermore, as the natural resources becoming very expensive, sustainability is becoming an important aspect to manufacturers in India and around the globe. The economics of manufacturing is less labour incentive, consumes less material and also produces small amount of waste and can use new materials that are light in weight. It is used in various industries like engineering and construction, automotive, aerospace, industrial design and medical industries, education and consumer products also includes the fabrication of electronic components such as transistors, conductive and photovoltaic structures, mechanical actuators and sensors, moreover more materials are gradually available to fabricate parts and thereby reducing the usage of machines [2, 3]. Hence, RP technologies have tremendous potential to assist modern manufacturing globally and in India.

1.3 Types of 3D Printing Technology

A few technologies are commercially available methods of fabricating prototypes, others are quickly becoming industrially viable forms of production manufacturing, and modern technologies are rapidly being developed there are several types of RP technology available. Some are discussed below.

1.3.1 Fused Deposition Modelling

Fused Deposition Modelling (FDM) is commercialized by Stratasys in Eden Prairie, Minnesota.) and is one of the most popularly known techniques in additive manufacturing exclusively used for modelling, prototyping and production purposes. In this process, a plastic or wax matter is ejected through a mini nozzle that follows the cross-sectional area of the object (layer by layer). These plastics are extruded as a semi-molten filament. The 3D CAD data is processed and the filament is deposited in layers eventually resulting in construction of the desired object. FDM makes use of two kind of materials to print such as modelling material which is used to construct the prototype and the support material acts as a platform. Material filaments are moved in the X and Y coordinates, resulting in deposition of material before the base moves down in the Z-direction and the corresponding layer begins. All movements are directed by software that provides a track for the nozzle to follow. The consumer breaks the support material away or dissolves it in detergent and water, and the element is ready to use. The materials that are prominently made used in this process are polyamide, polycarbonate, polyethylene, ABS, polypropylene and investment casting wax. This technique is a clean, effective and user-friendly 3D printing process (Fig. 1).

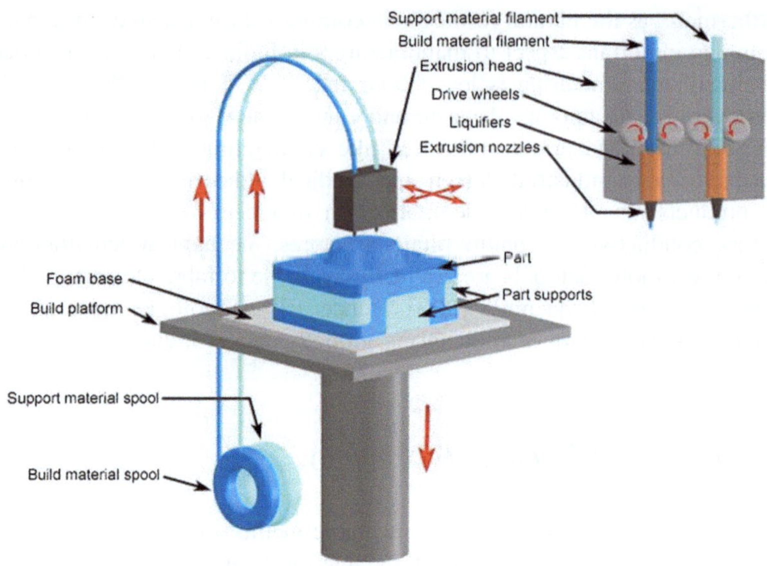

Fig. 1 A schematic diagram showing FDM printer

1.3.2 Stereo-Lithography

In 1986, SLA process is developed by 3D systems of Valencia, California, USA. The SLA rapid prototyping is the earliest technique to be incorporated in this field in the early 1980s and is still being dominantly used in the recent times. In this technique polymerization occurs with the introduction of low-power laser to reinforce photosensitive resin [3, 4]. A 3D CAD file is digitally sliced into a series of horizontal cross-sections which are parallel to each other and later is given to a Stereo-Lithography Apparatus (SLA) one at a time. The cross section of a bath of polymer resin is traced by a laser and further the cross section is made to solidify. The platform rises along the *z*-axis after one layer is completed, and additional layer of resin is coated on top of the cured layer (Fig. 2).

1.3.3 Selective Laser Sintering

The Selective laser sintering (SLS) is introduced to rapid prototyping by Carl Dekard and his colleagues at the university of Texas in Austin. SLS process is an useful manufacturing process based on the use of powder coated metal additives, a process used for rapid prototyping [4]. The surface of the build cylinder is spread by the thermoplastic powder assisted by a roller. Sintering occurs when fabrication chamber is maintained at a temperature slightly below the melting point of the powder and hence the laser raises the temperature (Fig. 3).

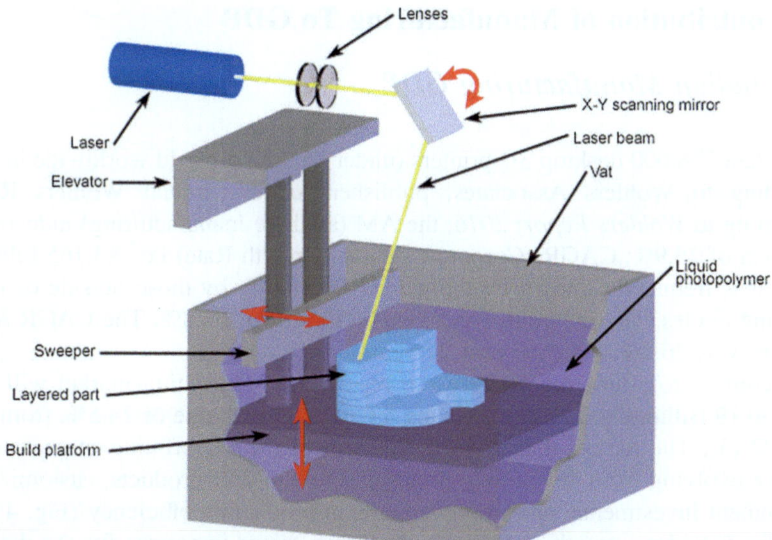

Fig. 2 A schematic diagram showing SLA printer

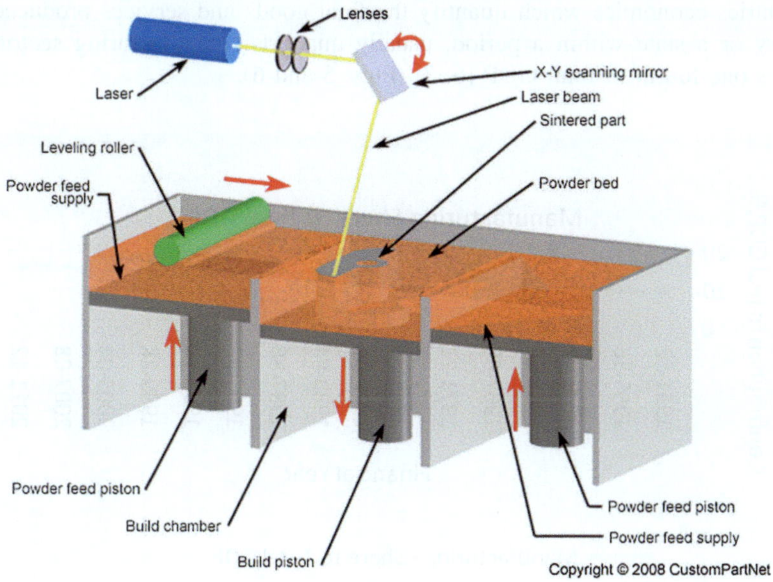

Fig. 3 Schematic diagram showing SLS printer

2 Contribution of Manufacturing To GDP

2.1 Indian Manufacturing GDP

More than 278,000 desktop 3D printers (under $5,000) are sold worldwide in 2015, according to Wohlers Associates, publishers of the annual Wohlers Report. According to *Wohlers Report 2016*, the AM (additive manufacturing) industry had a growth of 25.9%, CAGR (Corporate Annual Growth Rate) i.e., $5.165 billion in 2015. It is frequently known by the name 3D printing by those outside of manufacturing circles. The previous 3 years had a CAGR of 33.8%. The CAGR for the industry was 26.2%, over the past 27 years.

According to a new market research report, the 3D printing market will reach USD 30.19 Billion by 2022, i.e., having a CAGR growth rate of 28.5%, from 2016 to 2022 [5]. The factors that drive the growth of the 3D printing market are 3D printing evolving from developing prototypes to end-user products, customization, government investments, and improvements in production efficiency (Fig. 4).

India considers growth in the manufacturing sector important for the development of the economy. In the last decade, it is quite evident that there is a major shift in the composition of GDP with the contribution of manufacturing sector, GDP has recorded a continuous increase over the years [6]. GDP is a very vital component of a countries economics which quantify the final goods and services produced in a country or a state within a period, usually quarterly, manufacturing sector contributes one-fourth of total GDP [6, 7] (Figs. 5 and 6).

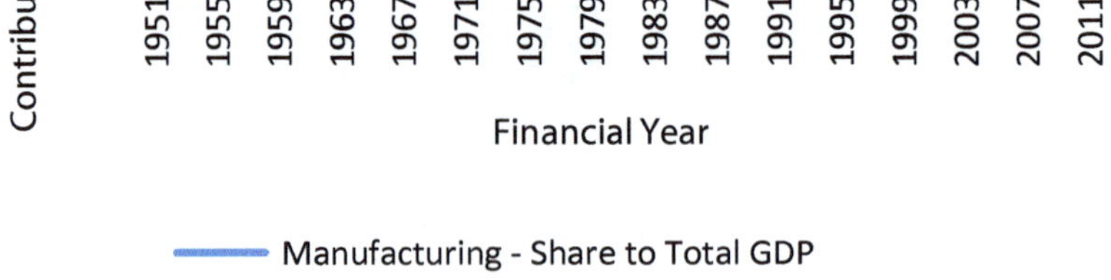

Fig. 4 Manufacturing sector contribution to real GDP over time

Fig. 5 Composition of manufacturing in GDP

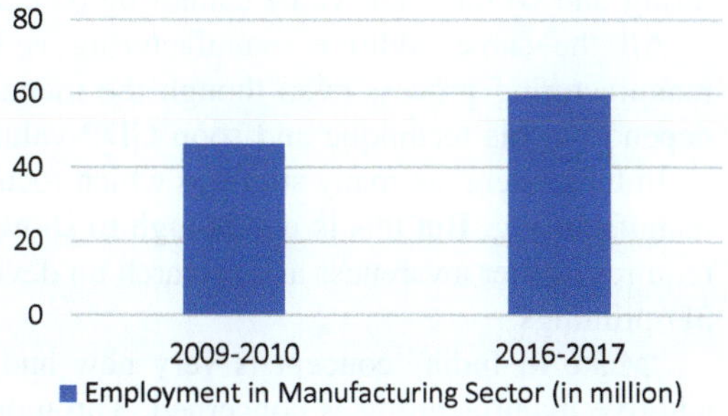

Fig. 6 Employment in manufacturing sector

2.2 Employment in Manufacturing Sector

In 2009–2010 the employment was 46.5 million. During 2016–2017 the employment increased by 22.51% that is 13.51 million. By this data we can say that employment is increasing every year. So by creating more opportunities through the introduction of a sustainable manufacturing technology employment can be increased which increases the production which in turn increases the GDP value [8].

The 3D printer or the AM processer can now be bought at affordable price as a result it can be used by most of the people. Economists reported that wastes could be reduced to a great extent as the accurate dimensions are noted and followed. The transporting charges of the parts can be controlled as they are digitally transferred to the required places. A Few parts are printed and are assembled by using traditional techniques, The number of parts to assemble reduces drastically, this makes production cost fall significantly. It also makes the supply chains simplified and shortened [5].

3 Effects of Additive Manufacturing on India

The additive manufacturing is a process in which a real physical object is created from a virtual 3D model. This technology is a rapidly evolving. This is used for different industries in India.

The possibility of adopting and integrating additive manufacturing technology into the manufacturing sector completely is not an easy task as it is still in research stage. If it is integrated, the sector could grow exponentially because it reduces the time to produce, It also saves a lot of raw material that is wasted for multiple unnecessary prototypes. In India additive manufacturing is just at the developing stage and is not used in large scale production. It is a long process in using this technique in all the sectors of manufacturing, since this process is still not used in industries. Therefore it is difficult to consider it as a separate sector of manufacturing and so the GDP value cannot be precisely evaluated.

All the same additive manufacturing is being used in small parts of huge manufacturing process even though the manufacturing process cannot completely depend on this technique and soon GDP value will find its place in India.

In India there are many start-ups which focus on development of 3D printing and manufacturing. But this is not enough to show a significant growth in GDP rate. It requires further awareness and research on development for efficient and low cost of 3D printings.

"Make in India" concept is very new and at the experimental stage as far as additive manufacturing is concerned. Through this it gives a better scope to make use of additive manufacturing techniques in the mass production of goods.

3.1 Applications

As mentioned above additive manufacturing can be used in different industries. They are mainly used for prototyping. Hence the word rapid prototyping is derived. This process has wide range of uses in architecture, construction, aerospace, dental and medical technology, fashion, food industry and so on. With the integration of 3D digitizer, 3D sensor, 3D scanner, the possibilities are endless.

Engineering: It is adopted in iterative designs, robust parts and precision prototyping. This could enhance designs and makes it easy for engineering teams to manage risks and understand the performance of their designs. Development cost and time are reduced with increase in quality and durability. It gives earlier performance insights in the development stage, where design changes can be made with less cost. 3D printing can enable earlier problem resolution, which can drastically reduce the costs associated with the product lifecycle.

Construction: In the construction industry, construction components or the entire buildings can be printed using this technology. The recent growth of building information modelling (BIM) may facilitate greater use of 3D printing. 3D printing will allow faster and more accurate and complex construction by lowering labour costs and wastage. It also enables construction in dangerous environments like space where humans cannot work safely.

The printer prints the building blocks from molten plastic or metal. The building blocks that are printed, are used to form component parts that can be arranged together like Lego to create a complete building.

Metal can be used and structures like bridges can be printed. Huge 3D printers are used to do this kind of jobs. Even the printing time is also very less, it is claimed that a house was printed in less than 20H.

3D printing in this industry is helpful in constructions in areas such as off-Earth habitats like on the Moon or Mars (Fig. 7).

Architecture: The 3D printer can be used for beautiful and durable models in order to showcase innovative designs [9]. Complicated designs can be made easily in less time (Fig. 8).

Defence: In military and defence manufacturing, customized equipment is vital and deadlines are non-negotiable. In such an industry 3D printing gives freedom for the government and manufactures, to design a single end-use part or to build a complex and precise prototypes. 3D printing enables design teams to rapidly produce high-quality precise and realistic prototypes, which has moving parts at relatively less cost when it compared to other traditional machining process. It can also reduce the outsourcing costs. The team can manufacture a part then and there when it is required. 3D printed parts has less weight and high strength which increase efficiency of the objects and reduces it carbon footprint.

During disasters, shelters can be printed on site. Special blend of cement is used in printing the walls of this shelters, which has strength greater than conventional walls.

Fig. 7 A view of 3D printed house

Fig. 8 A view of complicated architectural structure

Aerospace: Today, NASA has made such an advancement in aerospace that they have started testing rocket parts built by 3D printing and they may even use the same technology to build habitat in space. The achievements in aerospace industry using 3D printer:

- The Vector-R satellite launch vehicle has been successfully sent towards space with an essential 3D printed engine part for the first time.
- A group of students known as the MIT Rocket Team has successfully fired a fully 3D printed rocket.
- Aerojet Rocketdyne has reported two significant milestones have been reached in the development of the 3D printed AR1 rocket engine.

Aviation: In aviation industry, this technology has the capability of reducing aircraft weight, at the same time it increases customization and the overall construction efficiency. The FAA, i.e. Federal Aviation Administration, on a positive note cleared GE Aviation's first 3D printed part to fly, as of March 2015, Boeing has delivered more than 20,000 non-metallic additive manufactured parts that are on airplanes. Recently Air Bus has printed a 13 ft. aircraft, which weighs around 29 kg this aircraft is assembled using 50 different 3D printed parts. It is named as Test of High-tech Objective in reality (THOR). If any part is damaged, it can be easily replaced by printing that part again. The same technique can be implemented for a large size aircraft also.

Electronic Industries: Electronics can be used in prototyping as well as direct manufacturing of antennas, interconnects, PCBs and other components. This technology is used for electronic package application. The printed electronics market is well established and will see significant growth over coming years that will surpass the whole 3D-Printing market [10]. Recently optimec, has demonstrated the ability to print electronic circuits over 3D electronic packaged surfaces down to feature sizes of 10 μm at a reasonable time. Optimec and Stratasys undertook a joint venture to investigate a combination of Fused Deposition Modelling and Aerosol Jet to develop packages for sensor applications.

Biomaterials: We can use 3D printer to dispense biological materials. It is done by positioning biomaterials and living cells layer by layer. By using this technology different bio-parts like tissue, bones, heart valves can be printed [11]. In near future we can even print organs like kidneys also.

Bones: we can use hydroxyapatite a calcium mineral similar to one found in bone (It is mixed with polymers) to print bones [12].

Tissues: An integrated tissue–organ printer (ITOP) is capable of fabricating stable, human-scale tissue constructs of any shape. Mechanical stability is gained by printing cell-laden hydrogels together which uses biodegradable polymers in integrated patterns and anchored are on sacrificial hydrogels. Even organs can be printed [13]. Regenerative medicine applies 3D printing technology to make tissues and organs that are suitable for transplantation [11].

Different bio products printed till date are tissues, Tissues with blood vessels, Low–Cost Prosthetic PartsDrugs, Tailor-made sensors, Medical Models, Bone, Heart Valve, Ear cartilage, Cranium Replacement, Synthetic skin, liver, kidney. Hundreds of thousands of people worldwide are waiting for an organ donor, this technology could transform their lives (Figs. 9 and 10).

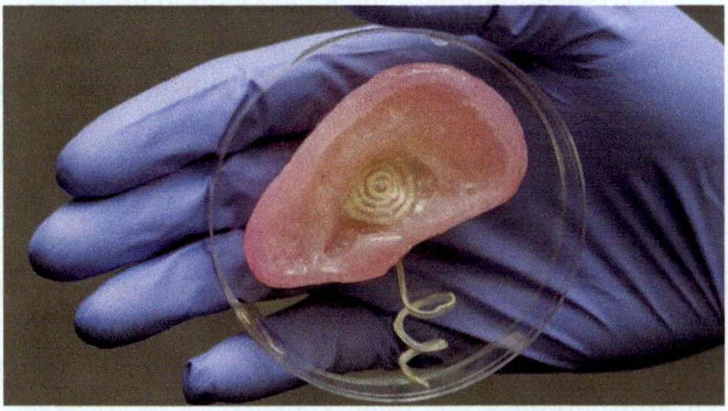

Fig. 9 A view of 3D printed ear cartilage

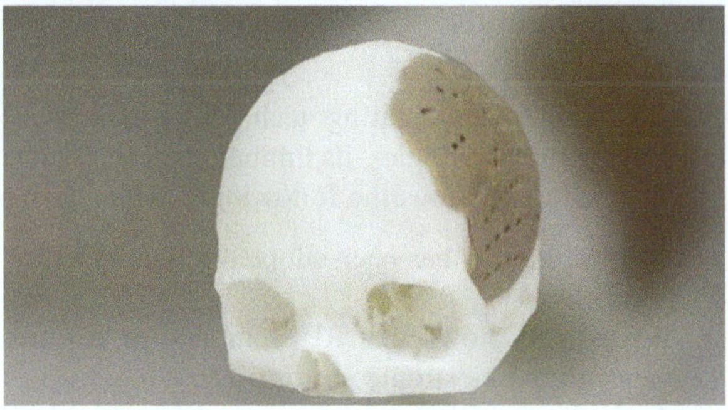

Fig. 10 A view Cranium replacement

4 Future Project Scope

The AM technology have been available in the market for over 30 years, we recently started to witness their more widespread adoption. Future scope of AM processes has been widely increasing due to the decrease in cost and development of direct metal technologies. We are able to visualize a disruption in the process of designing and manufacturing the products. The 3D printing requires high energy sources and complex scanning algorithms and devices, so it focuses on reducing complexity and cost to establish this technique in bulk.

Additive manufacturing will spread wide in next few years first through two paths, i.e. classrooms and then through homes. Many researchers are finding potential uses for bio-printing. Researchers from AFIRM and NIH are now trying to produce artificial kidneys, viable human tissues and simple organs for transplantations. These are very much helpful to treat soldier's wounds. The licences for use of 3D bioprinters in food and drug administrations are still at pending. In future these printers are used to print spare parts for battlefields, business requirements and even for the engineers or designers to show up their ideas.

The transportation of the manufactured products increases carbon foot print. This can be reduced by sending the designs around the world, rather than shipping the products. Finally, the products are printed using the design in the place where they are nearer to use. This could mainly help the army or security sources to easily manufacture as they need spare parts. In space required tools can be printed in the space stations itself. In future even houses can be printed on mars while colonizing. The AM process is eco-friendly. Material is added layer by layer so only the required amount of material is used. So there is no waste produced during the process. The Economist predicts that this technology will have a great impact on manufacturing, since 3D printing will rewrite economies of scale by making it as cheap to produce one item as many.

5 Conclusions

The additive manufacturing technology has been discussed in detail, Various processing methodologies, its limitations and advantages were also discussed. From the detailed discussions the following conclusions are drawn and presented below:

- AM technology has been adopted to manufacture a large number of engineering components.
- It was also shown that RP technologies are most useful in the medical, aeronautical and electronic sectors.
- A significant GDP growth was achieved under Indian economical conditions. Moreover, conventional and widely used manufacturing technologies are replaced with AM technologies.

- An adoption of RP technologies for manufacturing a of large number of engineering products would be suitably considered for the "Make in India" program, India's most important technological innovative programme.

Acknowledgements The authors would like to acknowledge principals of CMRIT Bangalore and PESIT-BSC for giving us an opportunity to work in this area.

References

1. D.A. Arcos-Novillo, D. Güemes-Castorena, Development of an additive manufacturing technology scenario for opportunity identification. Futures **20**, 1–15 (2017)
2. R. Schmid, S. Kalpakjian, *Manufacturing Engineering and Technology* (Pearson Prentice Hall, Upper Saddle River, 2006)
3. A.K. Kamrani, E.A. Nasr, *Rapid Prototyping: Theory and Practice,* Vol. 6 (Springer, Berlin, 2006)
4. J.P. Kruth et al., Lasers and materials in selective laser sintering, in *Proceeding of 3rd Laser Assisted Near shape Engineering Conference (LANE-2001)*, Erlangen, Germany: Lehrstuhl fur Fertigungstechnologie, pp. 3–24 (2001)
5. https://www.forbes.com/sites/tjmccue/2016/04/25/wohlers-report-2016-3d-printer-industry-surpassed-5-1-billion/#22a8be9c19a0
6. https://data.gov.in/catalog/gdp-india-and-major-sectors-economy-share-each-sector-gdp-and-growth-rate-gdp-and-other
7. https://data.gov.in/catalog/manufacturing-gdp-sector-and-employment-projections
8. A. Abeliansky, I. Martínez-Zarzoso, K. Prettner, The impact of 3D printing on trade and FDI. cege, Number 262–October 2015
9. I. Hager, A. Golonka, R. Putanowicz, 3D printing of buildings and building components as the future of sustainable construction? in *International Conference on Ecology and new Building materials and products*, ICEBMP 2016
10. C. Bailey, S. Stoyanov, T. Tilford, G. Tourloukis, *3D-Printing And Electronic Packaging* (University of Greenwich, London, 2016)
11. S.V Murphy, A. Atala, 3D bioprinting of tissues and organs. Nature America, Inc. (2014)
12. V. Mironov et al., Organ printing: computer-aided jet-based 3D tissue engineering. Trends Biotechnol. **21**(4), 157–161 (2003)
13. S. Bose, S. Vahabzadeh, A. Bandyopadhyay, Bone tissue engineering using 3D printing. Materials Today, **16**(12), 496–504 (2013). *Additive Manufacturing*, Atlantic Council, Washington, DC (2011)
14. G.N. Levy, R. Schindel, J.-P. Kruth, Rapid manufacturing and rapid tooling with layer manufacturing (LM) technologies, state of the art and future perspectives. Cirp Ann. Manuf. Technol. **52**(2), 589–609 (2003)
15. C.M. Cheah et al., Rapid sheet metal manufacturing, part 2: direct rapid tooling. Int. J. Adv. Manuf. Technol. **19**, 510–515 (2002)
16. I. Gheorghe, *Microtehnologii avansate prin prototipare rapidă cu sinterizare selectivă cu laser* (Editura CEFIN, Bucureşti, 2010)

Optimization of the Print Quality by Controlling the Process Parameters on 3D Printing Machine

R. Devicharan and Raghav Garg

Abstract Fused Deposition Modelling is one of the most widely used Additive Manufacturing technologies for various engineering applications. The present work investigated the aesthetics of the cube. In FDM process of printing, one of the major problems faced in printing the objects, specifically the cylindrical objects is the non-sticking of the preliminary layers of the molten filament on to the hot bed. There are other serious problems which occur while printing they are under extrusion, over extrusion of filament from heated printing nozzle, gaps occurring in the top layers of the printed object, stringing or oozing of the extra filament from the heated nozzle, overheating of print bed, Layer shifting, layer separation and splitting, grinding filament, clogged extruder, extruder stopping in the middle of printing process, gaps between infill and outline, etc. In this study, the above-mentioned problems are observed and identified through Pareto analysis that on the machine, the non-sticking of the preliminary layers to the hot bed occurred more, thus solutions are provided to avoid this problem which in turn has helped to reduce various other problems occurring while printing the 3D objects.

1 Introduction

Fused Deposition Modelling (FDM) is one of the most widely used Additive Manufacturing (AM) technologies for producing components for various applications. In this method 3-D modelling of the product is made by using Computer Aided Design (CAD) software. The geometric data is further converted in the form of slices to generate the print path [1]. The products are manufactured by depositing the heated material layer by layer. Advantages of this technology are lower cost,

R. Devicharan (✉) · R. Garg
Printing and Media Engineering Department, Manipal Institute of Technology, Manipal University, Manipal, India
e-mail: devicharan.r@manipal.edu

L. J. Kumar et al. (eds.), *3D Printing and Additive Manufacturing Technologies*,
https://doi.org/10.1007/978-981-13-0305-0_16

shorter production time and ability to print objects of any complex shape [2]. Potential applications of this technology are in the field of automobile, fashion designing, architectural models and medical implants [3]. This technology helps to designers to demonstrate their ideas quickly at lower cost and any required modifications can be incorporated in the design [4]. Most widely available AM technologies are direct metal deposition, selective laser sintering, stereolithography, digital light processing, fused deposition modelling, electronic beam melting, laminated object manufacturing. These processes differ mainly in the type of the raw material used in the fabrication process 3D objects [5]. Due to its ability to produce complex geometrical shapes accurately in a relatively shorter duration, FDM is widely used in the industries for prototyping small scale and small size applications [6]. Figure 1 shows typical FDM process. In this process the material is heated to plastic state in the heating liquefier head and then selectively deposited on to the hot bed. The path of deposition is controlled by three-axis computer numerical controller.

A wide range of plastic material are available in the market which include the ABS Acrylonitrile Butadiene Styrene, PLA Poly Lactic Acid, Nylon Polyamide, PET Polyethylene Terephthalate, TPU Thermoplastic Poly Urethane, etc.

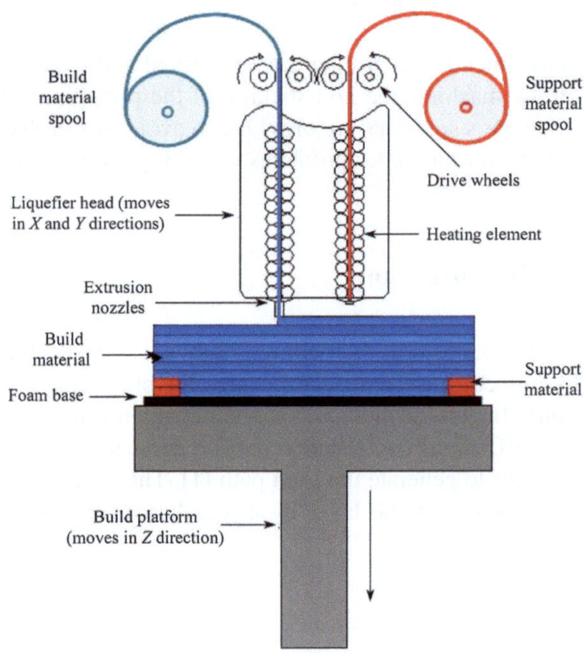

Fig. 1 Working of the FDM process of printing

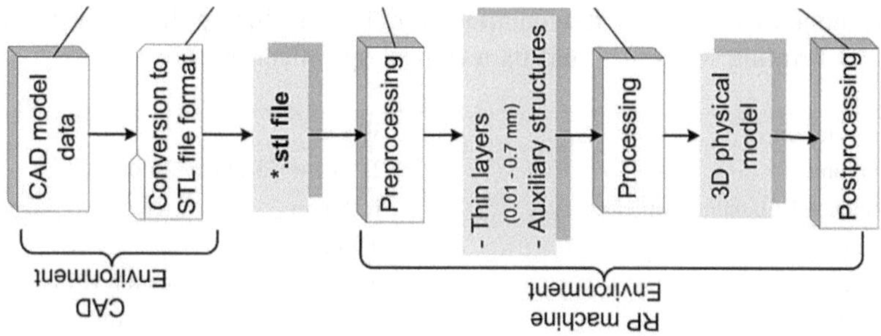

Fig. 2 Workflow followed to print the sample on the 3D printing press

2 Methodology

The methodology followed in the present work is shown in the form of flow chart as shown in Fig. 2.

2.1 *Work Material*

ABS is a theromoplastic resin which melts at 200 °C (392 °F) on the FDM machine, it is derived from acrylonitrile, butadiene and styrene. Ammonia and propylene are used to produce the Acrylonitrile and butadiene which is a petroleum hydrocarbon produced by butane. Finally the styrene monomers are derived from benzene and ethylene from coal. The features of ABS are it is medium strength, good chemical resistance, electroplatable, formability, high impact strength, tensile strength and stiffness and also excellent ductility. It is mainly used to make light, rigid, molded parts such as the piping, musical instruments, prototypes, enclosures, etc.

In this study ABS is utilized and a compendious review of FDM process, its technical problems and solutions and has put light on some research gap on which further research and development works can be directed to make this technology deliver products with high level of accuracy, quality and desired properties at the fastest speed possible.

2.2 *Experimental Set Up*

In this study "Pronterface" is the printer interface is utilized. This software helps to control the position of the printer head to position the molten filament which is deposited in thin layer form. The FDM machine is set according to the technical specifications mentioned by the manufacturer. The machine utilized for the study is "Julia 3D Printing Machine" and its technical specifications are

Layer resolution	80 μm
Build volume	21 cm × 25 cm × 26 cm
Recommended filament diameter	1.75 mm
Nozzle diameter	0.5 mm
Print technology	Fused filament fabrication (FFF)
Frame dimension *X Y Z*	48 mm × 47 mm × 42 mm
Operation nozzle temperature: 180–260	220 °C
Software	Pronterface

3 Model data

3.1 Process Parameters

Bead (road width)	0.200
Air gap	0
Model build temperature	230°°C
Raster orientation	90°
Sample size	$L \times B \times H = 1$ in.

It mainly included the process parameters such as the

Bead (Road Width): this is the thickness or the road at which the nozzle of the FDM printer will deposit the molten plastic filament. It vary from 0.012 to 0.396 for the T12 nozzle which is installed on the FDM machine.
Air Gap: Indicates the space available between the beads of the FDM material. Generally the default is zero which means that the beads just touch each other. It can be altered to provide a positive or a negative gap which means the beads may or may not touch each other lines. This will directly influence the strength of the object printed on the FDM machine. The time taken to print the object is exponentially varied according to this specification.
Model Build Temperature: This is the temperature of the heating element utilized to heat the material used for printing. It has a control over the molten material as it is extruded from the nozzle. 230° is set to melt the ABS material this helps the ABS to stick to the Kapton tape, PET tape put on the hot bed of the 3D press.
Raster Orientation: This is the direction in which the molten filament is drawn layer by layer on to the printing bed. It is set at 90° in this study.

The factors such as the envelope temperature (The temperature of the air around the part) is not considered in this study. The process followed to conduct this study

Fig. 3 Top view the cube samples printed on the 3D printing machine

is shown in Fig. 2, given below and the sample which are printed on the machine are shown in Fig. 3. The specification of the printed samples are Length × Breadth × Height = 1 in. Cube.

4 Results and Discussion

Figure 4 represents the Pareto analysis done to present the problems identified on the 3D printing machine. It is observed that at the initial stages "Print not sticking to the bed" (Refer Fig. 3) is a major problem which has occurred 96 times when printing 100 numbers of test samples. Thus the cause for this problem is identified and fixed by applying suitable solutions.

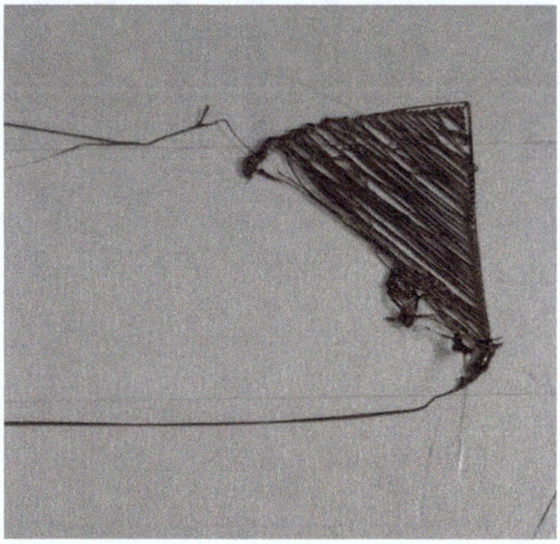

Fig. 4 Image displaying the problem—print not sticking to the bed

The major causes for the above-mentioned problem are

1. Uneven levels of the build platform
2. The nozzle starts too far away from the bed
3. First layer printing too fast
4. Lower bed temperature settings
5. Absence of Kapton Tape
6. The Brims and Rafts (Figs. 5 and 6)

The initial layers of the print need to be of good quality as they form the foundation of the complete sample. Due to the variation of the position of the screws or knobs which determine the levelness of the print bed. This problem is occurring. The screws and the knobs were checked and tightened and made sure that the plate was level and the print nozzle comes in contact with the print bed equally on all points of the bed.

The correct height relative to the build platform is achieved by locating the extruder in the perfect distance away from the build plate not too far, not too close. This is made using the printer interface "Pronterface" used to control the machine at the beginning of the printing operation. Each layer of the part is usually around 0.2 mm thick, so minor adjustments of 0.005 was made to obtain the desired results.

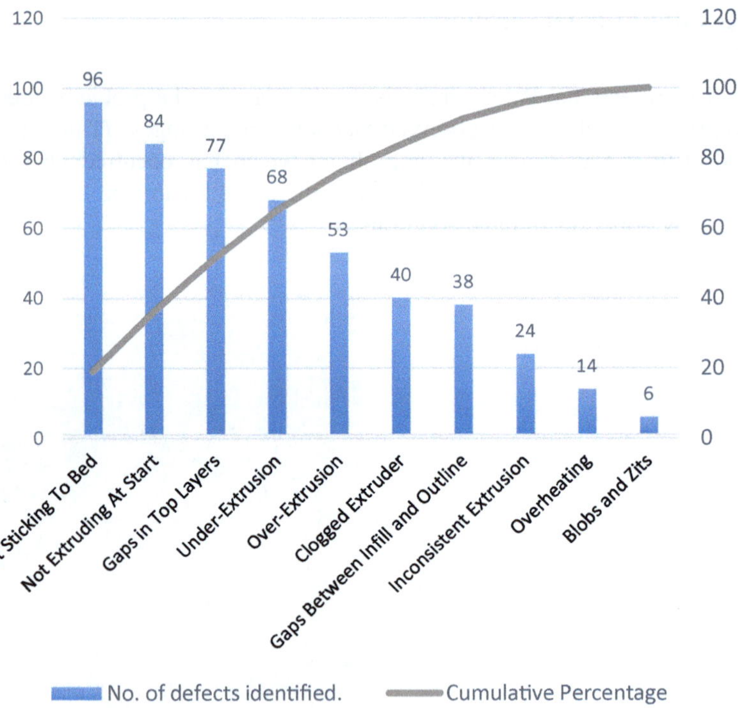

Fig. 5 Problems identified on the machine before troubleshooting

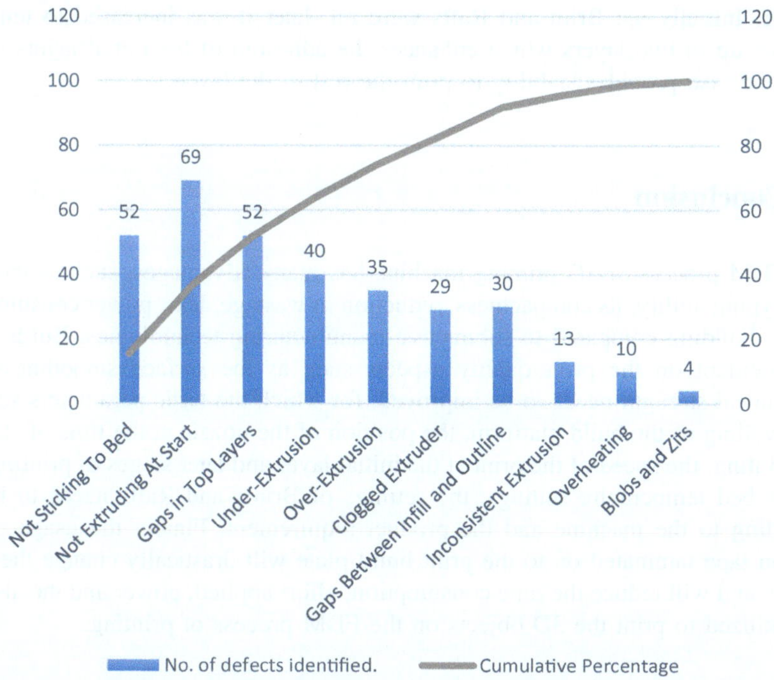

Fig. 6 Problems identified on the machine after the troubleshooting

The printing speed of initial layers was as same as the rest of the layers of the sample which was of 150 mm/s. Thus a lot of failures occurred. Later the speed was reduced to 50% that is to 75 mm/s which resulted in slower deposition and good adhesion was observed between the print bed and the initial layers of the print.

The molten ABS coming from the heated nozzle needs to be deposited on to the print bed which is heated up to 100–120 °C, if the bed temperature is lower, the initial layers of print does not stick to the bed. The cooling fan also needs to be switched off while printing initial layers to avoid to sudden cooling of the initial layers and by giving time for gradual cooling thus reducing the damage of the printed sample.

Initially the printer build plate was not laminated with any plastic film nor was given any special coating or treatment to enhance the receptivity and adhesion of the initial print layers, later the Kapton tape (Polyimide film) which are heat resistant are laminated on to the build plate and the plate is heated to 100–120 °C before printing, this has solved the problem of the non-adhesion of the initial layers of the print to the print bed.

The Brim and Rafts are the supporting extra layers which can be created around the sample to be printed to enhance the support for the initial layers of the print. Note: These can be removed easily without any damage to the actual printed sample. Initially the Brim and Rafts were nil, later it was increased to ten skirt outlines up to two layers which enhances the adhesion of the initial layers of the print and also provides stability to print the rest of the layers.

5 Conclusion

The FDM process of 3D printing machine has many advantages, such as the easy prototyping ability, its compactness, reduction in wastage, least power consumption in part building compared to subtractive manufacturing technologies. But a lot of enhancement on the print quality aspects such as the surface smoothness and mechanical strength needs to be improved, for which the basic parameters such as the levelling of the build platform, the position of the nozzle at the time of starting the printing, the speed of the print at the initial layer and later stages of printing, the printer bed temperature settings, the settings of Brims and Rafts needs to be set according to the machine and the product requirement. Finally the usage of the Kapton tape laminated on to the print build plate will drastically change the print quality and will reduce the time consumption, effort applied, power and the also the cost utilized to print the 3D objects on the FDM process of printing.

References

1. V. Vijayaraghavan, A. Garg, J.S.L. Lam, Process characterization of 3D-printed FDM components using improved evolutionary computational approach. Int. J. Adv. Manuf. Technol. (2014)
2. O. Lužanin, D. Movrin, M. Plančak, Effect of layer thickness, deposition angle, and infill on maximum flexural force in FDM-built specimens. J. Technol. Plast. **39**(1), 49–58 (2014)
3. A.K. Sood, R.K. Ohdar, S.S. Mahapatra, Parametric appraisal of mechanical property of fused deposition modelling processed parts. Mater. Des. **31**(1), 287–295 (2010)
4. O.A. Mohamed, S.H. Masood, J.L. Bhowmik, Optimization of fused deposition modeling process parameters: a review of current research and future prospects. Adv. Manuf., 42–53 (2015)
5. G.C. Onwubolu, F. Rayegani, Characterization and optimization of mechanical properties of ABS parts manufactured by the fused deposition modelling process. Int. J. Manuf. Eng. **11**(1), 13–15 (2014)
6. C.C. Kai, L.K. Fai, *3D Printing and Additive Manufacturing*, 4th edn. (World Scientific Publishing Co. Pte. Ltd., 2014), Chap. 1, pp. 2–17

A Review on Current State of Art of Bioprinting

Devarsh Vyas and Divya Udyawar

Abstract Bioprinting, an extension of traditional 3-D printing is the computer-aided additive manufacturing of cells, tissues, and scaffolds to create organs. It has emerged as the most innovative solution to the healthcare catastrophe of organ shortage and transplantation. Bioprinting makes use of rapid prototyping (RP) technology to print cells and biomaterials individually or in tandem, one layer over the other, producing 3-D tissue-like structures which can be reorganized and regrouped together to form vascularized organs. This review paper, sheds light on the current state of the art of bioprinting technology, various bioprinters used and focuses on the potential applications of bioprinted tissues in regenerative medicine. Challenges faced, limitations and future prospects of the technology have also been presented.

Keywords Bioprinting · Bio-additive manufacturing · Tissue engineering
Organ transplantation

1 Introduction

Organ shortage has emerged as a serious crisis despite an increase in willing donors [1]. In the last 10 years, the number of patients requiring an organ has doubled, while the number of transplants has barely increased. This is now a public health crisis. The solution to this problem, similar to the solutions to other complex engineering challenges, requires long-term solutions by manufacturing living

D. Vyas
Biomedical Engineering Department, Thadomal Shahani Engineering College,
Mumbai University, Mumbai, India
e-mail: devarshvyas94@gmail.com

D. Udyawar (✉)
Biotechnology Engineering Department, Thadomal Shahani Engineering College,
Mumbai University, Mumbai, India
e-mail: divyau008@gmail.com

L. J. Kumar et al. (eds.), *3D Printing and Additive Manufacturing Technologies*,
https://doi.org/10.1007/978-981-13-0305-0_17

organs from a person's own cells [1]. Thus various in vitro methods are being researched upon to tackle this problem. These methods involve the development of autologous cells derived from the patient and using them as a fundamental cell source to develop tissues or organs for transplantation. These 3D tissue analogues can be manufactured by incorporating native cells with appropriate biocompatible materials using an accurate fabrication operation [2].

Bioprinting is three-dimensional manufacturing of biological tissue and organs through the layering of living cells. It can be defined as the use of material science and production techniques to manufacture biological constructs with cell viability and function kept intact [2]. The process involves designing of the organ which is done by computer-aided design softwares. The actual printing is carried out through a bioprinter using the software design. The bioprinted construct then goes through tissue remodelling and maturation in a bioreactor, which escalates tissue maturation [3].

Regenerative medicine has been paying enormous attention to bioprinting due to its ability to manufacture organs from native cells. Even though bio-additive manufacturing of an entire organ, which can be transplanted, is yet to be achieved, this technology is pacing up and could soon solve the crisis of organ shortage [2].

2 3D Bioprinting Approaches

3D bioprinting is based onthree principle approaches: biomimicry, autonomous self-assembly and mini-tissue building blocks [4]. A detailed discussion of the three approaches is mentioned below:

Biomimicry: The main aim of this approach is to create fabricated structures similar to the natural structure found in the tissues and organs in our body. Biomimicry requires replication of the shape, framework, and the microenvironment of tissues.

Biomimicry has a potential application in creating indistinguishable cellular and extracellular structures of organs. For this approach to be successful, the replication of biological tissues on a micro scale is necessary. Hence it is necessary to understand the microenvironment, including the arrangement of gradients of various factors, functional and supporting cell types, and makeup of the extra cellular matrix and nature of the biological forces in the microenvironment [4].

Autonomous self-assembly: This approach depends on the process of embryonic organ development then further replication of the tissues using the former as a prototype. The early cellular components of a developing tissue produce their own components and have a self-governed organization to generate the desired biological micro-architecture and application. This approach completely depends on the cell as the primary driver of histogenesis, further influencing the constitution, location and properties of the tissue [4].

Mini tissues: The third approach of bioprinting is called mini tissues. It is a combination of the two approaches mentioned above. These mini tissues are the structural and functional unit of a tissue. They are manufactured and arranged into a larger construct by rational design, self-assembly or a combination of the two. This approach of Bioprinting makes use of two techniques. The first strategy involves self-governed assembly of cell spheres which restructure into a macro-tissue using the native design as a guide. The second strategy, includes designing accurate, high quality reproduction of a tissue and allowing them to assemble themselves into a functional tissue.

3 The Process

The essential steps in the bioprinting process are imaging and design, selection of materials and cells, and printing of the tissue or organ.

Figure 1 mentioned underneath shows the process of bio-additive manufacture of Human Ear.

The steps involved in the process of Bioprinting are briefly explained below

- Imaging of the damaged tissue is used as a fundamental to design bioprinted tissues. Tissue engineers make use of non-invasive imaging techniques to study the 3D structure of a specific organ and gain information about its function at different levels in the body. The most commonly used technologies are computed tomography (CT) and magnetic resonance imaging (MRI). Computer-aided design and computer-aided manufacturing (CAD-CAM) softwares and mathematical modelling are also used to collect and compute the intricate tomographic and architectural data for tissues [4].
- The three approaches mentioned above, are used individually or in consolidation depending on the organ/tissue being produced.
- The selection of materials and cell source is crucial and depends on the tissue being formed and its role in the body. Most commonly used materials include synthetic or natural polymers and decellularized ECM [4]. Cell sources can be allogeneic or autologous, such as differentiated cells, pluripotent stem cells or multipotent stem cells.
- The components selected for the manufacture of the bioprinted tissue are then integrated in one of the mentioned bioprinting systems.
- Some tissues may require a span of maturation before transplantation in a bioreactor. After the post processing stage, the tissues can be used for in vitro or transplantation purposes.

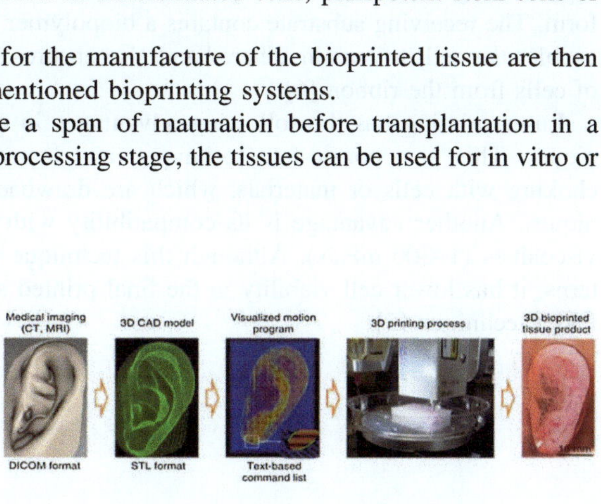

Fig. 1 Process of bio-additive manufacture of Human Ear

4 Bioprinting Techniques

Currently based on the concept of printing for each technique, bioprinting is broadly categorized as: (1) laser based, (2) inkjet based, or (3) extrusion based. All three techniques are widely used and have their own positives and negatives. A detailed explanation on each of the three techniques is given below:

Inkjet-Based Bioprinting: Printers based on this technique are also known as drop-on-demand printers because of its ability to reproduce digital information by printing small bioink drops onto a known location in a appropriate substrate. These printers have many biological and nonbiological applications. The print material is bioink, and the z-axis movements is controlled by an electronic system [2]. The inkjet technique can be driven by a thermal source, hence also being called Thermal printers. The thermal printers produces heat from the electricity which causes the printhead to produce pressure pulses that pushes droplets out of the nozzle. The confined heating in thermal printers can range from 200 to 300 °C, but for a very short duration of 2 μs; thus, the average temperature increases only by 4–10 °C hence not causing an impact on the viability of cells [5].

Advantages of this technique is economical and its versatility. Another advantage being ability to produce high-resolution structures (20–100 μm). The nozzle diameter and hence the size and deposition of bioinks can be controlled electronically. This allows the deposition of the suitable cells with required density and enables high cell viability and function after printing. Drawbacks of this technique are lack of accurate directionality and dimension control of droplet, thermal and mechanical stress applied on cells and biomaterials, repeated nozzle block, and inconsistent cell encapsulation [5].

Laser printing: This technique was first introduced in 1999 by Oddeetal to process 2-D cell patterning [1]. This technique has three parts: a pulsed laser source, a ribbon coated with liquid biological materials that are deposited on the metal film, and a receiving substrate. The lasers expose the ribbon to radiation, allowing the liquid biological materials to vapourize and reach the receiving substrate in droplet form. The receiving substrate contains a biopolymer or cell culture medium which enables the cells to remain adhered and also ensures sustained growth after transfer of cells from the ribbon [6].

Laser printing has the following advantages in comparison to the other techniques. This is a nozzle-free technique, and hence have no issues like nozzle choking with cells or materials, which are drawbacks of other bioprinting techniques. Another advantage is its compatibility with a broad range of biomaterial viscosities (1–300 mPa/s). Although this technique produces high-resolution patterns, it has lower cell viability in the final printed structure in comparison to the inkjet technique [5].

Microextrusion: Microextrusion 3D bioprinters produce a prolonged bead chain of the material that is deposited in 2D with a syringe nozzle in which the viscous material gets squash out under constant pressure. This layer acts as a base for the next layer, while the stage or microextrusion head is moved along the *z*-axis, ultimately leading to a 3D structure [5]. Extrusion-based bioprinting is a blend of a fluid-dispensing system and an automated robotic system.

Piston-driven deposition normally provides more direct control over the flow of bioink from the nozzle. Screw-driven systems give spatial control and are favourable for depositing bioinks with greater viscosities [7].

This technique is comparatively easier to construct and is economical. The only drawback is materials with high viscosity can be extruded. Printing complex structures through this method might require more time, but, the microstructure is well constructed [2] (Figs. 2 and 3).

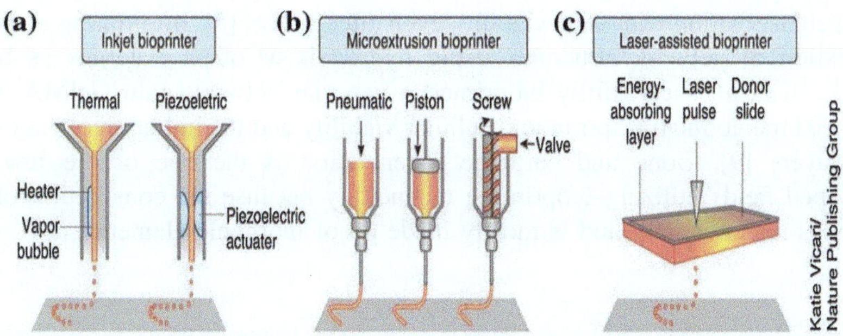

Fig. 2 Mechanism of an inkjet bioprinter, microextrusion bioprinter and laser-assisted bioprinter

Category	Inkjet	Microextrusion	Laser printing
Material viscosity (mPa/s)	Low (3–12)	High (30–6×10^7)	1–300
Gelation/cross-linking methods	Chemical or photo-cross-linking, temperature	Chemical or photo-cross-linking, temperature	Chemical or photo-cross-linking
Print speed	1–10,000 droplets/s	10–50 μm/s	200–1600 mm/s
Resolution	50–300 μm wide droplets	100 μm to 1 mm wide	50 μm
Accuracy	Medium	Medium–low	High
Cell viability	>85%	40%–80%	>85%
Cell density	10^6–10^7 cells/mL	High: cell spheroids	10^6–10^7 cells/mL
Biomaterials used	Hydrogels, fibrin, agar, collagen, alginate	Hyaluronic acid, gelatin, alginate, collagen, fibrin	Hydrogels, nano-hydroxyapatite
Mechanical/structural integrity	Low	High	Low
Fabrication time	Medium	Short	Long
Scalability	Yes	Yes	Limited
Cost	Low	Medium	High
Example applications	Skin,[22,53] vascular,[65,66,69] cartilage[46,78,79]	Trachea,[38,71] cardiac valve[75–77]	Skin[54,55]

Fig. 3 Table comparing the three techniques of bioprinting on different parameters

5 Applications

Bioprinting techniques makes it easier to print cells in constructs in precise locations. This is necessary for mimicking native tissue structure [2]. Many bioprinting approaches have been shown to trigger vascularization of scaffolds for tissue engineering application [2]. Another potential upcoming application of bioprinting is in situ printing, where living organs can be printed within the body during surgeries. Presently, it has already been tested for repairing and to reinstate external organs such as skin, where the damaged part is packed with multiple different cells, including human keratinocytes and fibroblast, with stratified zones throughout the wound bed [1]. In most of the skin injuries, tissue-engineered substitutes are trending as the current treatment choices. Unlike conventional tissue engineering techniques, 3D bioprinting of skin is more accurate and takes into consideration subtle interactions between two cells as well ascell–matrix interactions and precise cell layer positioning [5]. This newfangled technology has also allowed internal stem cell deposition with 90% viability even after 3 days [8]. Bioprinting allows the fabrication of network structures using hydrogels or other materials as bioink. Bertassoni et al. successfully bioprinted a vascular network with GelMA, which enhanced metabolic transportation, cellular viability and the initiation of endothelial monolayers [9]. Bone and cartilage regeneration is the one of the few fully developed fields utilizing bioprinting technology because the constitution of hard tissues is undemanding and is mostly made up of inorganic elements.

6 Future Prospects

Bioprinting can act as a revolutionary breakthrough in the future of healthcare majorly due to the reasons mentioned underneath

1. It can untangle the crucial issue of donor shortage in organ transplantations.
2. CAD tissue/organ models makes it easier for the scientific community to accept the technology.
3. It can vanquish the immune rejection symptoms caused by allograft tissue/organ transplantation.
4. It serves as an economical option thus tending to a larger population, with reverse engineering and personal customization techniques [5].

Despite its remarkable applications which can serve as a major breakthrough in healthcare, the technology is in its initiation phase and has a number of complications and drawbacks linked to it.

- Cell viability, feasibility and biocompatibility of the processes.
- Cell density, cytotoxicity, bioprintability and solidification mechanical and chemical properties, affordability and multitude and long-term functionality of bioinks [10].
- Compactness, accuracy, capability and motion speed, availability and accessability in the market, user-friendliness, asepticity and affordability of the bioprinters.

References

1. I.T. Ozbolat, Y. Yu, Bioprinting towards organ fabrication: challenges and future trends. IEEE Trans. Biomed. Eng. **60**(3) (2013)
2. D. Sundaramurthi, S. Rauf, C.A.E. Hauser, 3D bioprinting technology for regenerative medicine applications. Int. J. Bioprinting **2** (2016)
3. A.B. Dababneh, I.T. Ozbolat, Bioprinting technology: a current state-of-the-art review. https://doi.org/10.1115/1.4028512
4. S.V. Murphy, A. Atala, 3D bioprinting of tissues and organs. Nat. Biotechnol.
5. G. Dai, V.K. Lee, Three dimensional bio-printing and tissue fabrication: prospects for drug discovery and regenerative medicine. Dove Press J. (2015)
6. J. Li, M. Chen, X. Fan, H. Zhou, Recent advances in bioprinting techniques: approaches, applications and future prospects. J. Transl. Med.
7. A. Panwar, L.P. Tan, Current status of bioinks for micro-extrusion-based 3D bioprinting. Molecules (2016)
8. M. Lei, X. Wang, Biodegradable polymers and stem cells for bioprinting. Molecules (2016)
9. J.M. Lee, S.L. Sing, E.Y.S. Tan, W.Y. Yeong, Bioprinting in cardiovascular tissue engineering. Whoice Publications (2016)
10. S.-S.D. Carter, P.F. Costa, C. Vaquette, S. Ivanovski, D.W. Hutmacher, J. Malda, Additive biomanufacturing: an advanced approach for periodontal tissue regeneration. Ann. Biomed. Eng. (2016). https://doi.org/10.1007/s10439-016-1687-2
11. S.A. Irvine, S.S. Venkatraman, Bioprinting and differentiation of stem cells. Molecules (2016)
12. C. Mézel, A. Souquet, L. Hallo, F. Guillemot, Bioprinting by laser-induced forward transfer for tissue engineering applications: jet formation modelling. IOP Publishing (2010)
13. M. Nakamura, S. Iwanaga, C. Henmi, K. Arai, Y. Nishiyama, Biomatrices and biomaterials for future developments of bioprinting and biofabrication. IOP Publishing (2010)
14. D.T.J. Klein, J. Malda, P.J. Bartolo, D.W. Hutmacher, Additive manufacturing of tissues and organs. Prog. Polym. Sci. (2012)
15. K.F. Long, Solid freeform fabrication of three-dimensional scaffolds for engineering replacement tissues and organs. Biomaterials (2003)
16. F. Guillemot, V. Mironov, M. Nakamura, Bioprinting is coming of age: report from the International Conference on Bioprinting and Biofabrication in Bordeaux. IOP Publication (2010)

A Turnkey Manufacturing Solution for Customized Insoles Using Material Extrusion Process

Jia-Chang Wang, Hitesh Dommati and Jung Cheng

Abstract This paper discusses the development of a complete turnkey system to measure, design, and manufacture customized insoles. An optical foot measuring technology has been developed and used to detect the dimensions as well as the plantar pressure distribution of the foot pair instantly. A 3D CAD model of the insole is automatically generated followed by simple adjustments based on a certified medical professional's diagnosis. The generated pair of customized insole models are joined together by their edges to form a single piece for easy manufacturing and easily separable once fabricated. The fabrication is done by Material Extrusion (ME) process using PETG (a modified version of PET) material in a custom built machine. In this paper, it is discussed in detail about the Foot Measurement Apparatus (FMA), Design approach, and Fabrication using Additive Manufacturing (AM) process.

Keywords Customization · Footwear insoles · 3D CAD · Digitization Material extrusion process · Additive manufacturing

J.-C. Wang
Department of Mechanical Engineering, National Taipei University of Technology, Taipei, Taiwan
e-mail: jcw@mail.ntut.edu.tw

J.-C. Wang
Additive Manufacturing Center for Mass Customization Production, National Taipei University of Technology, Taipei, Taiwan

H. Dommati (✉) · J. Cheng
College of Mechanical & Electrical Engineering, National Taipei University of Technology, Taipei, Taiwan
e-mail: iamhitesh89@gmail.com

J. Cheng
e-mail: fogooxx@gmail.com

L. J. Kumar et al. (eds.), *3D Printing and Additive Manufacturing Technologies*,
https://doi.org/10.1007/978-981-13-0305-0_18

1 Introduction

Additive Manufacturing (AM) is a layer-based manufacturing process of components and prototypes used in product development and manufacturing at optimized cost with reduced time. AM process allows making components in variety of materials without requirement of tooling, assembly lines and supply chains. The AM can produce products with complex geometry in batch production or unit production at low cost. The AM process disrupts value chains and create new supply chain. The recent developments includes in personalized footwear for rheumatoid arthritis individuals, shoe insoles for diabetic foot problems, lower limb abnormalities, for sportsmen or anyone who want to become more active [1–3]. Combining the 3D scanning technology for plantar digitization, CAD systems for insole design, and AM system for manufacturing, a complete production process can be implemented to produce personalized products.

The footwear soles are essential to not only absorb the pressure and shocks which causes foot disorders but also, they are designed to offer enhanced comfort and quality protection to feet. Footwear insoles come in wide varieties including orthotic insoles which offers cushioning for support and avoid extra unnecessary movements. They are also used in treatment of a wide range of foot conditions including plantar fasciitis, knee pains, flat foot and Achilles tendon pain among other conditions. AM process has good potential in footwear industry and is already being used for prototyping of shoe insoles. Few industries started commercializing their 3D printed shoes which are not yet widely released into market.

The novelty of this paper lies in providing a turnkey solution for insole manufacturing using AM process and the technique to fabricate them by only printing the shear portion reduces the production time. This is possible by integrating AM technology with foot scanning system and CAD system to fetch, design, and manipulate the foot data, an insole that perfectly fits to the customer's foot is produced at very economic integration process. The produced insole can be directly utilized for manufacturing the rest of the shoe. There are various AM methods to fabricate footwear insoles. But, the most economic process to produce any part within a reasonable amount of time at a footwear shop would be by Material Extrusion (ME) process. Fused Deposition Modeling (FDM) is one of the ME process and mostly uses thermoplastic material. It has an advantage of using variety of raw materials and molders. Different parameters like raster width, slicing height, air gaps among other parameters were investigated and successfully fabricated various functional components [4–6]. Though the surface roughness obtained using ME process is comparably poor but, the post-processing is not necessary unless aesthetical appearances are highly required. The mechanical properties of materials are completely obtainable using ME process.

2 Insole Personalization and Manufacturing Process

An insole is required to provide a perfect fit to the feet within the footwear. Problems like diabetes, flat foot, plantar fasciitis, etc., were investigated and found that by using personalized soles in the footwear one can avoid troubles in harsh walking environments, straighten the neutral position [7, 8]. The footwear soles comes in two designs. One is removable and other is fixed ones. The fixed ones are time consuming to manufacture by using traditional methods. The most commonly used personalized footwear soles are the removable ones. But, by implementing modern methodologies, it is possible to make the insoles that can be fixed or removable. The Insoles are personalized and manufactured by following four steps as shown in Fig. 1:

(1) Foot Digitization using Foot Measurement System
(2) Foot Diagnosis
(3) Insole design using CAD tools
(4) Fabrication of insoles using ME process.

2.1 *Foot Digitization Using Foot Measurement System*

The foot data can be captured using modern methods like 3D scanning. But, processing 3D scanned data is a time consuming process and requires skilled person to edit the obtained 3D model. In this project, a Foot Measurement System (FMS) is developed. The apparatus consists of an optical measuring system and optical foot pressure measuring system.

(a) *Optical Measuring System (OMS)*: The optical measuring system scans and records the data of the plantar section of the feet. It is as showing in Fig. 2.

It consists of a line laser pointer, a camera sensor, a liner motion setup. The laser pointer shoots a laser line on to the plantar portion of the feet. The camera sensor captures the *X*, *Y* and *Z* axis data of foot profile. The laser pointer and camera are fixed on the linear motion setup which is belt-driven by a DC motor and supported on a linear rail. Using OMS the 3D CAD model of the foot profile is obtained.

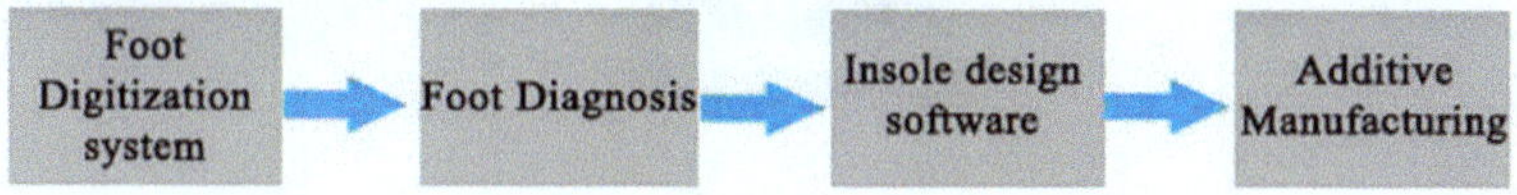

Fig. 1 Flowchart for Insole personalization and manufacturing

Fig. 2 Optical measuring system

Fig. 3 Foot pressure measuring system

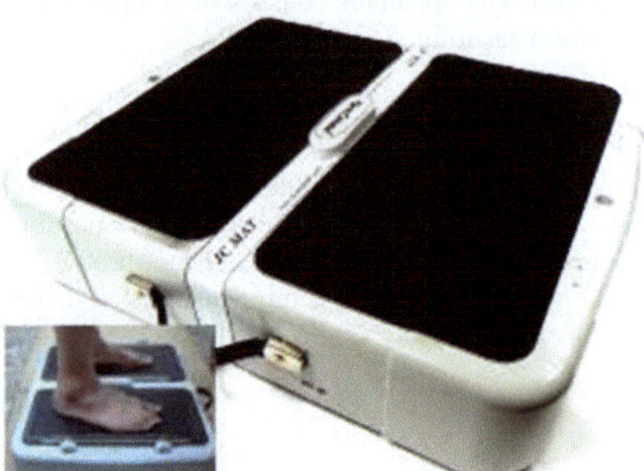

(b) *Foot pressure measuring system*: To know a person's neutral position, it is necessary to know the pressure points of the static and dynamic posture of the person. A foot pressure measuring system as shown in Fig. 3 is developed to measure the person's normal and abnormal pressure points of the plantar section [9]. This system consists of a specially made elastic surface and a setup to capture image and process it. This system also enables to analyze the health of person's default position.

A computer interface is required to visualize the generated 3D model and pressure points of the feet. The visualization of the 3D model of the foot is a relift surface of the plane *X* and *Y* parameters along with the *Z* axis guidelines as shown in Fig. 4. The pressure points data obtained is in the form of green and red colors as shown in Fig. 5. The green color region is the normal pressure area and the red color region is the portion where high pressure is applied. This data can help a

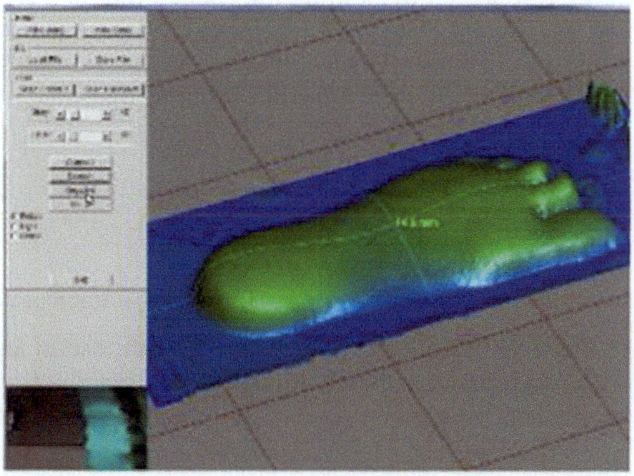

Fig. 4 Foot 3D model

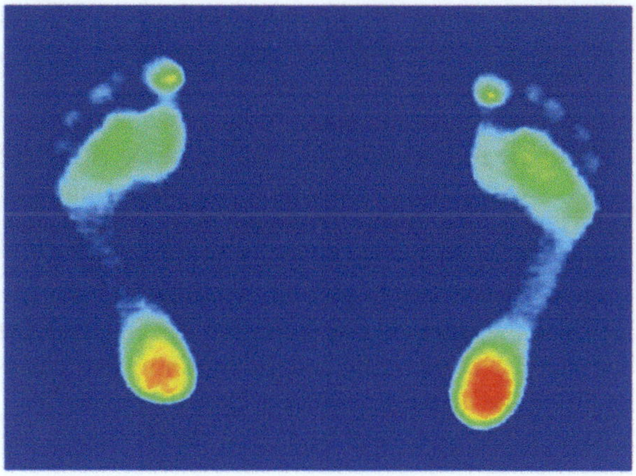

Fig. 5 Foot pressure data

certified medical professional to analyze and provide recommendations to correct the posture.

2.2 Foot Diagnosis

The foot pressure data indicates the possible foot problem or deformity. The medical professional checks the gait of the client and the amount of deformity for precise prescription of insoles. The most common procedure starts with prone process followed by holding the client's foot in a subtalar-neutral position to check the varus/valgus angle of rearfoot and forefoot. Finally, the medical professional can give a prescription according to the value of deformity and the type of gait. The figure-of-4 lower limb posture shown in Fig. 6 and the subtalar-neutral position is shown in Fig. 7 [10].

2.3 Insole Design Software

A software has been developed for the purpose to analyze the 3D shape of the feet data and pressure regions. In the insole design software, the first step is to create a total contact design. A designer requires to edit the 3D model of the insole surface to create a perfect fit to the person's foot shape. Such that, the insole provides an even distribution of foot pressure and provides comfort while wearing. The even distribution is as shown in the lower left window in Fig. 8. Secondly, the designer requires to set proper embedded wedges which is the important step in intervention protocol to correct person's biomechanical deficits. Figure 9 shows the wedges embedded and elevated local pressure points.

After providing wedges, the design software automatically smoothens the insole surface to remove the noise of the row data. From the obtained insole 3D model, 25% of the foot starting from the tip of toe is flattened, because, this portion of the feet is a common portion which does not require personalization of the insole. The rest 75% of the insole 3D model only is subjected to personalization. After flattening of the forefront region of the insole models, both insoles are faced against each other and joined together to form one single piece and added a flat bottom cover to the front side of the insole. By reducing the size of the insole to only the size of 75% of total foot and the adjoining two pieces with a flatten bottom plate to improve the manufacturability (Fig. 10).

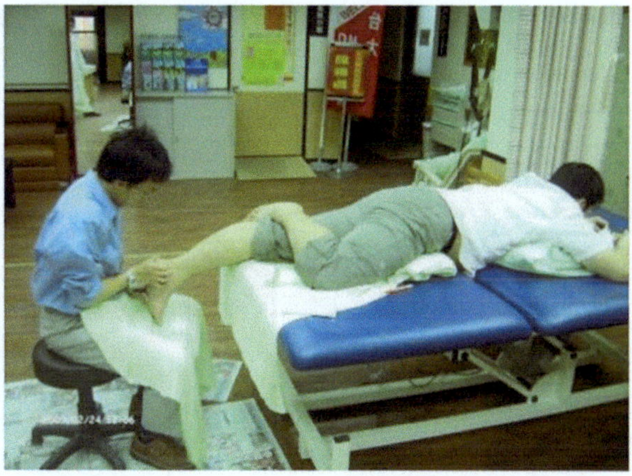

Fig. 6 Prone and figure-of-4 lower limb posture

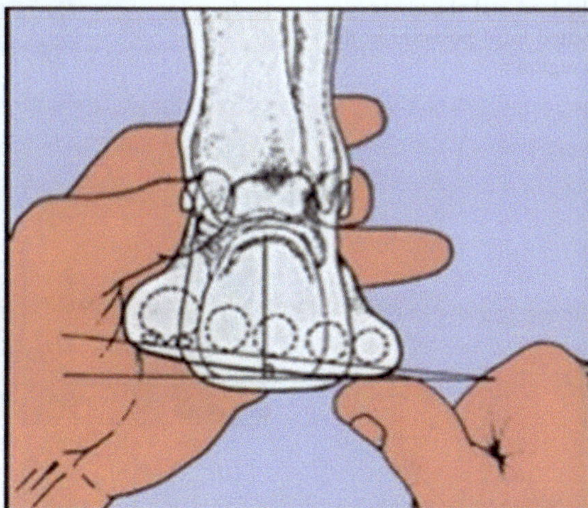

Fig. 7 Subtalar-neutral position [10]

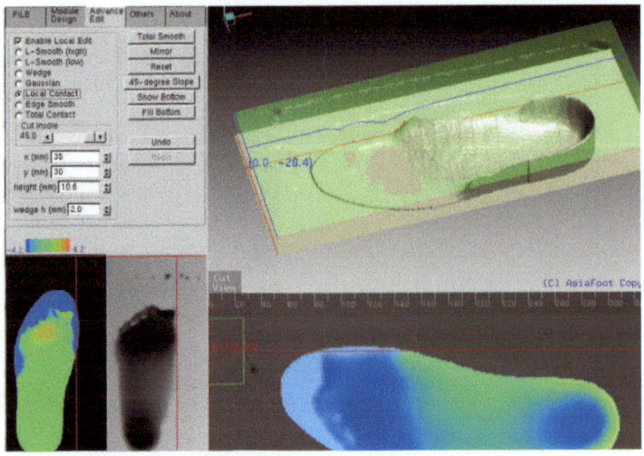

Fig. 8 Total contact process: green region showing even distribution of foot pressure

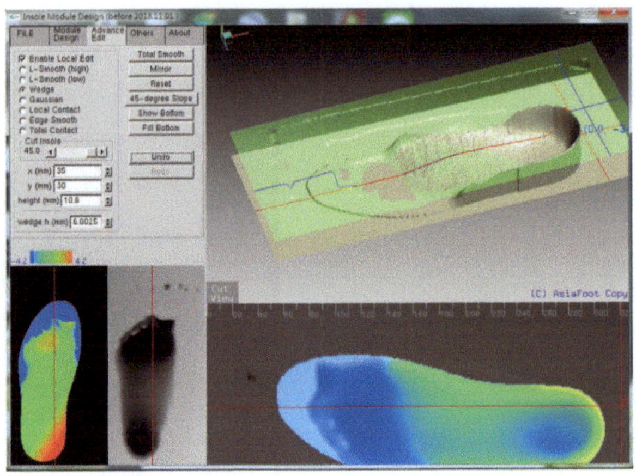

Fig. 9 Biomechanical wedge embedded and shown as elevated local pressure in the red region

2.4 Additive Manufacturing of Insoles

The .STL file obtained after the conversion step, is a single file vase type design which has two insoles combined and the bottom of the structure is a flat surface. The inside of the structure is hollow to save the plantar shape. The top of the structure at the adjoining it has a gradual slope curve which doesn't require any support structures while printing. A converted and ready to print .STL file is depicted in Fig. 11. Having no support requiring, the contour needed by the extruder to travel is reduced which makes the printing time faster. An ME type 3D printer in general retracts the filament where there is no printing object during the contour and extrudes when required. By deduction of two individual insoles to one the filament retraction and extrusion time is reduced as well as the surface quality is well maintained.

Fig. 10 Conversion of two individual insole files to one single file after automatically cutting the 25% of forefront portion

Fig. 11 Converted .STL file

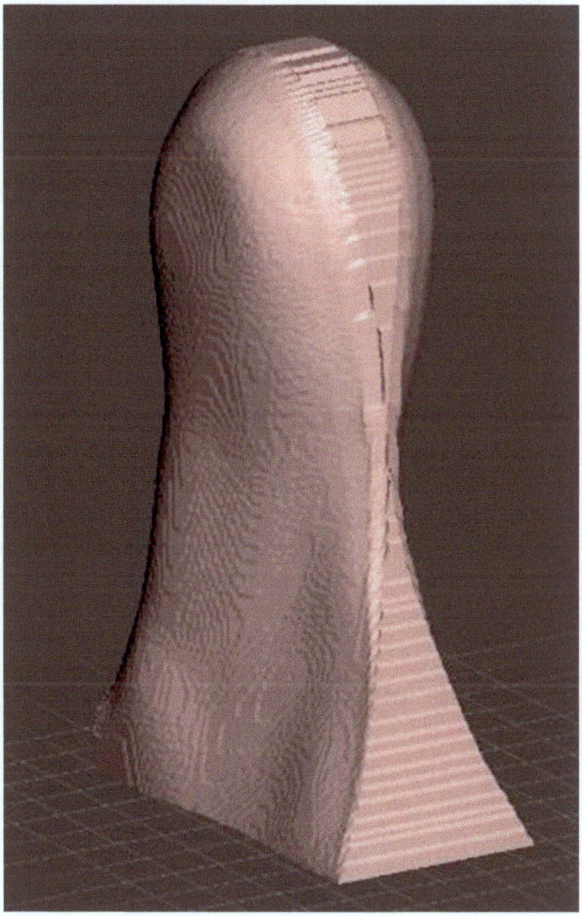

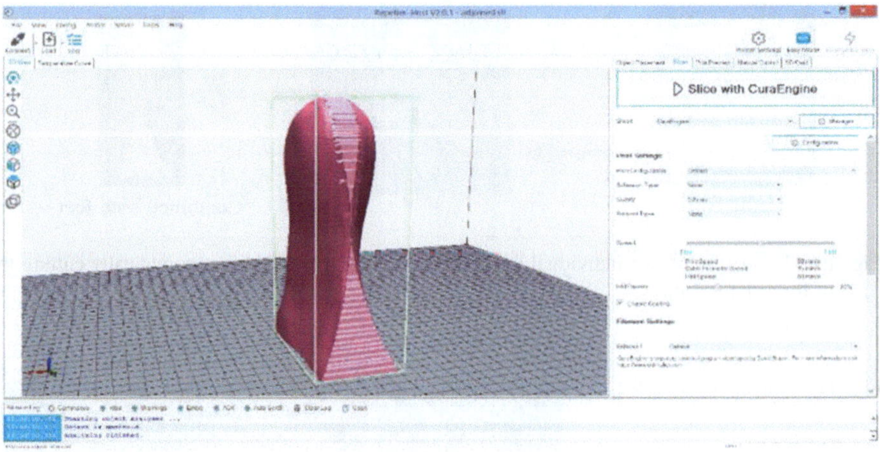

Fig. 12 Slicing interface to slice the .STL file

The filament material chosen in this project to fabricate insoles is PETG material. Even though the most common filament materials used in FDM process are PLA and ABS, the applications of PETG is more effective than PLA and ABS. Because, PLA material isn't suitable for the applications which requires less flexibility under stressed conditions where it tends to break. And, ABS is not suitable for the enclosed heated condition where it emits a sharp toxic smell. Hence, recognizing the properties of PETG like moisture bearing and high mechanical strength, it is chosen to be used as insole support material above which a soft cushioning material is arranged. The ready to print 3D model of the insole file is sliced using slicing software and fed the data to the 3D printer directly. During slicing there is no support structure required to print these models because of no overhangs. Figure 9 shows the converted .STL file which is ready to 3D print and Fig. 12 shows the slicing interface.

3 Results and Evaluation Criteria

Figure 13 shows the fabricated insoles using ME process. The fabricated model is an insole pair stitched together. It is required to separate them into two equal halves by carefully cutting on the joining line. Once separated, the edges have to be smoothened as shown in Fig. 14. The final parts are the insoles of three-fourth the size of foot made out of PETG material. Above the surface of the printed insoles an EVA foam sheet is glued. The foam sheet adapts the shape of the insole structure and completes the forefront portion of the insole along with sponginess. This can be directly used in footwear. The final parts are perfect fit and in total contact of the person's plantar surface of feet. The plantar pressure is evenly distributed causing

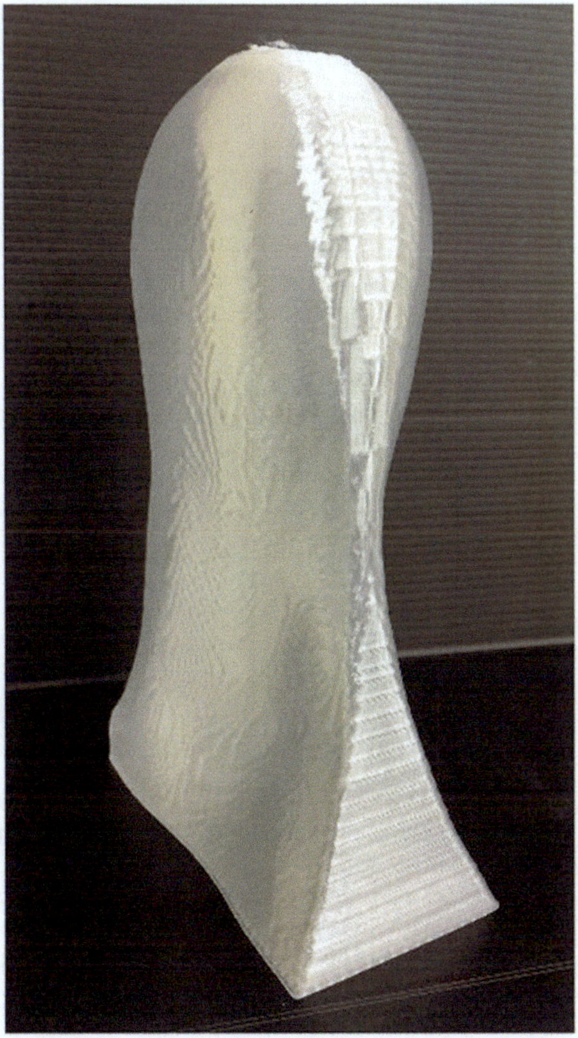

Fig. 13 3D printed insole pair

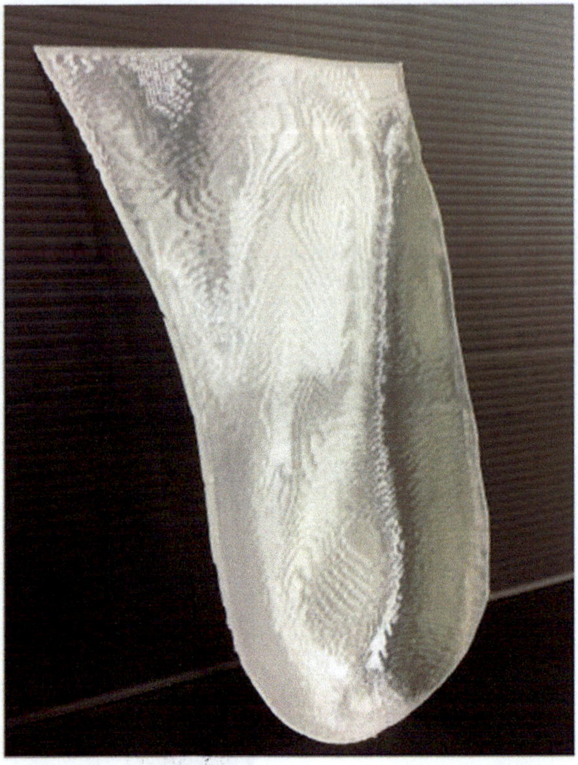

Fig. 14 Separated and post processed insole

no stresses at one particular region. This helps for the people having diabetes foot or plantar fasciitis.

The insoles pair obtained is hard material which can be used by gluing a plane EVA (Ethylene-Vinyl Acetate Copolymer) foam pad which is cut according to the plantar foot shape. This foam sheet once glued on to the insole body, it acts as a skin that contacts with the plantar surface of the foot. The completed version of the insole is as shown in Fig. 15.

Evaluation: The fabricated insoles are used practically by the related person in real time to check the actual foot dimensions, the discomforts caused, among other biomechanical parameters including varus and valgus foot angle correction and leg length discrepancy. The varus, valgus foot angle correction and the leg length discrepancy are verified by a certified medical profession. The personalization of the insoles is done by also considering a professional's recommendation. If the person's requirement is not for a medical purpose there requires no medical professional's analysis. The personalized insoles can be directly used in the footwear. For non-medical use, the insoles are evaluated by checking the total contact of the insole body with the scanned model of the plantar surface of the person's foot in real time. In normal cases, the plantar faces of both feet are symmetric. If found to

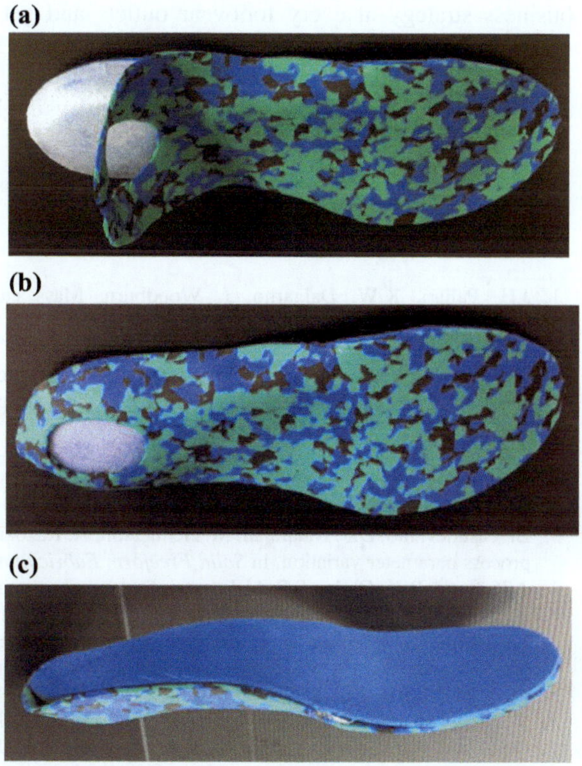

Fig. 15 **a** The EVA foam pad glued to the bottom insole body, **b** the bottom view of the insole, **c** EVA foam on top side of insole body for better cushioning

be non-symmetric, by following scanning and evaluation criteria it can be recognized and modify the designs at ease if necessary.

4 Conclusions

The traditional method consumes at least two working days for a skilled person. By following the proposed digital turnkey process, a pair of personalized insoles can be generated within 90 min. To minimize the cost of fabrication, ME process is adapted in this turnkey process. The time it takes to scan one foot is around 10–15 s. And, to generate a 3D model including wedges and corrections it takes 5–6 min depending upon the number of corrections. To 3D print the model, it takes around 60 min on an FDM 3D printer depending upon the size of the insole. In this project all the processes are fully integrated. A complete User Interface (UI) is developed to scan, manipulate the 3D model and combine the two separate insoles into one and finally 3D print directly. The obtained results had always proved right in the total contact surface evaluation. They are able to evenly distribute the foot pressure. This modern method of personalization can be taken to a subsidiary business strategy at every footwear outlets and get the fabrication done at ease either at the outlet itself or at a fabrication center. The efficiency of this turnkey process is very high as there requires no skilled technician to fabricate the insole and the repeatability of the machines are always high comparable to humans.

References

1. J.H. Pallari, K.W. Dalgarno, J. Woodburn, Mass customization of foot orthoses for rheumatoid arthritis using selective laser sintering. IEEE Trans. Biomed. Eng. **57**(7), 1750–1756 (2010)
2. S.P. Sun, Y.J. Chou, C.C. Sue, Classification and mass production technique for three-quarter shoe insoles using non-weight-bearing plantar shapes. Appl. Ergon. **40**(4), 630–635 (2009)
3. A.S. Salles, D.E. Gyi, Delivering personalised insoles to the high street using additive manufacturing. Int. J. Comput. Integr. Manuf. **26**(5), 386–400 (2013)
4. M. Srivastava, Some studies on layout of generative manufacturing processes for functional components. Ph.D. thesis, Delhi University, 2015
5. B. Vasudevarao, D.P. Natarajan, M. Henderson, A. Razdan, Sensitivity of RP surface finish to process parameter variation, in *Solid Freeform Fabrication Proceedings* (2000), pp. 251–258
6. A.K. Sood, R.K. Ohdar, S.S. Mahapatra, Parametric appraisal of mechanical property of fused deposition modelling processed parts. Mater. Des. **31**(1), 287–295 (2010)
7. J.S. Paton, E.A. Stenhouse, G. Bruce, D. Zahra, R.B. Jones, A comparison of customised and prefabricated insoles to reduce risk factors for neuropathic diabetic foot ulceration: a participant-blinded randomised controlled trial. J. Foot Ankle Res. **5**(1), 31 (2012)
8. S. James, Flat feet treatment: a holistic approach. Available from: http://www.somastruct.com/flat-feet-treatment-a-holistic-approach. Accessed 15 July 2014
9. J.C. Wang, Foot pressure measurement devices, components and their measurement methods. Patent CN102283657, 21 Dec 2011
10. D. Shavelson, A closer look at neoteric biomechanics. Podiatry Today **9**, 234–241 (2007)

Parameter Optimization for Polyamide in Selective Laser Sintering Based on Mechanical Behavior

B. Karthick Raja, R. Jegan Pravin Raja, K. Karan, R. Soundararajan and P. Ashokavarthanan

Abstract Selective laser sintering (SLS) permits the fast, flexible, cost-efficient, and easy production of prototypes of required shape and size by using powder-based material. The physical prototype for design verification and working analysis is done directly on CAD tools. In SLS process optimization of influenced parameters will contribute to save time and material. In this study, optimal SLS process parameters by changing the layer thickness, bed temperature, and part orientation for hardness and ultimate tensile strength for the intended specimen by using Polyamide and also evaluate the mechanical behavior by using Vickers hardness tester and Ultimate tensile machine. The tests were conducted conferring to the Taguchi design of L9 orthogonal array at various combinations of process parameters and statistical optimization technique. Analysis Of Variance (ANOVA) was used to determine the optimal levels and percentage of influence of each parameter. The results postulate that the bed temperature is the main key factor followed by the part orientation and layer thickness for optimal value of the hardness and ultimate tensile strength of the SLS processed. This optimized value serves as a data base for the industries.

Keywords SLS process parameter · Polyamide · Mechanical behavior Optimization

1 Introduction

Manufacturing processes are radically challenged worldwide by economic, socio-political, and technological dynamics that have a tremendous impact on enterprise behavior in the market [1]. Prototyping (RP) technique is one among the

B. Karthick Raja (✉) · R. Jegan Pravin Raja · K. Karan · R. Soundararajan
P. Ashokavarthanan
Department of Mechanical Engineering, Sri Krishna College of Engineering & Technology, Coimbatore, India
e-mail: rajakarthick304@gmail.com

L. J. Kumar et al. (eds.), *3D Printing and Additive Manufacturing Technologies*,
https://doi.org/10.1007/978-981-13-0305-0_19

manufacturing deeds that leads in reduction of product development time [2]. Rapid prototyping of mechanical components uses techniques that makes quick, low-volume production of prototype parts, often with several iterations [3]. Recent developments in additive manufacturing (AM), a construction technique where a three dimensional object is created through the build-up of thin layers of a base material, have resulted in the commercialization and simplification of what is commonly known as 3D printing (3DP) [4]. Some of the successfully commercialized solid freeform fabrication processes include Stereo lithography (SL), Fused Deposition Modeling (FDM), Selective Laser Sintering (SLS), Laminated Object Manufacturing (LOM), Three Dimensional Printing (3D-P) and Laser Engineered Net Shaping (LENS) [5]. One of the important techniques is stereo-lithography [6]. Rapid prototyping in general, is more flexible and can readily provide accommodation changes in product design as compared to conventional method of casting, molding, or machining [7]. The application of RP has been shown to greatly shorten the design-manufacturing cycle time hence dropping the cost of product and increasing competitiveness [8]. Almost all RP process are based on a layered manufacturing methodology in which objects are built as a series of horizontal cross-sections, each one being formed individually from the related raw materials and bonded to preceding layers until the object is completed [9]. The RP technology also needs several supporting technologies such as solid modeling, database management, and electronic networking for data exchange [10]. RP covers a range of technologies that build physical 3D objects from 3D data sources output by solid modeling CAD systems, 3D laser scanners, and CT/MRI scanners [11]. The future is clear as RP is now becoming a key gen that reduces product growth time for faster building of physical prototypes, tooling, and models [12]. The technology is currently conquered by "additive" processes developed by companies such as 3D systems, Stratasys, EOS, ZCorp and Object Geometries. These build models by linking a range of materials like plastics, resins, sheet metal, etc.,—layer by layer using horizontal cross-sections of the computer model to run the RP equipment [13]. The best mechanical behavior obtained is produced of SGC machines and followed by SLA, LOM, FDM and SLS [14]. The RP machines are continuously being advanced to improve the mechanical behaviors without sacrificing build speed [15]. Taguchi method is one of the best optimizing technique for optimizing process parameters [16]. Part quality of the prototype in the RP technique mainly depends upon the parameters such as hatch cure depth, layer thickness, orientation, laser power, temperature, hatch spacing, etc. [17]. Thus an effort was made to study and optimization of process parameters prevailing the system which are linked with developed part characteristics on hardness and UTS by using the Taguchi method. Based on the above facts, a statistical method based on the ANOVA techniques was espoused in this present work to determine the degree of significance of each process parameter on the hardness, ultimate tensile strength of SLS process by using Polyamide (nylon 12).

2 Experimentation

2.1 Preparation of Test Samples by Using SLS Process

The foremost experimental device adopted comprises, Sinter station 2500+ SLS Selective Laser Sintering Machine where it uses the Nylon—Duraform PA (Polyamide) to construct geometrical shape of the specimen. The experimental setup and specimen is shown in Fig. 1.

The experimental construction material adopted is Nylon—Duraform PA (Polyamide) as powder. The experimental specimens are constructed based on L9 OA. The STL format is created by PRO-E Creo 2.0 and sends to the Sinter station 2500+ SLS Selective Laser Sintering Machine.

The several preprocessing stages such as STL verification, Layer thickness, Bed Temperature and Part orientation, building inner structure form and deposition way are to be followed by the layer slicing process to create the building path. Construction of mechanical behavior includes the Larger-the-better for both hardness and ultimate tensile strength. The Sinter station 2500+ machine can build parts of size 381 × 330 × 457 mm. On the other hand, the desktop RP machines are intended to construct scaled down specimens within a size of 200 × 200 × 200 mm. For larger parts, an alternative way is to make several parts individually and adhesive them together. In this study, the Nylon 12 (polyamide) was used as the base material. Due to its high elongation and high abrasion resistance properties, it is commercially used in electronic and automobile industries. The standard specimen size is 165 mm length, 19 mm outer width, 13 mm inner width, 0.6 mm thickness was fabricated by using SLS process.

The hardness of the specimens was tested in Vickers Hardness Testing Machine. The Vickers Hardness Testing Machine is shown on Fig. 2. From this testing the hardness values was taken and the mean value is tabulated. The universal testing

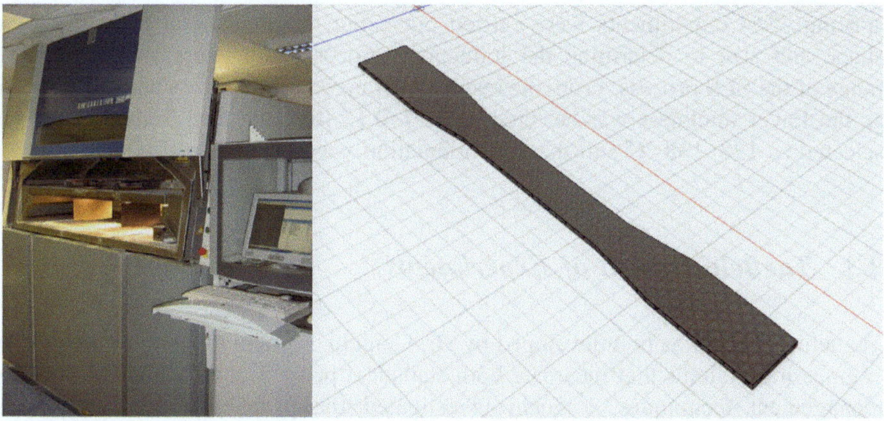

Fig. 1 Experimental setup with CAD model

Fig. 2 Vickers hardness testing machine

machine was employed to determine the ultimate tensile strength of the specimens. The tensile testing machine with specimens is shown in the Fig. 3. From the tensile test experiment, the ultimate tensile strength of all the specimens was noted and the mean value is taken.

2.2 Selection of Factors and Their Levels of SLS Process

In the current study, three process parameters, i.e., Layer thickness, Bed Temperature, and Part orientation were considered. The chemical formula for nylon 12 is $(C_{12}H_{23}NO)_n$ used in this study. Among various SLS process parameters like hatch cure depth, layer thickness, orientation, laser power, bed temperature, hatch spacing, Test experiments were carried out using nylon material for a deserved shape and size to determine the influenced process parameters and operational range of the SLS process. From the results, the range of the process parameters such as the Layer thickness was selected as 0.07–0.15 mm, the Bed Temperature was selected as 128–138 °C and the Part orientation was selected as 0°–90°.

2.3 Taguchi Design of Experiments

Taguchi method was first developed by Dr. Genichi Taguchi mention to the quality engineering systems that incarnate both statistical process control and new quality management techniques. Taguchi investigated the problems with a statistical approach and proposed that the engineering optimization of a process should be

Fig. 3 Ultimate tensile strength machine

carried out in three phase approaches namely the system design, the parameter design, and the tolerance design. Taguchi method uses OA from the DOE theory to learn large number of variables with a small number of trials. The OA decrease the number of alignments to be studied. The results from this analysis will be true within the range of the experimental area defined by the control factors. Taguchi has renewed their use by providing tabulated sets of standard orthogonal arrays and corresponding linear graphs. These trial results are then reformed into a signal-to-noise (*S/N*) ratio, as a measure of quality characteristics. The three categories of quality features in the analysis of the *S/N* ratio are: the lower-the-better, the larger-the better, and the nominal-the-better. The *S/N* ratio for each parameter is worked out based on *S/N* analysis. From the analysis of parameters, larger the *S/N* ratio, larger the quality characteristics and better properties. Moreover, ANOVA can be used to determine the significance of each process parameter on quality features. The involved process\parameters optimization there are, to determine the quality characteristic to be optimized and to identify the noise factors, control factors and test environments. Then to design and construct the matrix experiment also to analyzes the data and determine the optimum levels for control factors. Finally, to calculate the performance.

2.4 Assigned Orthogonal Array

Tests have been carried out using Taguchi's L9 Orthogonal Array (OA) experimental design which consists of 9 combinations of layer thickness, bed

(a)

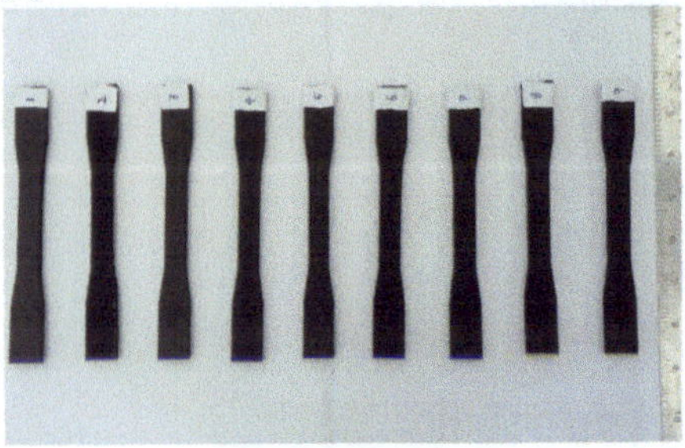

(b)

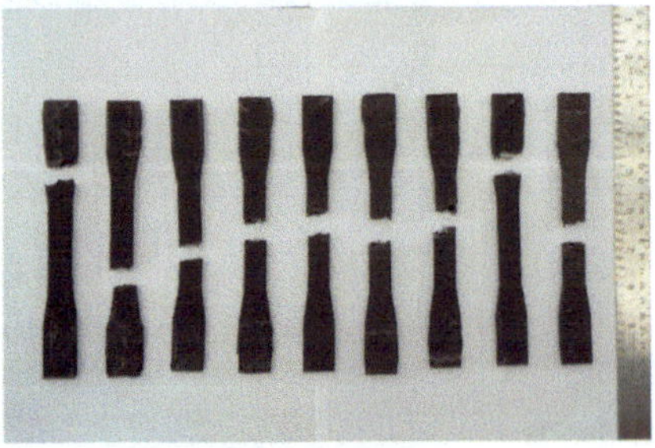

Fig. 4 **a** Standard specimens before testing, **b** Standard specimens after testing

temperature and part orientation. It considers three process parameters (without interaction) to be varied in three discrete levels. As per Taguchi design, two degrees of freedom was allocated to each parameters. This gives a total of 6 DOF for three process parameters. As a result, we have a total of 8 DOF for the factors for the present tests. The nearest three-level OA accessible, sustaining the criterion of selecting the OA is L9.

2.5 *Design of Experiments with Results*

Fabricated nine specimens are taken for checking the dimensional accuracy like hardness and ultimate tensile strength which are to be done by Vickers Hardness Testing Machine and UTS machine is shown in Fig. 4 and average value was noted in the given Table 1.

Table 1 Acquired data for SLS process parameter on responses

S. No.	LT (mm)	BT (°C)	PO (°)	*H*1 (VHN)	*H*2 (VHN)	*H*-mean (VHN)	UTS1 (N/mm^2)	UTS2 (N/mm^2)	UTS-mean (N/mm^2)
1	0.07	128	0	77	76	76.5	46	45	45.5
2	0.07	133	45	82	82	82	49	50	49.5
3	0.07	138	90	73	72	72.5	43	42	42.5
4	0.11	128	90	79	78	78.5	47	46	46.5
5	0.11	133	45	82	81	81.5	49	48	485
6	0.11	138	0	70	71	70.5	42	42	42
7	0.15	128	45	85	84	84.5	50	49	49.5
8	0.15	133	0	75	74	84.5	45	46	45.5
9	0.15	138	90	73	73	73	43	43	43

3 Results and Discussion

3.1 Taguchi Analysis of Hardness (H) on SLS Process

From this experiment we conclude that the hardness gradually increases from layer thickness from 0.07 to 0.15 mm. When considering bed temperature, the hardness decreases gradually from 128 to 138 °C. When considering part orientation, the hardness increases from angle 0° to 45°. The optimum value of hardness is at 45° and the hardness value increases when angle increases. This is same in *S/N* ratio analysis. The bed temperature contributes about 50.06% in change of hardness when compared to layer thickness and part orientation and for *S/N* value the bed temperature contributes 50.48%. The optimal process parameters value for required hardness value in Nylon 12 material is 0.15 Layer thickness, 133 °C Bed temperature, 45° part orientation (Figs. 5 and 6; Tables 2 and 3).

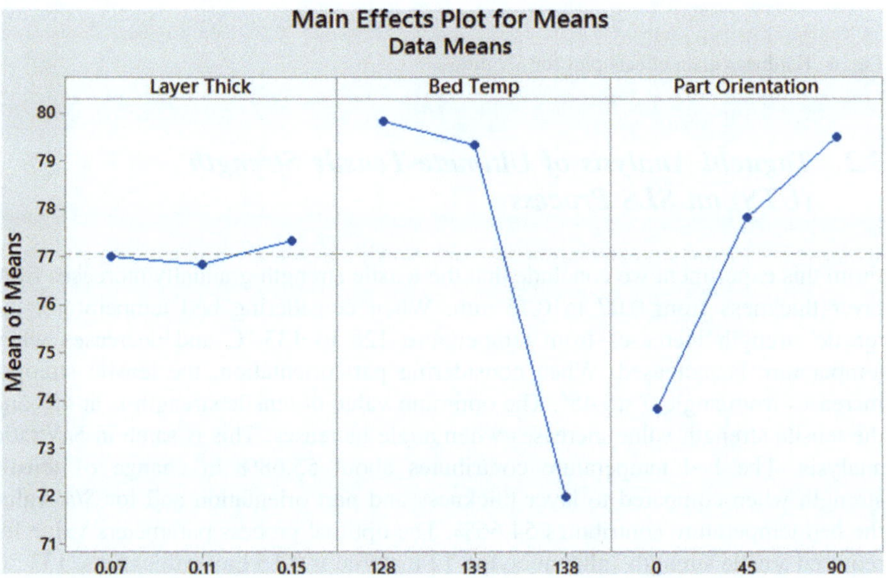

Fig. 5 Hardness main effects plot for means

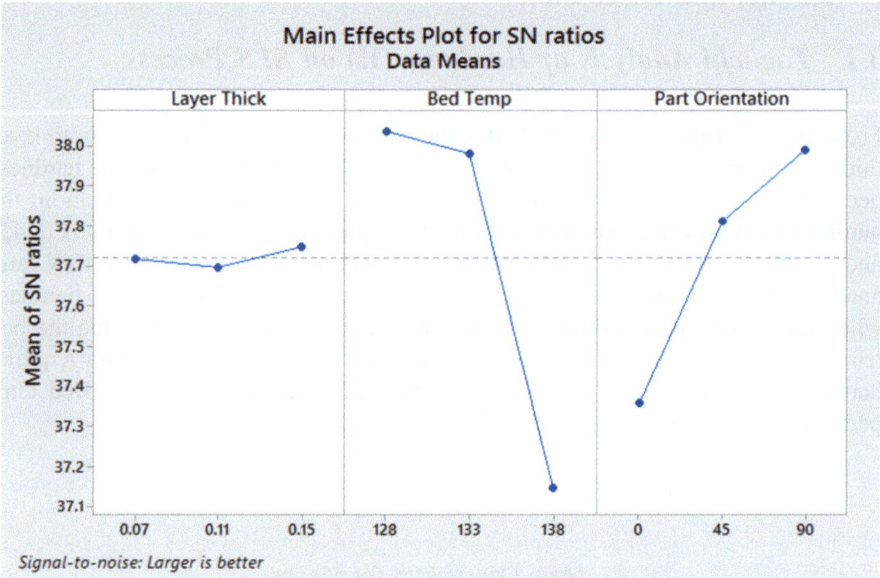

Fig. 6 Hardness main effects plot for *SN* ratio

3.2 *Taguchi Analysis of Ultimate Tensile Strength (UTS) on SLS Process*

From this experiment we conclude that the tensile strength gradually increases from layer thickness from 0.07 to 0.15 mm. When considering bed temperature, the tensile strength increases from temperature 128 to 133 °C and decreases when temperature is increased. When considering part orientation, the tensile strength increases from angle 0° to 45°. The optimum value of tensile strength is at 45° and the tensile strength value increases when angle increases. This is same in *S*/*N* ratio analysis. The bed temperature contributes about 55.68% in change of tensile strength when compared to layer thickness and part orientation and for *S*/*N* value the bed temperature contributes 54.66%. The optimal process parameters value for required tensile strength value in Nylon 12 material is 0.15 Layer thickness, 133 °C Bed temperature, 45° part orientation (Figs. 7 and 8; Tables 4 and 5).

Table 2 Analysis of variance for means

Source	DF	Seq SS	Adj SS	Adj MS	F	P	% contribution
Layer thick	2	0.000991	0.000991	0.000495	6.32	0.137	14.7
Bed temp	2	0.003402	0.003402	0.001701	21.71	0.044	50.06
Part orientation	2	0.002167	0.002167	0.001083	13.83	0.067	32.2
Residual error	2	0.000157	0.000157	0.000078			2.33
Total	8	0.006716					
$S = 3.321$	$R\text{-Sq} = 98.3\%$	$R\text{-Sq(adj)} = 95.3\%$					

Table 3 Analysis of variance for *SN* ratios

Source	DF	Seq SS	Adj SS	Adj MS	*F*	*P*	% contribution
Layer thick	2	0.000030	0.000030	0.000015	6.32	0.137	14.56
Bed temp	2	0.000104	0.000104	0.000052	21.69	0.044	50.48
Part orientation	2	0.000067	0.000067	0.000033	13.83	0.067	32.52
Residual error	2	0.000005	0.000005	0.000002			2.42
Total	8	2.38219					
$S = 0.3552$	*R*-Sq = 97.4%	*R*-Sq (adj) = 95.6%					

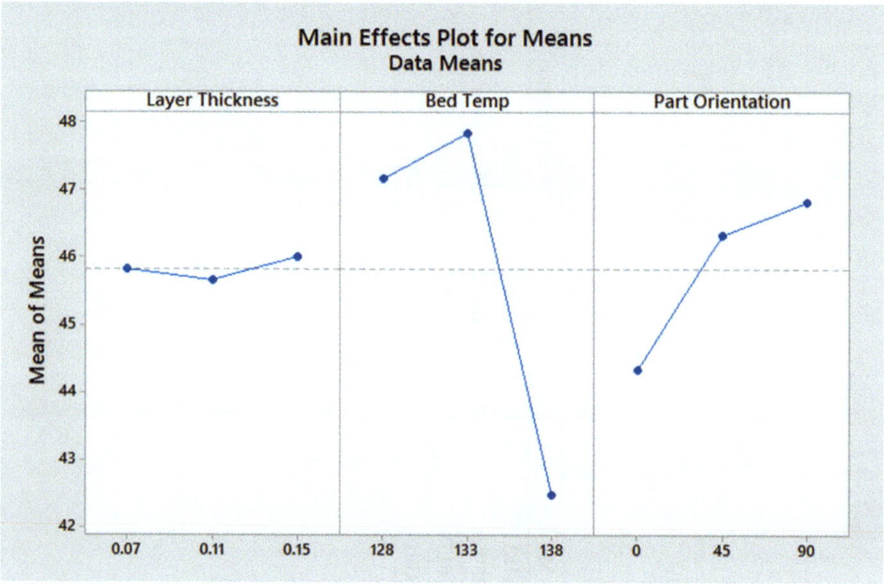

Fig. 7 Ultimate tensile strength main effects plot for means

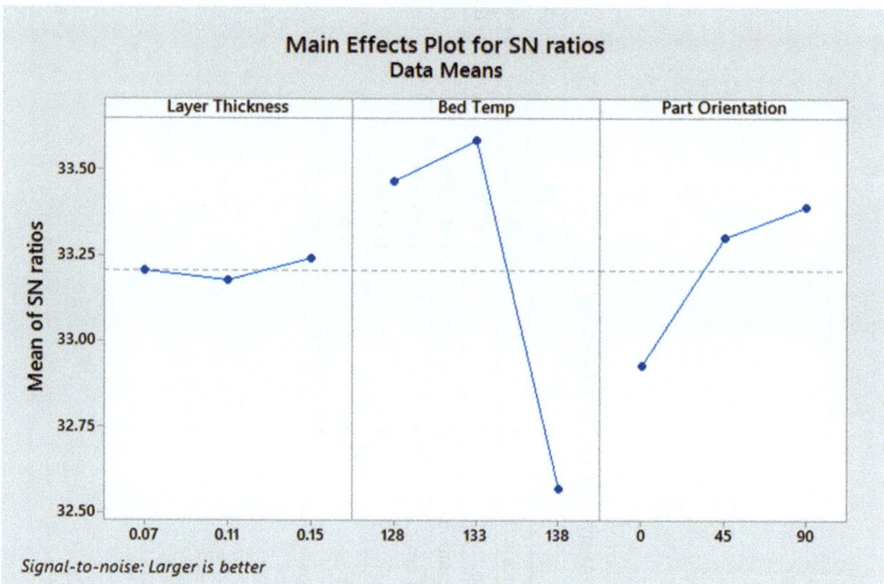

Fig. 8 Ultimate tensile strength main effects plot for means

Table 4 Analysis of variance for means

Source	DF	Seq SS	Adj SS	Adj MS	*F*	*P*	% contribution
Layer thickness	2	0.023622	0.023622	0.011811	3.91	0.204	16.87
Bed temp	2	0.077939	0.077939	0.038969	12.91	0.072	55.68
Part orientation	2	0.032356	0.032356	0.016178	5.36	0.157	23.11
Residual error	2	0.006039	0.006039	0.003019			4.31
Total	8	0.139956					
$S = 1.893$	R-Sq = 99.5%	R-Sq (adj) = 97.2%					

Table 5 Analysis of variance for *SN* ratios

Source	DF	Seq SS	Adj SS	Adj MS	*F*	*P*	% contribution
Layer thickness	2	0.7210	0.7210	0.36052	4.88	0.170	17.89
Bed temp	2	2.2029	2.2029	1.10144	14.92	0.063	54.66
Part orientation	2	0.9580	0.9580	0.47898	6.49	0.134	23.77
Residual error	2	0.1477	0.1477	0.07383			3.66
Total	8	4.0295					
$S = 0.3413$	*R*-Sq = 94.6%	*R*-Sq (adj) = 91.2%					

4 Conclusion

The result of SLS process variables such as Layer thickness, Bed temperature, and Part orientation on the hardness and ultimate tensile strength of Polyamide has been studied. The tests were conducted as per Taguchi parametric design concept and the results were analyzed using ANOVA technique. The following conclusions have been drawn from the current study:

(1) The SLS process parameters were optimized and the optimum values of the process parameters were found to be 0.15 Layer thickness, 133 °C and 45° part orientation.
(2) The ANOVA was carried out to observe the effect of process parameters on the Hardness: The bed temperature contributes larger about 50.06% followed by part orientation about 32.2% and layer thickness about 14.7%. For ultimate tensile strength the bed temperature contributes larger about 55.68% followed by part orientation about 23.11% and layer thickness 16.87% on Nylon 12 (polyamide).

References

1. J.D. Williams, C.R. Deckard, Rapid Prototyp. J. **4**, 90–100 (1998)
2. C.C. Wang, T.W. Lin, S.S. Hu, Rapid Prototyp. J. **13**(5), 304–315 (2007)
3. P. Hanzl, M. Zetek, T. Baksa, T. Kroupa, 25th DAAAM Int. Symp. Intell. Manuf. Autom. DAAAM **100**, 1405–1413 (2015)
4. A. Nizam, R.N. Gopal, L. Naing, A.B. Hakim, A.R. Samsudin, Dimensional accuracy of the skull models produced by rapid prototyping technology using stereolithography apparatus. Arch. Orofac. Sci. **1**, 60–66 (2006)
5. B. Kianian, S. Tavassoli, T.C. Larsson, O. Diegel, The adoption of additive manufacturing technology in Sweden. 3D Print. Add. Manuf. **2**(4), 152–158 (2015)
6. G. Bugeda, M. Cervera, G. Lombera, E. Onate, Numerical analysis of stereolithography processes using the finite element method. Rapid Prototyp. J. **1**, 13–23 (1995)
7. Y.T. Oh, G.D. Kim, A benchmark study on rapid prototyping processes and machines: quantitative comparisons of mechanical properties accuracy, roughness, speed, and material cost. J. Eng. Manuf. **222**, 201–215 (2008)
8. G. Cardano, R. Giannoccaro, A.D. Ludovico, E.L.J. Bohez, S.L. Campanelli, Statistical analysis of the stereo lithographic process to improve the accuracy. Comput. Aided Des. **39**, 80–86 (2007)
9. J.P. Kruth, X. Wang, T. Laoui, L. Froyen, Lasers and materials in selective laser sintering. Assem. Autom. **23**(4), 357–371 (2003)
10. B.H. Lee, J. Abdullah, Z.A. Khan, Optimization of rapid prototyping parameters for production of flexible ABS object. J. Mater. Process. Technol. **169**, 54–61 (2005)
11. P.M. Pandey, N. Venkata Reddy, S.G. Dhande, Part deposition orientation studies in layered manufacturing. J. Mater. Process. Technol. **185**, 125–131 (2007)
12. S.H. Choi, S. Samavedam, Modelling and optimisation of rapid prototyping. Comput. Ind. **47**, 39–53 (2002)
13. Moore JP, Williams CB. Fatigue characterization of 3D printed elastomer material, in *19th Annual International Solid Freeform Fabrication Symposium*, 2008, pp. 641–655

14. R. Anitha, S. Arunachalam, P. Radhakrishnan, Critical parameters influencing the quality of prototypes in fused deposition modeling. J. Mater. Process. Technol. **118**, 385–388 (2001)
15. S.O. Onuh, K.K.B. Hon, Optimising build parameters for improved surface finish in stereolithography. Int. J. Mach. Tools Manuf. **38**(4), 329–392 (1998)
16. K. Thrimurthulu, P.M. Pandey, N.V. Reddy, Optimum part deposition orientation in fused deposition modeling. Int. J. Mach. Tools Manuf. (in press)
17. P.M. Pandey, N.V. Reddy, S.G. Dhande, Real time adaptive slicing for fused deposition modeling. Int. J. Mach. Tools Manuf. **43**, 61–71 (2003)

Analysis of Adjacent Vertebrae Post Vertebroplasty

A. Kavitha, G. Sudhir, V. D. Deepak, M. Pavithra and V. Vallabhi

Abstract Vertebral Compression Fractures refer to the collapse of a vertebral body under excessive trauma to the spine. The surgical procedure to relieve pain and restore the normal height of the vertebral body is called vertebroplasty, in which a small amount of cement is injected into the spine. The problem with vertebroplasty is that the amount of cement injected is not accurate and leads to leakage into discs, which may develop into adjacent vertebral fractures. The current study aims at assessing the stress incurred on the adjacent vertebrae after vertebroplasty and the accurate cement levels required. The data in the form of CT images of the spine were acquired from a normal subject. A three-Dimensional model was generated by identifying the Regions of Interest (T11-L3), after performing volume and surface rendering techniques using Mimics and 3-MATIC. A force of 600 N was applied to the vertebrae for varying conditions of Bone Mineral Density (BMD), and the stress levels were calculated individually. Then, three fractures were induced at L1 and the corresponding 3-D models were generated. The stress levels on the fractured spine for forces of 600, 1200 and 1800 N were assessed. To assess the conditions after vertebroplasty, cement was injected in the fractured spine using Boolean Operations, which helped in optimizing the cement level. Results showed that the amount of cement required for the three cases were 4.5631, 5.5771 and 6.5849 mL respectively. Stress levels for the cement injected spine were analyzed, and were found to be much lesser than the stress incurred in the fractured cases with similar BMD. Among the adjacent vertebrae L2 was found to have a higher stress. Thus, this work seems to be clinically highly relevant in estimating the exact amounts of cements to be injected for different fracture cases thereby avoiding excessive cement leakage into the discs.

A. Kavitha (✉) · V. D. Deepak · M. Pavithra · V. Vallabhi
Centre for Healthcare Technologies, Department of Biomedical Engineering,
SSN College of Engineering, Chennai, India
e-mail: kavithaa@ssn.edu.in

G. Sudhir
Department of Orthopedics, Sri Ramachandra Medical College, Chennai, India

L. J. Kumar et al. (eds.), *3D Printing and Additive Manufacturing Technologies*,
https://doi.org/10.1007/978-981-13-0305-0_20

Keywords Vertebroplasty · Vertebral compression fracture · Mimics 3-matic · Finite element analysis · 3D modelling

1 Introduction

Vertebral compression fractures are caused as a result of major trauma to the spine. Fractures or dislocation of vertebra can cause bone fragments to collapse which in turn can damage the spinal nerves or spinal cord. Bone injuries range from mild muscle strains, to fractures and dislocations of the bony vertebrae, to unbearable spinal cord damage. While fractures heal with traditional cure, severe fractures may require surgery to reform the bones and restore the normal height of the vertebral body.

Compression fractures are very common in osteoporotic patients, or patients whose bones have been weakened by other chronic diseases. Such fractures occur due to combined bending forward and downward pressures on the spine. Fracture is induced when the vertebral body collapses and the anterior part forms a wedge shape. Wedge fractures are the most common, accounting for more than 50% of all the Vertebral Compression Fractures and are characterized by compression of the anterior segment of the vertebral body [1]. The most common procedure used to relieve pain caused by Vertebral Compression Fractures are the various vertebral augmentation techniques. All these surgical procedures involve injecting bone cement in the fractured vertebra to create an 'internal cast' inside the bone, with the goal of stabilizing the fracture and reducing the patient's pain [2, 3].

Among the augmentation techniques, vertebroplasty has proved to be the most efficient with the highest success rate. Vertebroplasty involves injection of bone cement into the deformed region of the spine under X-ray guidance with the aim of restoring the height of the vertebral structure and pain management. PMMA and Hydroxyapatite are the cements of choice. Though vertebroplasty helps in pain reduction, the injection of cement in the vertebral structure might induce a higher stress in the adjacent vertebrae and the discs. This is mainly attributed to the fact that the cement could be injected in excess which might lead to extravasation into the discs [4, 5].

Assessing the accurate stress values on the fractured and the adjacent vertebral bodies, concentrating on the contributing factors and calculating the right amount of cement required for each case of fracture might serve as a reference in clinical conditions of Vertebroplasty [6].

2 Methodology

The data used for 3D modelling was acquired in the form of CT scan images in DICOM format. The vertebral column scan of a healthy patient was obtained.

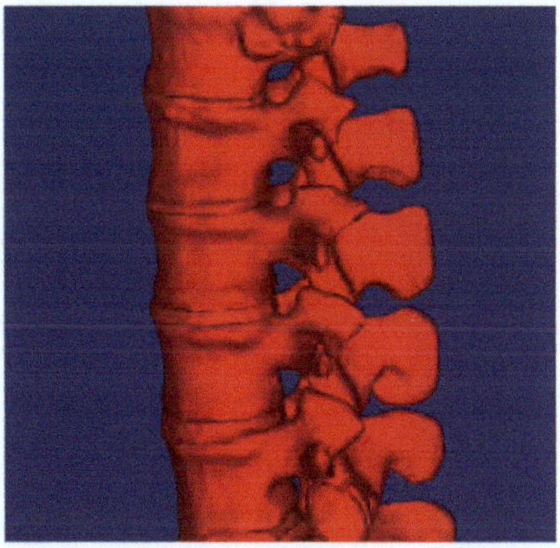

Fig. 1 3D model of T11-L3 spine

2.1 3D Modelling of Normal Vertebrae Using Mimics

The first procedure involved creating a 3D model of the vertebrae using Mimics. Mimics provides options for editing the masks in the axial, coronal and sagittal views. Segmentation of the vertebral body from the intervertebral discs has been done by identifying the fine details separating the bone and the soft tissues. Image segmentation was performed by considering the characteristics of the CT. The pixels having CT numbers in the threshold range were considered as bone tissue and collected in a segmentation mask. The segmentation has been done in axial view and copied to the other views using appropriate options. By segmenting, the regions required for this model were separated out, using multiple slice edit tool and the floating pixels were removed using region growing tool. The T11-L3 region was selected as the region of interest. Figure 1 represents the 3D model of the T11-L3 region of the spine used for analysis.

2.2 Meshing in 3-Matic

Reduction of the model needs to be performed for eliminating unnecessary curves and lines. The non-manifold assembly was created using the appropriate tool. An inspection scene was created and surface meshing was performed by adjusting the optimal edge length of the triangle. Volume meshing was also performed in the similar manner. Figure 2 shows the model of the vertebra after meshing done in 3-matic.

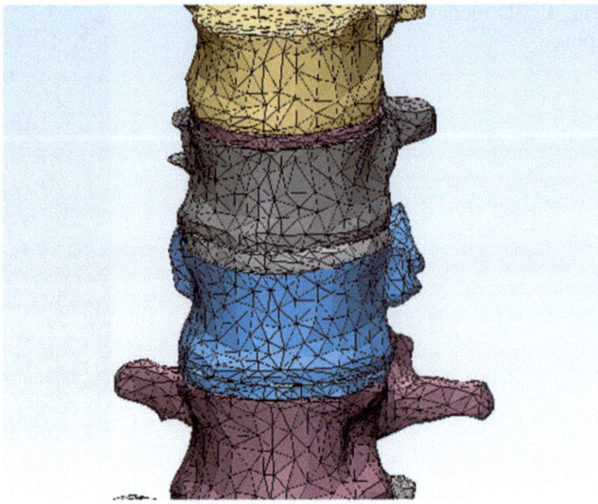

Fig. 2 Meshing of normal vertebra in 3-matic

2.3 Stress Analysis of Normal Vertebrae Using ANSYS

The ANSYS workflow included two modules. The FE Modeller module was used to define the skin geometry and mesh shape. The final mesh was checked for errors using mesh diagnosis. The Static Structural module was used to link the mesh from the FE Modeller to the analysis module. The material properties were defined under the knowledge base, materials were assigned, compression support was given to L3, and an axial force of 600 N was applied on the T11 vertebra [7]. The final mesh was solved and the results were evaluated.

2.4 3D Modelling of Fractured Vertebrae Using Mimics

In the L1 region, which is most susceptible to Vertebral Compression Fractures (VCF), the mask of the normal vertebra was edited to induce a fracture. Three levels of compression fractures in the shape of wedges were created and separate 3D models were created for the same. The model of the fractured L1 vertebra is shown in Fig. 3. Smoothening and meshing operations were carried out in a way similar to that of the normal vertebrae.

2.5 Stress Analysis of Fractured Vertebrae Using ANSYS

The modules used were similar to that of the stress analysis of normal vertebrae. The material properties were defined separately for the fractured vertebrae and the adjacent vertebrae, materials were assigned and the stress values were evaluated. Figure 4 represents the stress analysis performed on the fractured L1 vertebra.

Fig. 3 Fractured L1 vertebra

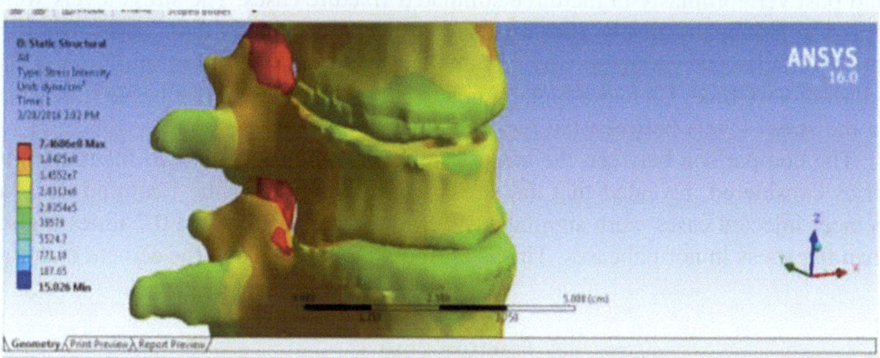

Fig. 4 Stress on fractured L1 vertebra

2.6 *3D Modelling of Cement Injected Vertebrae Using Mimics*

To simulate the condition of vertebroplasty, the masks of the fractured vertebrae were edited using MIMICS. A separate mask for the cement was created and by using a subtraction operation between the normal and fractured vertebrae, the accurate area that has to be filled with cement was assessed. Depending on the area to be filled, the accurate amount of cement required for each case of fracture was evaluated and filled. Smoothening and meshing operations were carried out in a way similar to that of the normal vertebrae.

2.7 *Stress Analysis of Cement Injected Vertebrae Using ANSYS*

The operations similar to that of stress analysis in normal vertebrae were performed. An additional density, elastic modulus and Poisson's ratio values corresponding to Poly (methyl methacrylate) (PMMA) and Hydroxyapatite (HA) cements were introduced to compare the difference in stress values [8–10].

3 Results and Discussions

Vertebroplasty was performed for the three cases of fracture with two different cements. Poly (methyl Methacrylate) (PMMA) and Hydroxyapatite (HA) were the cements of choice. Figures 5, 6 and 7 represent the stress levels in the vertebra pre and post vertebroplasty in the three simulated fracture cases. The stress levels were found to reduce drastically after injection of the cement in all three cases. But, the stress on adjacent vertebrae was higher than in the normal cases, and among the adjacent vertebrae, L2 had a higher stress value. There was a negligible difference in the stress levels between the two cements used.

The comparison between the stress level distributions on the two intervertebral discs considered, revealed that the stress reduction between the fractured and the cement injected cases were significant. But, the stress levels were 0.7 times higher than the stress in normal cases. This is attributed to the fact that the cement material

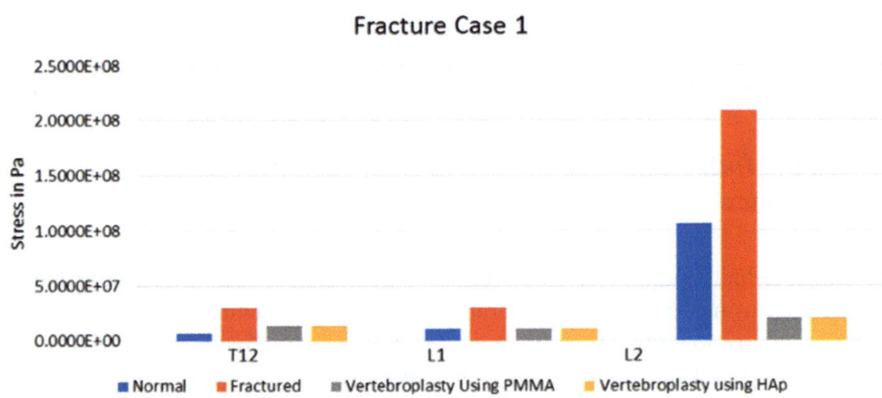

Fig. 5 Stress levels pre- and post vertebroplasty in fracture case 1

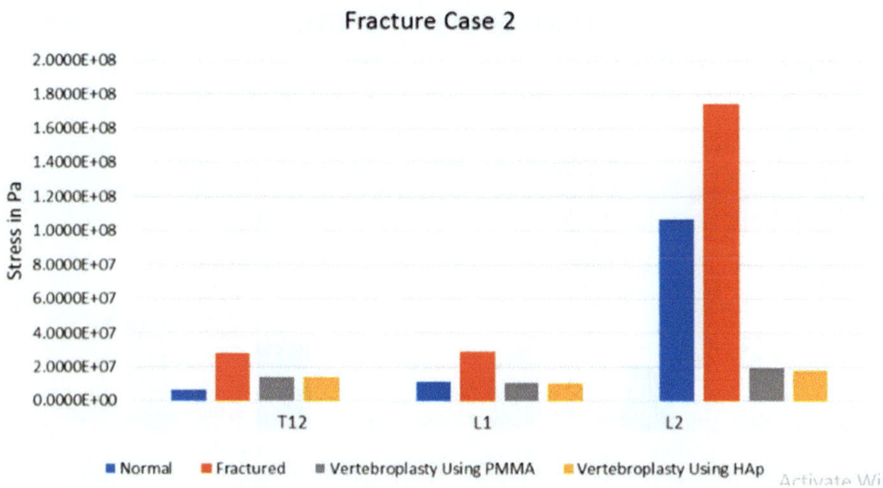

Fig. 6 Stress levels pre- and post vertebroplasty in fracture case 2

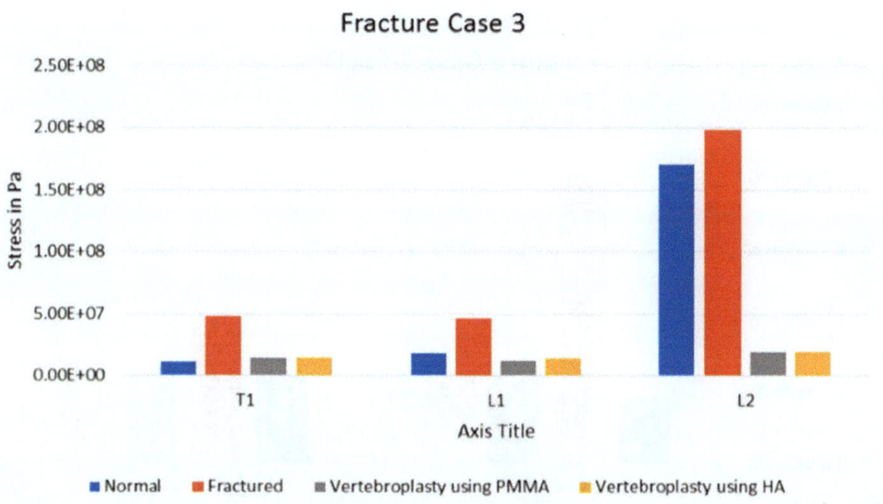

Fig. 7 Stress levels pre- and post vertebroplasty in fracture case 3

came in direct contact with the disc, which has a major liquid composition, making the disc stiffer. Figures 8, 9 and 10 show the stress levels in the disc pre and post vertebroplasty in the three fracture cases. Among the T12-L1 and L1-L2 discs, the higher stress concentration was in the T12-L1 disc as it was the interface between the normal vertebra and the region of the vertebra that faced maximum degradation.

The amount of cement injected for the three fracture cases increased 1.2 times with each case. In case 1, the volume of cement injected was 4.5631 mL, which

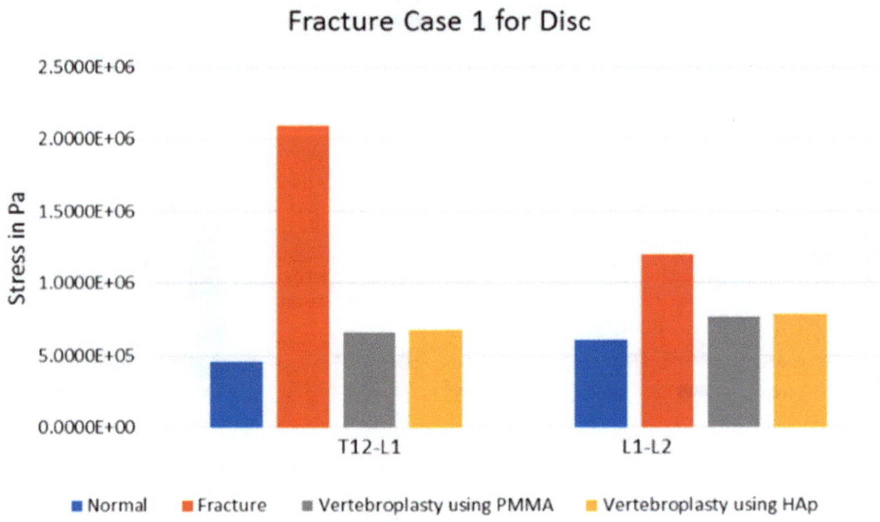

Fig. 8 Stress levels in discs pre- and post vertebroplasty in fracture case 1

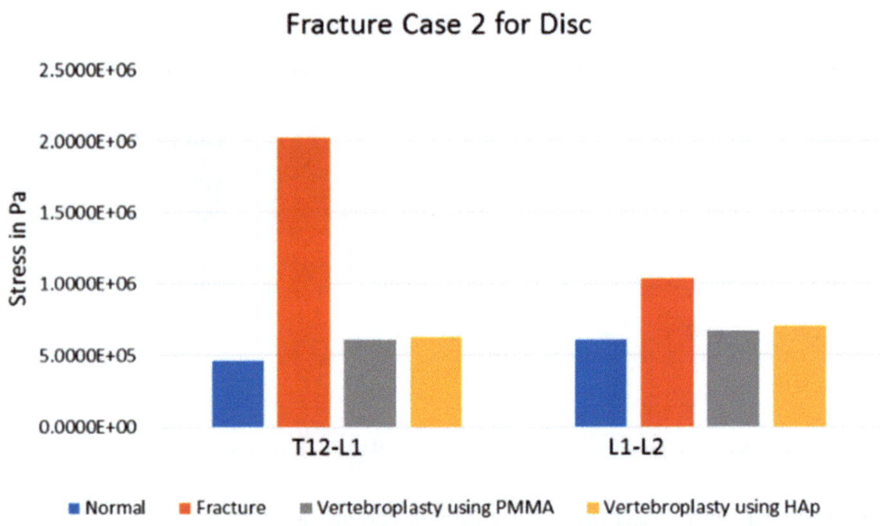

Fig. 9 Stress levels in discs pre- and post vertebroplasty in fracture case 2

increased to 5.5771 mL in case 2. In case 3, the volume was calculated to be 6.5847 mL. While the normal practice in vertebroplasty surgeries is to inject a volume above 5 mL, on an analysis with three cases, 4–7 mL seemed to be the optimal amount.

The volume parameter proved to be a crucial factor in determining the stress on adjacent vertebrae. Therefore, by making use of Boolean Operations in the

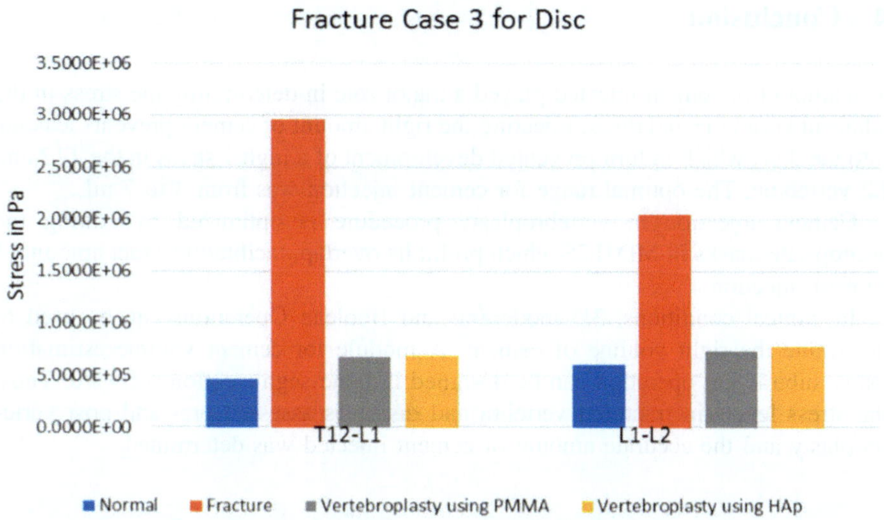

Fig. 10 Stress levels in discs pre- and post vertebroplasty in fracture case 3

Fig. 11 Cement injection in L1

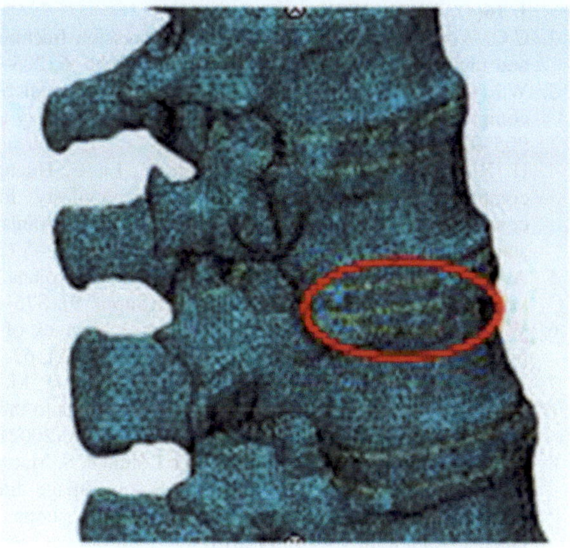

prediction of cement volume requirements, an optimal value can be obtained. This accuracy helps prevent excessive stress on the adjacent vertebrae and tissues, while also preventing cement leakage into the disc. The injection of cement done in L1 is shown in Fig. 11.

4 Conclusion

The amount of cement injected played a major role in determining the stress in the adjacent vertebrae and discs. Injecting the right amount of cement prevents leakage into the disc, which in turn prevented development of a higher stress in the T12 and L2 vertebrae. The optimal range for cement injection was from 4 to 7 mL.

Cement injection in vertebroplasty procedure is optimized by editing the appropriate masks in MIMICS which prohibits overlap, facilitating exact amount of cement injection.

In clinical conditions, 3D modelling and Boolean Operations can be used to determine the right volume of cement. A module for cement volume estimation using subtraction operation can be designed in bone segmentation software. Thus, the stress levels in fractured vertebra and disc was assessed pre- and post vertebroplasty and the accurate amount of cement injected was determined.

References

1. D. Alexandru, W. So, Evaluation and management of vertebral compression fractures. Perm. J. **16**(4), 46–51 (2012)
2. C.C. Wong, M.J. McGirt, Vertebral compression fractures: a review of current management and multimodal therapy. J. Multidiscip. Healthc. **6**, 205–214 (2013)
3. W.J. Choe, Y.-J. Rho, Y.I. Chun, Risk factors predicting the new symptomatic vertebral compression fractures after percutaneous vertebroplasty or kyphoplasty. Eur. Spine J. **21**(5), 905–911 (2012)
4. H. Xiao, J. Yang, X. Feng, P. Chen, Y. Li, C. Huang, Y. Liang, H. Chen, Comparing complications of vertebroplasty and kyphoplasty for treating osteoporotic vertebral compression fractures: a meta-analysis of the randomized and non-randomized controlled studies. Eur. J. Orthop. Surg. Traumatol. **25**(1), 77–85 (2015)
5. A. Chotivichit, E. Korwutthikulrangsri, A. Churojana, D. Songsaeng, Complications in vertebroplasty. J. Med. Assoc. Thail. **95**(Suppl. 9), S75–S81 (2012)
6. W. Yi, Z. Ling, Influence of degenerative changes of intervertebral disc on its material properties and pathology. Chin. J. Traumatol. **15**(2), 67–76 (2016)
7. M. Ito, A. Nishida, A. Koga, S. Ikeda, A. Shiraishi, M. Uetani, K. Hayashi, T. Nakamura, Contribution of trabecular and cortical components to the mechanical properties of bone and their regulating parameters. Bone **31**(3), 351–358 (2002)
8. H. Ikeda, M. Nishio, S. Matsuoka, B.D. Lohman, S. Matsushita, Y. Ogawa, S. Hamaguchi, Y. Nakajima, A. Kojima, Y. Torii, Y. Sasao, Image findings following vertebroplasty in osteoporotic vertebral compression fractures: bone healing and sagittal alignment. Open J. Radiol. **3**, 152–158 (2013)
9. E. Carlsson et al., In vitro and in vivo response to low-modulus PMMA-based bone cement. Biomed. Res. Int. **2015**, 409–418 (2015)
10. J.-M. Kim et al., Effect of bone cement volume and stiffness on occurrences of adjacent vertebral fractures after vertebroplasty. J. Korean Neurosurg. Soc. **52**(5), 435–440 (2012)

Design and Processing of Functionally Graded Material: Review and Current Status of Research

Manu Srivastava, Sandeep Rathee, Sachin Maheshwari and T. K. Kundra

Abstract Functionally graded materials (FGMs) are superiorly engineered as well as natural materials with customized properties. These materials offer a variety of advantages over conventional materials in specific engineering applications. High strength, improved ductility, superior mechanical properties and enhanced surface properties are some advantages of FGMs when compared to homogeneous materials of the same type. Engineered FGMs have intrigued numerous researchers in recent years. These materials were conceived as thermal obstruction materials for various critical applications. These are increasingly being employed for numerous conventional as well as advanced applications. Many methods are utilized in developing FGMs possessing specific advantages and disadvantages. This article reviews current trends and developments of functionally graded materials (FGMs). Techniques to attain gradation especially structural and functional are emphasized in the current work. A few real life illustrations are discussed to present a glimpse of different FGM fabrication strategies to the readers.

Keywords Functionally graded materials · Classification · Composites
Fabrication · Applications

M. Srivastava (✉)
SGT University, Gurgaon, India
e-mail: manyash@gmail.com

S. Rathee · S. Maheshwari
Netaji Subhas Institute of Technology, New Delhi, India

T. K. Kundra
Indian Institute of Technology Delhi, New Delhi, India

L. J. Kumar et al. (eds.), *3D Printing and Additive Manufacturing Technologies*,
https://doi.org/10.1007/978-981-13-0305-0_21

1 Introduction

The advent of functionally graded materials (FGM) led to the development of innovatively engineered tailor-made materials possessing customized properties [1, 2]. This concept basically advanced almost four decades ago (in early 70s) [3, 4]. Several methods were suggested earlier also but substantial investigations commenced only after 80s. Japanese researchers came up with novel FGM fabrication to accomplish temperature change of around 1000 K across 10 mm thick hypersonic space plane component dimension [5, 6]. Initially FGMs were fabricated by giving structural gradation in early 70s [3, 4]. Continuous control of texture was proposed in 1985 to derive improvement of properties [7]. This texture control was found of huge value and also led to the development of concept of material ingredients [8]. In 1986, this new class of materials having graded structures came to be known as functionally gradient materials which was finally refined to functionally graded materials (FGMs) in 1995 [9].

FGMs are synthesized in specific manner to suit the need of end user. Fundamental or elementary unit for defining FGM is termed as elemental or material ingredient [10] and includes its chemistry, physics, biology and orientation [5, 11, 12]. Its chemistry includes the organic, inorganic, metal, ceramic, alloy, etc., nature of FGM. Physics of FGM includes its properties like dipole, magnetic, potential state, etc. Its biology includes its cellular, tissue or molecular levels. Its orientation includes its granular, needle, platelet states, etc. These elementary units also form the basis of property gradation in FGMs. Pores and their location in a FGM is also of crucial value and may significantly affect properties like insulation, strength, stress relaxation, etc.

Excellent examples of FGMs can be seen in nature. Various natural load carrying units undergo adaptation corresponding to need. Examples include stems, hard tissues, human bones, etc. [13, 14]. These basically have three load sustenance mechanisms: first, by microstructural changes (refer Fig. 1a); second, by shape/size changes (refer Fig. 1b) and third (refer Fig. 1c), by combination of first two mechanisms. In fact, these natural tissues are the inspirations behind many beautiful and innovative designs [15]. Most of the biological organizations are complex but highly effective and have substantial engineering potential if properly understood. Bamboo and bone tissues are two most important examples extensively researched in the past [16]. It is a proven fact that bamboo cells are able to sense at least some external stimulus. Also, bone cells exhibit piezoelectricity when subjected to stress. Roots of plants, stems of trees, palm tree leaves, mollusc shells, etc., are numerous examples of naturally existing FGMs. The behaviour of these natural facts can be utilized to achieve FGMs possessing newer and structurally efficient intelligent components with graded properties.

FGMs are a special class of material where the material properties are continuously varying and clear interfacial boundaries are not present in general (Refer Fig. 2a) [1, 8]. However, some cases of FGMs have been reported where discrete boundaries are present (Fig. 2b) [17, 18]. Sometimes, the property variation can

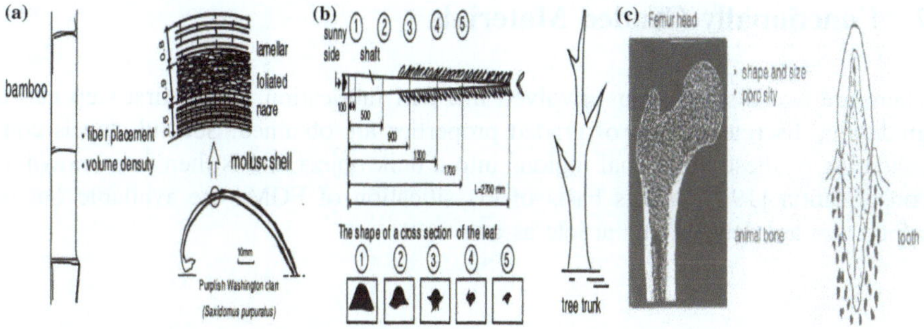

Fig. 1 Natural load sustenance systems [9, 12]

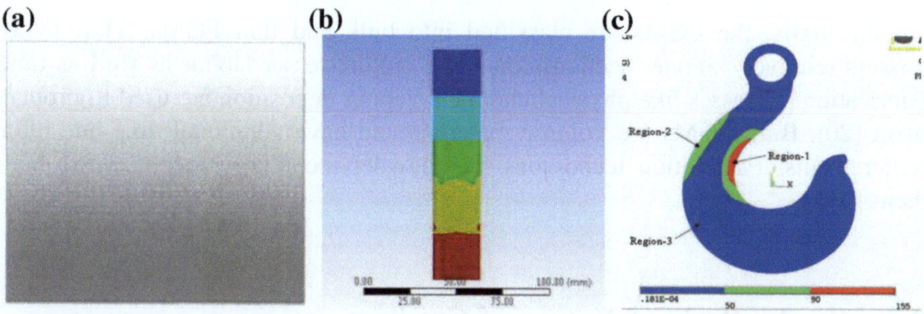

Fig. 2 Gradation in FGM structures. a Continuously varying microstructures [9]. b Variations with discrete properties in steps [18]. c Variations confined to small regions [17]

also be confined to a small region and not globally in FGMs (Fig. 2c). Gradation in structures initiated from composites and polymers.

This technology is still in development stage despite enormous quantum of work done. This can majorly be attributed to the diversity of the elementary unit for defining FGM. Design and modelling of FGM require substantial research owing to their heterogeneous nature. Systematization of the fabrication techniques and applications of FGM is a pressing requirement to fully explore the versatility these materials.

Techniques to process FGMs has been the primary focus of research of majority of the researchers in the recent past. However, fabrication approaches have rarely been reviewed. Therefore, an attempt has been made to review various methods to fabricate tailored materials on the basis of the underlying fabrication techniques with the help of practical examples.

2 Functionally Graded Materials

There are two discrete steps involved in FGM fabrication. In the first step called gradation, discrete regions of graded properties are obtained. Second step is consolidating of these individual regions into a bulk object and is therefore known as consolidation [19]. Various basis of classification of FGMs are available out of which two are dealt in this article as below:

2.1 On the Basis of Amount of Volume in Which Graded Properties Are Present

On this basis, the FGMs are classified into bulk and thin FGMs. Thin FGMs possess relatively thinner sections/coats and require lesser labour as well as time. Fabrication processes like physical/chemical vapour deposition are used to produce them [20]. Bulk FGMs are volume materials and have additional time and effort requirements. Fabrication techniques like powder metallurgy are used to obtain them [21].

2.2 On the Basis of Nature of Gradient

Gradation process involves constitution, homogenization and segregation processes. Constitution basically refers to the building up of discrete graded structures from the individual components of the FGM. While homogenizing the sharp boundaries at gradations need to be eliminated to give rise to specific gradients by transportation of material using some externalized fields. In segregating, the homogeneous material gets transformed into FGM. The gradients are thus are important parameter during FGM fabrication and form an important basis of FGM classification [19, 22]. On this basis, FGMs are classed into four distinct categories namely fraction, shape, orientation and size/material gradient types.

2.3 Achievement of Graded Structures

As mentioned in Sect. 2.1 of this article, gradients are formed during the second phase of gradation process. These gradients are the heart of FGM and this can be achieved in many ways. One way is by microstructural variation since microstructures play a predominant role in the material properties. In FGMs, varying microstructures are present at least in one dimension with the objective to obtain better functional property. Tremendous research has been accomplished in the direction of improvement of material properties by controlling the microstructures and vice versa [23]. Phase, interface and defect are the three elements of microstructure. Phases affect properties like volume fraction, size, shape, etc. Interfaces affect boundary areas, continuity, etc. Defects affect number density, electronic nature, orientations, etc. Some plausible relationship of microstructure with material properties and its processing is required in order to utilize microstructural gradients in a material. Key parameters to characterize microstructure of FGM are: (1) crystal parameters which include composition and crystal structure; (2) geometric parameters which include nomenclature, fractional volume, boundary area density, specific boundary area, size, shape, contiguity and some special features. Advancements in this field have enabled researchers to allow control of microstructure of the materials well before actual fabrications. Another way can be through compositional variation. However, this kind of variation is sometimes followed with other gradients also [9]. Another probable basis of introducing gradients in FGMs are porosity and pore size, volume content of phases and granular sizes [19], melt processing techniques like centrifugal casting, sedimentation casting, controlled mould filling [24], etc.

3 Overview of Techniques Utilized for FGM Fabrication

Corresponding to the extensive scope of FGMs in engineering applications, much attention is focused in development of FGMs since last two decades. Lot of techniques have been utilized in fabrication of FGMs. Each fabrication technique has its own pros and cons and are utilized based on their suitability for particular type of material and applications. In this section, these fabrication techniques are briefly outlined. To simplify, these techniques can be classed into three major categories viz. liquid-phase-based processes, gas-phase-based processes and solid-phase-based processes [22]. Liquid-based processes mainly include centrifugal casting [25–27], tape casting [28, 29], slip casting [30], gel casting [31, 32], chemical solution deposition techniques [33], laser technique [34, 35], etc.

Gas-based processes mainly include chemical vapour deposition [36, 37] and thermal spray deposition techniques [38, 39], etc.

Majority of FGM research pursued in field of FGMs is focused on ceramic-/metal-based systems [19, 22]. In case of metal matrix, FGMs developed by liquid and gas phase techniques, problem of deleterious phases, undesirable reactions and defects formation arises due to involvement of elevated temperature also, inadequate microstructural refinement takes place. These issues can be avoided in solid-state techniques [40, 41]. Due to these reasons, more emphasize is given to solid state processing techniques to fabricate metal matrix FGMs during last decade. Solid state processes mainly include spark plasma sintering [42, 43], friction stir processing [22, 23, 44], etc.

FGM fabricated via conventional manufacturing methods have relatively simpler geometry, restricted design freedom and therefore lack advanced functionality required for satisfying application requirements. Advancements and complexity in 3D periodic cellular structure which is close to natural structures and are fabricated by control of density and strength can be obtained by additive manufacturing. AM additionally permits conceptualization and creation of FGMs possessing lighter weights that are close to optimal design instead of compromising owing to manufacturability restrictions. Discussion on FGMs developed by additive manufacturing is presented in the Sect. 5.

4 Additive Manufacturing

Additive manufacturing is an evolving technique which forms a versatile pillar of manufacturing advancement and design visualization. It is known by different key names like additive (layer/digital) manufacturing, layered (based/oriented) manufacturing, rapid (prototyping and tooling and manufacturing) technology, digital (fabrication/mock up), direct (tooling and manufacturing) technology, 3D (printing/modelling) technology, desktop, freeform, generative, on-demand manufacturing. It forms a critical pillar of worldwide manufacturing together with conventional manufacturing techniques (both substractive and formative). Standardization in the field of AM is still limited to a few standards, example VDI3404 by VDIsociety and F2792 09e1/F2792/, also called *standard terminology for additive manufacturing technologies by ASTM* [45]. Economically, AM predicts enormous growth and around $3 billion asset size by 2016 [46]. Despite emerging as a forbearer amongst innovations in manufacturing technology, AM still faces six-fold challenges. These are shown in Fig. 3.

The process chain for AM is based on general working principle of this technology and can be summarized as: it initiates with obtaining 3-dimensional data corresponding to the physical model via some suitable technique. This is followed by slicing of the obtained models uniformly using suitable techniques. This data is

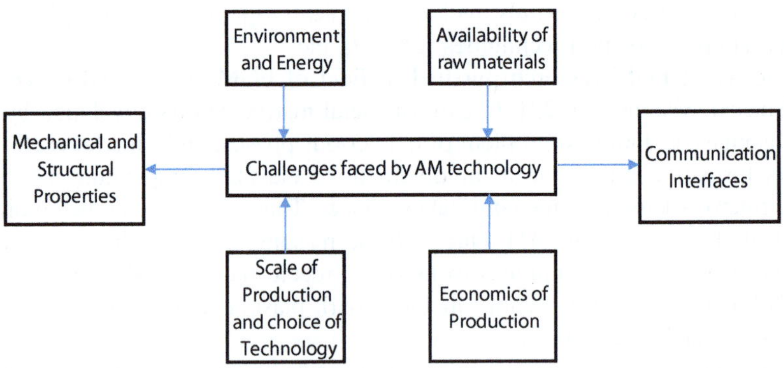

Fig. 3 Challenges faced by AM technology

inputted to the AM modeller which processes the data to build the part in layers with suitable process parameters and set up by the designer. The individual layers are bonded together by process specific techniques. Though the general trend in AM is addition of material in layered fashion but sometimes there can be variations like subtractive additive manufacturing or adding material in voxels rather than in layers (as in ballistic particle manufacture) [45, 47, 48]. Also, it should be clearly understood that while AM is based on 3D input, it deals with 2D layers only during actual manufacturing. It is independent of the design process which in turn facilitates its introduction at any point of the product cycle.

In general, there are six major deciding variables in commercialization of AM techniques. These include economics of production, production quantity, environmental issues, underlying manufacturing cost, choice of raw materials and structural efficiency of manufactured components [18, 49, 50]. The choice of AM process depends upon the application specific selection criterion [51, 52]. Same component can be fabricated using different AM modellers.

Addition of material in layers gives rise to inherent defects like anisotropy [53], stair stepping effect [45], reduced lateral strength [17], communication interfaces [54], limited raw materials since most materials are proprietary [18], etc.

5 Development of FGMs via Additive Manufacturing

Additive manufacturing or rapid manufacturing has been extensively used for the production of FGMs owing to the tremendous flexibility of conversion of design ideas into tangible components [17, 18]. A new term functionally graded rapid prototyping (FGRP) is used for the science of fabricating FGMs via AM techniques. FGRP is fundamentally an innovative approach and technological skeleton that enables control of material property gradation in functional applications [55]. Various AM techniques have been extensively researched for FGM designs. Amongst these, fused deposition process (FDM) and selective laser sintering (SLS) are found more suitable for polymeric components. On the contrary, laser engineered net shaping (LENS), selective laser melting (SLM), electron beam melting (EBM), etc. have been used for metallic components.

Components fabricated via these approaches find extensive applications ranging from biologically inspired design for making carpal or beast like skin [55], structural components and their analysis [17, 18, 56], variable density concrete foams [57], etc.

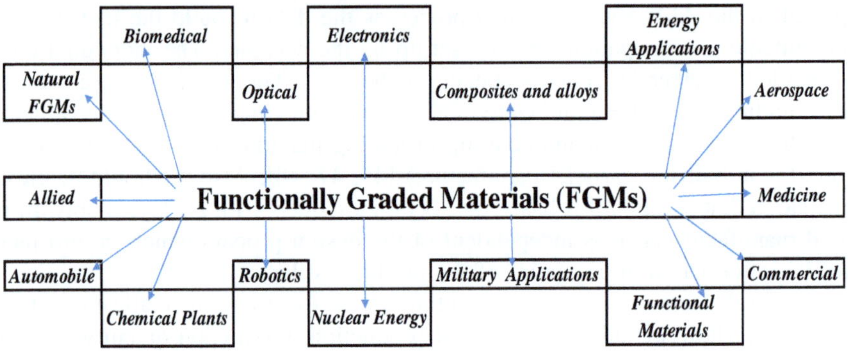

Fig. 4 Application areas of FGMs

6 Applications of FGMs

FGMs are excellent engineering composite materials owing to their specially tailored properties. These are widely utilized in large space station, shuttle, aircraft, automotive and many more [55]. These are found to reduce thermal stresses and can resist super high temperatures. They can be applied for weight reduction applications, applications requiring severe thermal gradients, applications requiring high corrosion, fatigue resistance, micro hardness, surface hardness, load bearing capabilities, malleability, toughness, etc. [56, 58–60]. There is practically no aspect of life where FGMs cannot be applied to impart desirable functions [8]. Figure 4 depicts the different aspects of FGM applications.

Various applications of FGMs can be divided into the following classes:

(a) Defence Applications: FGM possess a distinctive ability to stop cracks from spreading further which can be utilized in defense where it can be used as an invasion resistive material utilized in amour plates and bullet-proof jackets. The recent development in the application of FGMs is weight reduction for army. Minimal weight weapons will be critical to any potential battles and boundary securities [61].
(b) Medicine and Bio-medical applications: As discussed earlier, tissues like bone and teeth are natural FGMs. Their replacement necessitates need of some synthetic/engineered FGM to be used as original bio-tissue. FGMs therefore find huge applications in dental and orthopedic applications [62–65]. Tailored nano HA/PVA gel fabricated by combination of additive manufacturing principles with freeze/thaw cycles technology can be a brilliant choice for repairing cartilages [66] owing to their enhanced bioactivity as well as mechanical characteristics.
(c) Optoelectronics: Another probable region of FGM function can be as a storage medium [66, 67].

(d) Energy: FGMs can be effectively utilized to convert energy. These can function as thermal obstructions and coats on turbine blades [68, 69].
(e) Aerospace: Ability to sustain appreciably high temperature make these FGMs a feasible option of improving structural performance in bodies of highly critical applications like aerospace, space jets, etc. Their application domain can be further broadened by incorporating improvements in the fabrication and processing techniques [70, 71].
(f) Automobile industry: FGMs are used for manufacturing multiple engine and other automotive parts which includes racer car braking systems, advanced components to absorb shocks and so on.
(g) Marine applications: Many marine components are FGMs which can include pipe system, dive cylinders, etc.
(h) Commercial Applications: Apart from the applications stated above, there are numerous commercial applications of FGMs produced via different fabrication routes. A few examples include tool inserts, laptop cover, aviator frame, instruments in music industry, etc.

Numerous alternative applications of the graded materials are available which have their stems in manufacturing, tooling, nuclear industries, tribological applications, etc. [72–76].

7 Future Trends

Numerous research studies on FGMs have opened avenues of newer opportunities in the field of FGMs. This can partly be attributed to their varied engineering applications and partly to the boundless tailoring approaches that can be applied to achieve these versatile materials. These materials have the capability to be engineering marvels and revolutionize the entire manufacturing and materials spectra. Work is under progress to aim at optimal production of these materials specially in terms of time and cost factors. Additive manufacturing is one of the most technically advanced and economical ways of FGM fabrication. However, lots of work need to be accomplished before the roadblocks of FGM fabrication via AM can be economically done. This specially includes the modeller availability and its cost, material compatibility with various AM techniques and issues related to communication strategies.

8 Concluding Summary

Following conclusions can be drawn from this review:

1. FGMs are exquisite class of graded materials with unique tailored properties which are designed and fabricated for specific functional requirements.

2. Nature is the best FGM fabricator and gives exemplary examples. Synthetic FGMs have still a long way to go to match the perfection of naturally existing FGMs.
3. Cost and technology are the two main inhibitors in FGM fabrication.
4. AM is one of the most versatile and economical methods of FGM production. However, full scale production of FGMs by AM is still a distant dream.
5. The repeatability of fabrication processes and the authenticity of FGM performance are still debatable. Controlling quality is also a major challenge of this area.
6. An insight into application areas of FGMs suggests that sensible research put in this field will be worth the effort owing to the vastness of FGM application.

References

1. J.J. Sobczak, L.B.L. Drenchev, Metal Based Functionally Graded Materials (2009), p. 74
2. R.K. Joshi, S. Alwarappan, M. Yoshimura, V. Sahajwalla, Y. Nishina, Graphene oxide: the new membrane material. Appl. Mater. Today **1**(1), 1–12 (2015)
3. M. Shen, M.B. Bever, Gradients in polymeric materials. J. Mater. Sci. **7**, 741–746 (1972)
4. M.B. Bever, P.E. Duwez, Gradients in composite materials. Mater. Sci. Eng. **10**, 1–8 (1972)
5. T. Hirai, R. Watanabe, M. Niino, The functional gradient materials. J. Jap. Soc. Compos. Mat. **13**, 257–264
6. Fundamental study on relaxation of thermal stress for high temperature material by tailoring the graded structure (1992)
7. M. Niino et al., Fabrication of a high pressure thrust chamber by the CIP forming method. AIAA Pap. 84–1227 (1984)
8. R.M. Mahamood, E.T.A. Member, M. Shukla, S. Pityana, Functionally graded material : an overview, **3**, 2–6 (2012)
9. R.G. Ford, Y. Miyamoto, W.A. Kaysser, B.H. Rabin, A. Kawasaki, Functionally graded materials design, processing and applications (1999)
10. Y. Miyamoto, W.A. Kaysser, B.H. Rabin, A. Kawasaki, R.G. Ford, Functionally graded materials: design, processing and applications (1999)
11. A. Mortensen, S. Suresh, Fundamentals/Functionally Graded Materials. 10 M Commun. Ltd (1998)
12. A. Mortensen, S. Suresh, Functionally graded metals and metal-ceramic composites: Part I processing. Int. Mater. Rev. **40**(6), 239–265
13. C. Mattheck, S. Burkhardt, A new method of structural shape optimization based on biological growth. Int. J. Fat. **12**(3), 185–190 (1990)
14. J.C. Koch, The laws of bone architecture. Am. J Anat. **21**, 177–198
15. J. Gottron, K.A. Harries, Q. Xu, Creep behaviour of bamboo. Constr. Build. Mater. **66**, 79–88 (2014)
16. E.C.N. Silva, M.C. Walters, G.H. Paulino, Modeling bamboo as a functionally graded material: lessons for the analysis of affordable materials. J. Mater. Sci. **41**(21), 6991–7004 (2006)
17. P.K. Jain, P.M. Pandey, P.V.M. Rao, Tailoring material properties in layered manufacturing. Mater. Des. **31**(7), 3490–3498 (2010)
18. M. Srivastava, S. Maheshwari, T.K. Kundra, Virtual modelling and simulation of functionally graded material component using FDM technique. Mater. Today Proc. **2**(4–5), 3471–3480 (2015)

19. B. Kieback, A. Neubrand, H. Riedel, Processing techniques for functionally graded materials. Mater. Sci. Eng. A **362**, 81–105 (2003)
20. G.R. Palmese, J.K. Sutter, M. Ivosevic, R. Knight, S.R. Kalidindi, Solid particle erosion resistance of thermally sprayed functionally graded coatings for polymer matrix composites, Surf. Coat. Technol. (2005)
21. G.E. Knoppers, J.W. Gunnink, J. Hout Van den, W.P. Wliet Van, The reality of functionally graded material products. TNO Sci. Ind. Netherl. 38–43
22. M. Naebe, K. Shirvanimoghaddam, Functionally graded materials: a review of fabrication and properties. Appl. Mater. Today **5**, 223–245 (2016)
23. S. Rathee, Maheshwari, A.N. Siddiquee, Issues and strategies in composite fabrication via friction stir processing: a review, Mater. Manuf. Process. (2017)
24. B. Kieback, A. Neubrand, H. Riedel, Processing techniques for functionally graded materials. Mater. Sci. Eng., A **362**(1–2), 81–105 (2003)
25. A.G. Arsha, E. Jayakumar, T.P.D. Rajan, V. Antony, B.C. Pai, Design and fabrication of functionally graded in-situ aluminium composites for automotive pistons. Mater. Des. **88**, 1201–1209 (2015)
26. Y. Watanabe, N. Yamanaka, Y. Fukui, Control of composition gradient in a metal-ceramic functionally graded material manufactured by the centrifugal method. Compos. Part A Appl. Sci. Manuf. **29**(5), 595–601 (1998)
27. T. Prabhu, Processing and properties evaluation of functionally continuous graded 7075 Al alloy/SiC composites. Arch. Civ. Mech. Eng. **17**(1), 20–31 (2017)
28. J.-G. Yeo, Y.-G. Jung, S.-C. Choi, Design and microstructure of ZrO2/SUS316 functionally graded materials by tape casting. Mater. Lett. **37**(6), 304–311 (1998)
29. D. Hotza, P. Greil, Review: aqueous tape casting of ceramic powders. Mater. Sci. Eng., A **202** (1), 206–217 (1995)
30. T. Katayama, S. Sukenaga, N. Saito, H. Kagata, K. in Nakashima, Fabrication of Al_2O_3-W functionally graded materials by slipcasting method, in *3rd International Congress on Ceramics, ICC 2011, IOP Conference Series: Materials Science and Engineering, Osaka, Japan, J. Andertová et al., Functional gradient alumina ceramic materials*, vol. 97, 2011
31. L. García-Gancedo, S.M. Olhero, F.J. Alves, J.M. Ferreira, C.E. Demoré, S. Cochran, T.W. Button, Application of gel-casting to the fabrication of 1–3 piezoelectric ceramic–polymer composites for high-frequency ultrasound devices. J. Micromech. Microeng. **22**(12), 125001 (2012)
32. C. Tallon, G.V. Franks, Recent trends in shape forming from colloidal processing: a review. J. Ceram. Soc. Japan **119**(1387), 147–160 (2011)
33. R. Slowak, S. Hoffmann, R. Liedtke, R. Waser, Functional graded high-K $(Ba_{1-x}Sr_x)TiO_3$ thin films for capacitor structures with low temperature coefficient. Integr. Ferroelectr. **24**(1–4), 169–179 (1999)
34. I. Shishkovsky, F. Missemer, I. Smurov, Direct metal deposition of functional graded structures in Ti–Al system. Phys. Procedia **39**, 382–391 (2012)
35. J.M. Wilson, Y.C. Shin, Microstructure and wear properties of laser-deposited functionally graded Inconel 690 reinforced with TiC. Surf. Coat. Technol. **207**, 517–522 (2012)
36. M. Sasaki, T. Hirai, Thermal fatigue resistance of CVD SiC/C functionally gradient material. J. Eur. Ceram. Soc. **14**(3), 257–260 (1994)
37. M. Kawase, T. Tago, M. Kurosawa, H. Utsumi, K. Hashimoto, Chemical vapor infiltration and deposition to produce a silicon carbide–carbon functionally gradient material. Chem. Eng. Sci. **54**(15), 3327–3334 (1999)
38. V. Cannillo, L. Lusvarghi, A. Sola, Production and characterization of plasma-sprayed TiO_2–hydroxyapatite functionally graded coatings. J. Eur. Ceram. Soc. **28**(11), 2161–2169 (2008)
39. A.H. Pakseresht, A.H. Javadi, M. Nejati, K. Shirvanimoghaddam, E. Ghasali, R. Teimouri, Statistical analysis and multiobjective optimization of process parameters in plasma spraying of partially stabilized zirconia. Int. J. Adv. Manuf. Technol. **75**(5), 739–753 (2014)
40. S. Rathee, S. Maheshwari, A.N. Siddiquee, Issues and strategies in composite fabrication via friction stir processing : A review. **0**, 0–22

41. S. Rathee, S. Maheshwari, A.N. Siddiquee, M. Srivastava, Distribution of reinforcement particles in surface composite fabrication via friction stir processing : suitable strategy **0**, 0–8
42. J.J. Sobczak, L. Drenchev, Metallic functionally graded materials: a specific class of advanced composites. J. Mater. Sci. Technol. **29**(4), 297–316 (2013)
43. A. Strojny-Nędza, K. Pietrzak, W. Węglewski, The influence of Al_2O_3 powder morphology on the properties of Cu-Al_2O_3 composites designed for functionally graded materials (FGM). J. Mater. Eng. Perform. **25**(8), 3173–3184 (2016)
44. J. Gandra et al., Functionally graded materials produced by friction stir processing. J. Mater. Process. Technol. **211**(11), 1659–1668 (2011)
45. A. Gebhardt, Understanding additive manufacturing (2012)
46. T. Wohlers, Additive manufacturing and 3D printing state of the industry (2012)
47. I. Gibson, D.W. Rosen, B. Stucker, Chapter 6 extrusion-based systems (2010)
48. M. Srivastava, Some studies on layout of generative manufacturing processes for functional components, Delhi University (2015)
49. I. Shiota, Y. Miyamoto, Functionally graded materials 1996. Vasa (1997)
50. S. Palanivel, H. Sidhar, R.S. Mishra, Friction stir additive manufacturing: route to high structural performance. JOM **67**(3), 616–621 (2015)
51. S. Srivastava, M. Maheshwari, S. Kundra, T.K. Rathee, Commercially available layered manufacturing techniques : a review, **7**(1), 31–46 (2016)
52. E.D. Herderick, Progress in additive manufacturing. JOM **67**(3), 580–581 (2015)
53. K. Senthilkumaran, P.M. Pandey, P.V.M. Rao, Influence of building strategies on the accuracy of parts in selective laser sintering. Mater. Des. **30**(8), 2946–2954 (2009)
54. A.S. Gogate, S.S. Pande, Intelligent layout planning for rapid prototyping. Int. J. Prod. Res. **46**(20), 5607–5631 (2008)
55. N. Oxman, S. Keating, E. Tsai, Functionally graded rapid prototyping (2008)
56. S. Keating, E. Tsai, N. Oxman, Biobeams: functionally graded rapid fabrication. in Mediated Matter, CRC Press (2016)
57. T. Tonyan, Mechanical behavior of cementitious foams (1991)
58. Y.S. Chan, G.H. Paulino, A.C. Fannjiang, Change of constitutive relations due to interaction between strain-gradient effect and material gradation. ASME J. Appl. Mech. **70**, 7–9 (2005)
59. R.H.L.C.Y. Tang, C.P. Tsui, Y.Q. Guo, B. Gao, Damage modelling of graded Ti-based composites using repeated unit cell approach. Int. J. Mater. Sci. **2**, 63–66 (2012)
60. P.G. Longmei Li, Q. Sun, C. Bellrehumeur, Composite modelling and analysis of FDM prototypes for design and fabrication of functionally graded parts. Rapid Prototyp. J. **4**, 14–25 (1998)
61. L. Lu, M. Chekroun, O. Abraham, V. Maupin, G. Villain, Mechanical properties estimation of functionally graded materials using surface waves recorded with a laser interferometer. NDT E Int. **44**(2), 169–177 (2011)
62. A. Sadollah, A. Bahreininejad, Optimum functionally gradient materials for dental implant using simulated annealing. University of Malaya (2012)
63. W. Pompe, H. Worch, M. Epple, W. Friess, M. Gelinsky, P. Greil, U. Hempel, D. Scharnweber, K. Schulte, Functionally graded materials for biomedical applications. Mater. Sci. Eng. A **362**, 40–60 (2003)
64. N. Ohata, S. Matsuo, F. Watari, Fabrication of functionally graded dental composite resin post and core by laser lithography and finite element analysis of its stress relaxation effect on tooth root. Dent. Mater J. **20**(4), 257–274 (2001)
65. F. Watari, A. Yokoyama, M. Omori, T. Hirai, H. Kondo, M. Uo, T. Kawasaki, Biocompatibility of materials and development to functionally graded implant for bio-medical application. Compos Sci Technol. **64**, 893–908 (2004)
66. I. Bharti, N. Gupta, K.M. Gupta, Novel applications of functionally graded nano optoelectronic and thermoelectric materials. Int. J. Mater. Mech. Manuf. **1**(3), 221–224 (2013)
67. T. Gürol, Finite element modeling of beams with functionally graded materials (2014)
68. E. Müller, Č. Drašar, J. Schilz, W.A. Kaysser, Functionally graded materials for sensor and energy applications. Mater. Sci. Eng. A **362**, 17–30 (2003)

69. M. Niino, K. Kisara, M. Mori, Feasibility study of FGM technology in space solar power systems (SPSS). Mater. Sci Forum **492**, 163–168 (2005)
70. T. Wallmersperger, C. Messe, D. Kliche, Modelling and simulation of heat transfer on FGM for hot structures in scramjet engines. J. Aircr. **6**, 472–474 (1969)
71. L. Marin, Numerical solution of the Cauchy problem for steady-state heat transfer in two dimensional functionally graded materials. Int. J. Solids Struct. **42**, 4338–4351 (2005)
72. J. Saifulnizan, Application of functionally graded materials for severe plastic deformation and smart materials (2012)
73. B. Woodward, M. Kashtalyan, Performance of functionally graded plates under localised transverse loading. Compos Struct (2012)
74. A. Kawasaki, R. Watanabe, Thermal fracture behavior of metal/ceramic functionally graded materials. Engng. Fract Mech. **69**, 1713–1728 (2002)
75. A. Xing, Z. Jun, H. Chuanzhen, Z. Jianhua, Development of an advanced ceramic tool material—functionally gradient cutting ceramics. Mater. Sci. Engng. A **248**, 125–131 (1998)
76. M. Malinina, T. Sammi, M.M. Gasik, Corrosion resistance of homogeneous and FGM coatings. Mater. Sci. Forum. **492–493**, 305–310 (2005)

Development of Electrical Discharge Machining (EDM) Electrode Using Fused Deposition Modeling (FDM)

Piyush Saxena and R. M. Metkar

Abstract Rapid prototyping (RP) is an innovating additive technology for creating functional prototypes and physical models directly from CAD model. Rapid Tooling (RT) is typically used to describe a process which either uses a Rapid Prototyping (RP) model as a pattern to create a mold quickly or uses the Rapid Prototyping process directly to fabricate a tool for a limited volume of prototypes. In this work, ABS electrode was metalized by electroless copper coating to make the RP electrode conductive. Applying a thin coating of copper to prototyped parts by electroless metallization has provided direct method from model to tool. Fused deposition modeling (FDM) process of rapid prototyping is employed to develop the electrode for electro-discharge machining (EDM). This revolutionary approach offer both designers and manufacturers attractive advantages of time compression and cost reduction. Time saving is of vital significance in production of EDM electrode for the fabrication of moulds and dies. The performance of these new type of electrode were compared with those obtained conventionally in term of, Surface Roughness (Ra), Tool Wear Rate (TWR), Working Time, Material Removal Rate (MRR) by using input parameters such as discharge Current, discharge Time, and discharge Voltage.

Keywords Rapid tooling (RT) · Fused deposition modeling (FDM)
Electro-discharge machining (EDM) · Metallization · Rapid prototyping

P. Saxena (✉) · R. M. Metkar
Mechanical Engineering Department, Government College of Engineering, Amaravti, Maharashtra, India
e-mail: piyushsaxena40@gmail.com

R. M. Metkar
e-mail: rajeshmetkar@yahoo.co.in

L. J. Kumar et al. (eds.), *3D Printing and Additive Manufacturing Technologies*,
https://doi.org/10.1007/978-981-13-0305-0_22

1 Introduction

Three-dimensional printing, also known as Additive Manufacturing (AM), refers to processes used to create a 3D solid object from a digital file model or another electronic data source such as an Additive Manufacturing File (AMF) [1].

The process builds objects by adding material in a layer by layer fashion to create a three-dimensional (3D) part, offering the benefit to produce any complex parts and lower cost compared to traditional manufacturing process [2].

Fused deposition modeling (FDM) is an additive manufacturing technology commonly used for modeling, prototyping, and production applications. It is one of the techniques used for 3D printing. FDM works on an "additive" principle by laying down material in layers; a plastic lament or metal wire is unwound from a coil and supplies material to produce a part [3].

Electrical discharge machining (EDM) is a well-established machining option for manufacturing geometrically complex parts or hard materials that are extremely difficult to machine by conventional machining processes. It is a non-conventional machining method. In electrical discharge machining process electrical energy is used to cut the material to final shape and size [4]. There is no mechanical pressure existing between work piece and electrode as there is no direct contact. Any type of conductive material can be machined using EDM irrespective of the hardness or toughness of the material.

2 Methodologies

In this experiment work, comparative study was carried out in between RP electrode and solid copper electrode for machining of mild steel. Both electrode having the same dimensions and geometry (i.e., 10 mm side and Quadrilateral square shape).

2.1 Electrode Preparation

Electrode 1 (RP Electrode): First of all a RP component of desired shape and dimension was produced on FDM RP machine uPrint SE Plus. Then electroless copper coating of 2 mm was done on the RP electrode of ABS material.

Electrode 2 (Solid Electrode): Machining of the solid copper rod is done on CNC machine to get the desired dimensions.

2.2 Experimental Set-up

uPrint SE Plus is used as rapid prototyping machine which made by US-based company "Strasys". The uPrint SE Plus 3D printer uses FDM Technology to build in real ABS plus thermoplastic, creating models and functional prototypes that are durable, stable, and pinpoint accurate [5].

The experimental investigations were carried out on a die-sinking EDM machine (SS 50 Spark Generator) of ELECRONICA ZNC (Figs. 1 and 2).

2.3 Design of Experiments

Experimental design was carried out by Taghuchi method for three input parameters with three levels, as given in Table 1.

Where current (I_p) is in ampere, Pulse on Time (T_{on}) is in microseconds and Pulse Duty Factor (t).

Fig. 1 uPrint SE plus machine

Fig. 2 Electronica ZNC die-sinking EDM

Table 1 Input parameters with their 3 levels

Parameter	Levels		
	1	2	3
Current (I_p)	6	8	10
Pulse on time (T_{on})	150	200	300
Pulse duty factor (t)	9	10	11

2.4 Data Collection

The initial weight of work piece and electrode was measured. The work material was mounted on the fixture table and clamped. The electrode was mounted on tool holder. After each machining operations, the work piece material electrode are taken out their weights are checked to find out the weight loss. Precision balance used to measure the weight of the work piece before and after the machining process is shown in Fig. 3.

2.5 Calculation of Responses

- Material Removal Rate (MRR)

  MRR can be calculated by using Eq. (1)

$$\mathrm{MRR} = \left(W_i - W_f\right)/\rho.T, \tag{1}$$

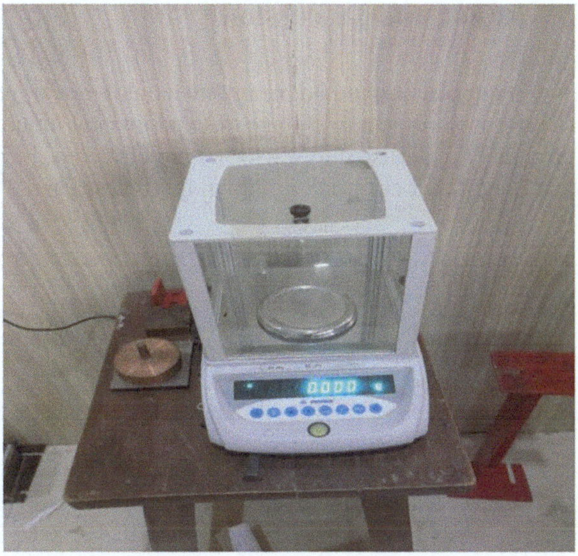

Fig. 3 Contech digital-balance weight measuring machine

where

W_i Weight of work piece before machining
W_f Weight of work piece after machining
ρ Density of work piece = 0.0078 gm/mm^3
T Machining time (min)

- Electrode Wear Rate (EWR)

  EWR can be calculated by using Eq. (1)

$$\text{EWR} = \left(E_i - E_f\right)/\rho.T, \tag{1}$$

where

E_i Weight of Electrode before machining
E_f Weight of Electrode after machining
ρ Density of the Electrode = 0.00896 gm/mm^3
T Machining time (min)

3 Result and Discussion

The change in output parameters material removal rate (MRR), Electrode wear rate (EWR) with respect to change of experimental input parameters Current (I_p), Pulse on time (T_{on}), Pulse duty factor (t) are observed. Experimental details for RP electrode and Solid electrode is shown in Tables 2 and 3 respectively (Figs. 4 and 5).

Table 2 Experimental details for RP elecctrode

Run No.	I_p	T_{on}	t	MRR	EWR
1	6	150	09	04.8264	0.017653
2	6	200	10	04.2672	0.016335
3	6	300	11	03.0029	0.004601
4	8	150	10	08.0763	0.022976
5	8	200	11	07.0504	0.023583
6	8	300	09	07.8296	0.005694
7	10	150	11	11.8463	0.016961
8	10	200	09	10.3325	0.073668
9	10	300	10	10.3813	0.059603

Table 3 Experimental details for solid electrode

Run No.	I_p	T_{on}	T	MRR	EWR
1	6	150	09	04.8013	0.02580
2	6	200	10	04.3832	0.01706
3	6	300	11	02.8945	0.00690
4	8	150	10	07.7878	0.07022
5	8	200	11	07.6998	0.01095
6	8	300	09	07.0064	0.01837
7	10	150	11	11.0090	0.03152
8	10	200	09	11.1594	0.03324
9	10	300	10	10.3523	0.08146

Fig. 4 Workpiece after experiment using square RP electrodes

Fig. 5 Workpiece after experiment by using square solid electrodes

4 Analysis

4.1 Mathematical Analysis

4.1.1 Regression Analysis

- Square RP Electrode

Regression Equation for MRR:

$$\begin{aligned} \text{MRR} = & -2.81 + 1.705\ \text{Current}\left(I_p\right) - 0.00694\ \text{Pulse On Time}(T_{\text{on}}) \\ & - 0.181\ \text{Duty Factor}(t) \end{aligned}$$

Regression Equation for EWR:

$$\begin{aligned} \text{EWR} = & \ 0.0382 + 0.00930\ \text{Current}\left(I_p\right) + 0.000003\ \text{Pulse On Time}(T_{\text{on}}) \\ & - 0.00864\ \text{Duty Factor}(t) \end{aligned}$$

- Square Solid Electrode

Regression Equation for MRR:

$$\begin{aligned} \text{MRR} = & -2.21 + 1.7035 \text{ Current}\left(I_p\right) - 0.00779 \text{ Pulse On Time}(T_{\text{on}}) \\ & - 0.227 \text{ Duty Factor}(t) \end{aligned}$$

Regression Equation for EWR:

$$\begin{aligned} \text{EWR} = & \ 0.091 + 0.00804 \text{ Current}\left(I_p\right) - 0.000018 \text{ Pulse On Time}(T_{\text{on}}) \\ & - 0.0047 \text{ Duty Factor}(t) \end{aligned}$$

4.1.2 Analysis of Variance (ANOVA)

Analysis of Variance (ANOVA) has been derived for each response to find the significant of each input factors. As per this technique, if the calculated P-value of the developed model does not exceed the standard tabulated value of P for a desired level of confidence (say 95%), then the model is considered to be satisfactory within the confidence limit. Analysis of resulted output was carried out by using Minitab v.16 Software (Tables 4, 5, 6 and 7).

Table 4 ANOVA table of MRR for RP electrode

Source	DF	Adj SS	Adj MS	*F*-value	*P*-value
Current (I_p)	2	69.8808	34.9404	63.12	0.016
Pulse on time (T_{on})	2	2.4768	1.2384	2.24	0.309
Duty factor (t)	2	0.2151	0.1076	0.19	0.837
Error	2	1.1071	0.5535		
Total	8	73.6798			

The *P*-value in this table, clearly indicates that the response MRR is highly influenced by current and least affected by duty factor

Table 5 ANOVA table for EWR for RP electrode

Source	DF	Adj SS	Adj MS	*F*-value	*P*-value
Current (I_p)	2	0.002472	0.001236	2.89	0.257
Pulse on time (T_{on})	2	0.000577	0.000289	0.67	0.597
Duty factor (t)	2	0.000621	0.000310	0.73	0.580
Error	2	0.000856	0.000428		
Total	8	0.004526			

The *P*-value in this table, clearly indicates that the response EWR is highly influenced by current and least affected by pulse on time

Table 6 ANOVA table of MRR for solid electrode

Source	DF	Adj SS	Adj MS	*F*-value	*P*-value
Current (I_p)	2	69.6522	34.8261	406.92	0.002
Pulse on time (T_{on})	2	2.2500	1.1250	13.15	0.071
Duty factor (t)	2	0.3226	0.1613	1.88	0.347
Error	2	0.1712	0.0856		
Total	8	72.3960			

The *P*-value in this table, clearly indicates that the response MRR is highly influenced by current and least affected by duty factor

Table 7 ANOVA table for EWR for solid electrode

Source	DF	Adj SS	Adj MS	*F*-value	*P*-value
Current (I_p)	2	0.001551	0.000776	3.05	0.247
Pulse on time (T_{on})	2	0.000766	0.000383	1.51	0.399
Duty factor (t)	2	0.002597	0.001299	5.11	0.164
Error	2	0.000508	0.000254		
Total	8	0.005423			

The *P*-value in this table, clearly indicates that the response EWR is highly influenced by duty factor and least affected by pulse on time

4.2 Comparison Between Experimental and Mathematical Analysis

4.2.1 RP Electrode

- Comparison of Material Removal Rate (EWR) (Table 8)
- Comparison of Material Removal Rate (MRR) (Table 9)

Table 8 Showing values obtained from experiment and regression equation for MRR

Run No.	MRR (experimental value)	MRR (value from regression equation)	|ERROR|
1	4.8264	4.750	0.07640
2	4.2672	4.222	0.04520
3	3.0029	3.347	0.34410
4	8.0763	7.979	0.09732
5	7.0504	7.451	0.40060
6	7.8296	7.119	0.71060
7	11.8463	11.208	0.63830
8	10.3325	11.223	0.89050
9	10.3813	10.348	0.03330

Table 9 Showing the values obtained from the experiment and the regression equation for EWR

Run No.	EWR (experimental value)	EWR (value from regression equation)	|ERROR|
1	0.017653	0.01669	0.000963
2	0.016335	0.00820	0.008135
3	0.004601	−0.00014	0.004741
4	0.022976	0.02665	0.003674
5	0.023583	0.01816	0.005423
6	0.005694	0.03574	0.030046
7	0.016961	0.03661	0.019649
8	0.073668	0.05404	0.019628
9	0.059603	0.04570	0.013903

4.2.2 Solid Electrode

- Comparison of Material Removal Rate (MRR) (Table 10)
- Comparison of Electrode Wear Rate (EWR) (Table 11)

Table 10 Showing values obtained from experiment and regression equation for MRR

Run No.	MRR (experimental value)	MRR (value from regression equation)	|ERROR|
1	04.8013	04.7995	0.0018
2	04.3832	04.1830	0.2002
3	02.8945	03.1770	0.2825
4	07.7878	07.9795	0.1917
5	07.6998	07.3630	0.3368
6	07.0064	07.0380	0.0316
7	11.0090	11.1595	0.1505
8	11.1594	11.2240	0.0646
9	10.3523	10.2180	0.1343

Table 11 Showing values obtained from experiment and regression equation for EWR

Run No.	EWR (experimental value)	EWR (value from regression equation)	|ERROR|
1	0.02580	0.02224	0.00356
2	0.01706	0.01664	0.00042
3	0.00690	0.01014	0.00324
4	0.07022	0.03362	0.03660
5	0.01095	0.02802	0.01707
6	0.01837	0.03562	0.01725
7	0.03152	0.04500	0.01348
8	0.03324	0.05350	0.02026
9	0.08146	0.04700	0.03446

5 Conclusions

This work investigated the direct production of EDM electrodes by means of the FDM technique. In addition, EDM experiments were conducted to evaluate the electrode performance in regards to material removal rate(MRR) and electrode wear rate(EWR) From the result and discussions the following conclusion can be drawn:

(i) The use of RP is especially useful for urgently replication worn-out parts that have a complex shape and are required one-off or in small numbers.
(ii) RP electrodes show good potential for use as EDM tools. EDM tools with complex shapes, that are not easily machined by traditional processes, have an alternative to look upon.
(iii) From the above work, it can be concluded that, RP electrode worked very similar to that conventional electrode in all aspects.
(iv) Hence it is feasible solution to replace the conventional electrode manufacturing process to evolutionary RP electrode process.

(v) The potential for application of copper coated RP models is currently limited by a number of factors. The efficiency of the RP electrodes depends on the quality of the electrode deposition copper coating and the EDM machine set-up.

(vi) Two problems are resolved by the proposed RP electrode:

- The excessive weight of conventional electrodes: The weight of the RP electrodes is quite low because of thin copper shell manufacturing (1–2 mm, no more, for the common parts).
- The difficulty in machining some complex geometries electrode: Geometries that are complicated to mill can be handled more easily.

Acknowledgements It gives me immense pleasure on bringing out the Project entitled "Development Of Electrical Discharge Machining (EDM) Electrode Using Fused Deposition Modeling (FDM)". I express my deep sense of gratitude and sincere regards to my respected guide Dr. R. M. Metkar, whose encouragement, valuable guidance and supervision really helped me in completing dissertation work. I express my sincere thanks to him for kind help and guidance. I would like to give my sincere thanks to my family, my friends, and well-wishers.

References

1. https://en.wikipedia.org/wiki/3D_printing. Accessed 21 oct 2016
2. S.K. Singh, Prof. R.K. Satankar, Investigating the process parameters of 3D printer extruder of Fused Deposition Modeling—A review. JETIR (ISSN-2349-5162) JETIR1710001 Journal, October (2017)
3. https://en.wikipedia.org/wiki/Fused_filament_fabrication. Accessed 21 oct 2016
4. N. Patel, Review on importance of electrodes in electrical discharge machining process. Int. J. Res. Aeronaut. Mech. Eng. **3**(10), 36–41 (2015)
5. http://www.stratasys.com/3d-printers/uprint-se-plus. Accessed 21 oct 2016

State of the Art of Powder Bed Fusion Additive Manufacturing: A Review

R. Verma and G. Kaushal

Abstract Premature failure of the components subjected to high temperature application has been a concern for last three decades now. Surface engineering is now an established approach to enhance component life for such application. A new approach; Additive manufacturing is gaining importance now a days. Additive Manufacturing, an incremental layer-by-layer manufacturing of components, has gained popularity as a possible option for producing fully dense metallic components in small span of time and creating its path in a constantly growing number of industries, such as aircrafts, military, automobile, medical, and architecture. The current review article includes the fundamentals of processes and introduces few metallic alloys which are presently available for layered manufacturing. The significance and prospective utility of the processes has been discussed with emphasis on mechanical properties like ultimate tensile strength and yield strength.

Keywords Additive manufacturing · Selective laser sintering · Electron beam melting · Mechanical properties

R. Verma (✉)
Mechanical Engineering Department, Thapar Polytechnic College, Patiala, Punjab, India
e-mail: rajan.r.verma@gmail.com

G. Kaushal
Mechanical Engineering Section, Yadavindra College of Engineering, Punjabi University, Guru Kashi Campus, Talwandi Sabo, Punjab, India
e-mail: gagankaushal@pbi.ac.in

L. J. Kumar et al. (eds.), *3D Printing and Additive Manufacturing Technologies*,
https://doi.org/10.1007/978-981-13-0305-0_23

1 Introduction

Materials having utilization for high temperature applications are subjected to different types of degradation such as high temperature corrosion, erosion-corrosion, overheating, solid particle abrasion, wear, etc. [1, 2]. Untimely collapsing of the components subjected to high temperature application has been a worry for last three decades now [3, 4]. Surface engineering is now an established approach to enhance component life for such application. A new approach; Additive manufacturing is gaining importance now a days. Additive Manufacturing, also famous as 3D printing, freeform fabrication, rapid prototyping or an incremental layer-by layer manufacturing of components, has recently attracted the attention of investigators worldwide for producing fully dense metallic components in small span of time. According to the ASTM standard F2792-10, additive manufacturing is the "process of joining materials to make objects from 3D model data, usually layer upon layer, as opposed to subtractive manufacturing methodologies, such as traditional machining". Other commonly used names of the process: additive fabrication, additive processes, additive techniques, additive layer manufacturing, layer manufacturing, and freeform fabrication [5]. Metal additive manufacturing (MAM) is creating its path into a continuously growing number of industries, such as aircrafts, military, automobile, energy, medical, medical, and architecture [6, 7].

All MAM parts start from 3D CAD solid model that completely illustrate the external geometry. MAM machine necessitates 3D CAD solid model of the component in the file format consisting of .STL extension. This .STL file becomes the root for calculation of the slices. The model is then sliced into number of cross-sectional layers of appropriate thickness with the help of the software. The data is then sent to the printing machine known as printer where the component is sintered layer by layer [8]. Support structures are used for the correct positioning of the part. During preprocessing stage, the supports can be made and inserted according to the need of the manufacturing of the component. After the completion of process, the whole component undergoes postprocessing treatments like polishing, machining, heat treatment as required but only after the assurance that all the supports have been removed. Hot isostatic pressing is also done to some particular parts to ensure part density.

MAM process can be categorized largely into two groups (Fig. 1),—Powder bed fusion additive manufacturing (PBF) and Directed energy deposition additive manufacturing (DED). In the powder bed fusion AM, the parts are manufactured by fusing each layer of metal powder with a heat source. In the Directed Energy Deposition AM, material in shape of powder, wire or sheets is added in layers and fused with a heat source. In this paper, an attempt has been made to describe working of laser and electron beam based powder bed fusion additive manufacturing and their respective applications.

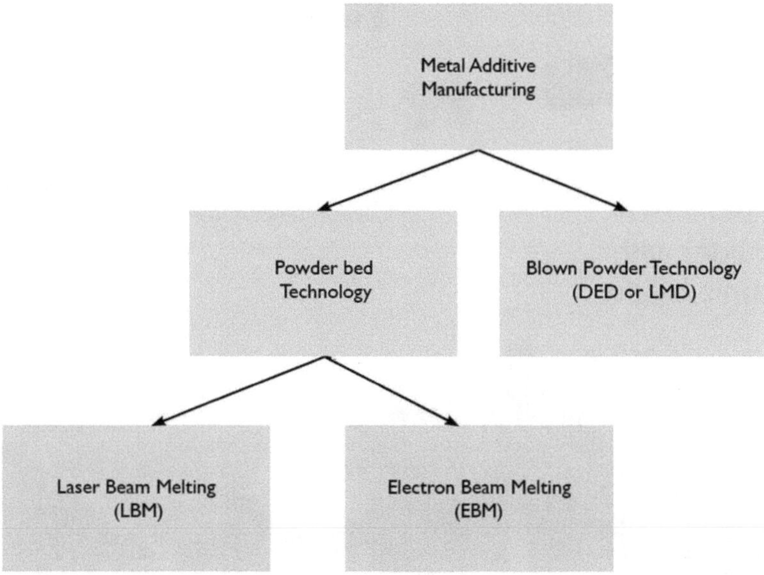

Fig. 1 Mapping of metal additive manufacturing processes

2 Powder Bed Fusion Additive Manufacturing (PBF)

The working area is enclosed and either filled with an inert gas or vacuum enclosed to offer protection to molten metal from reacting with the air. A roller spreads a layer of powder across build chamber. Afterwards, using the energy source (Laser or electron beam) melting of the metal powder takes place tracking the outline of each layer. The foundation layer is melted and then solidified into a solid state. When manufacturing of one layer has been completed, there is a downward motion of the piston in building chamber and upward movement of the piston in powder chamber. The upward movement is equivalent to thickness of the layer. A fresh powder layer is spread over the whole working area and whole process is recurred again until the complete part is built.

In laser based powder bed fusion additive manufacturing (Fig. 2), the powder is spread over the working platform with the help of blade or roller and this layer is melted with a laser beam. Then the lowering of platform equal to the thickness of the layer is done and again the powder layer is spread. This procedure is repeated till the whole component is manufactured. Generally Nd:YAG-laser and fiber laser is used for sintering of the powder.

In electron beam melting (Fig. 3), high power electron beam is used which is controlled by electromagnetic coils. The electron beam heats the whole powder bed for each layer build. The whole procedure occurs at high temperature and in vacuum.

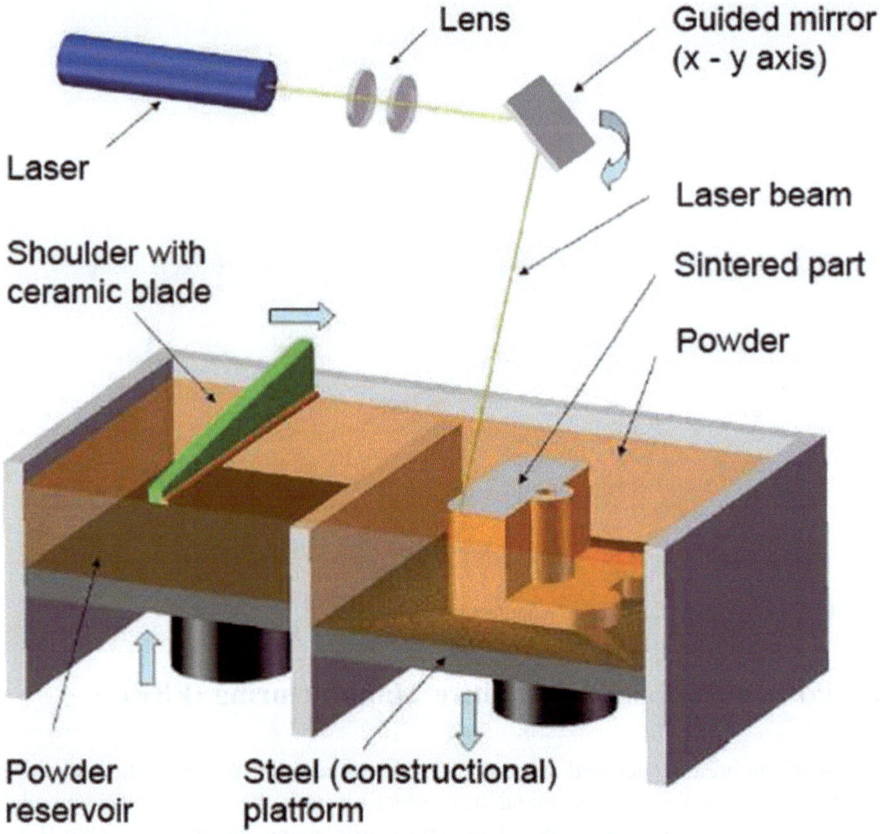

Fig. 2 Laser-based powder bed fusion additive manufacturing [20]

Fig. 3 Electron beam melting system [15]

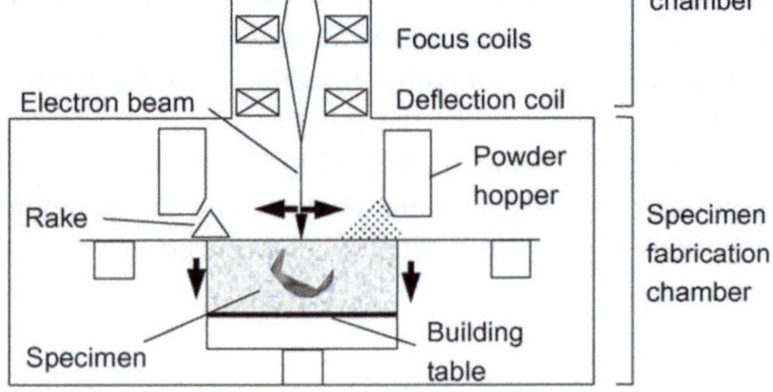

Selective laser sintering (SLS) or Direct metal laser sintering (DMLS) from EOS and Selective laser melting (SLM) from Renishaw and SLM solutions are a few famous PBF based technology which employ laser as supply of energy whereas Electron beam melting (EBM) from ARCAM is PBF based technology which make use of electron beam as supply of energy.

3 Some Recent Studies Related to Powder Bed Fusion Additive Manufacturing

Sun et al. [9] conducted an experiment with Ti6Al4V alloy powders using selective laser melting technique. The parameters of selective laser melting like laser power, scanning speed, powder thickness, hatching space and scanning strategy were chosen in order to achieve maximum density. Taguchi method was used for design of experiments. The outcomes were evaluated using ANOVA and the signal-to-noise (S/N) ratios by design-expert software for the optimization of the parameters. From the experiments, it was established that powder thickness was the most noteworthy influence in processing parameters on density and sample having density more than 95% was attained and the microstructure of the sample was mostly consisted of acicular martensite, α phase and β phase. The micro-hardness of produced sample was 492 $HV_{0.2}$.

Kruth et al. [10] proposed "energy density (E_d)" as an important processing characteristic for melting powdered material. This parameter represents the energy delivered by the laser beam to a volumetric unit of powder material and correlates some important laser and scan parameters.

$$E_d = \frac{P_{\text{Laser}}}{V_{\text{Scan}} \cdot S_{\text{Hatch}} \cdot t_{\text{Layer}}}, \tag{1}$$

where E_d is energy density, P_{Laser} is laser power, V_{Scan} is scan speed, S_{Hatch} is hatch spacing and t_{Layer} is layer thickness. From Equation, it can be seen that the energy density E_d is proportional to laser power, but reciprocal to scan speed, hatch spacing and layer thickness. The ratio of $P_{\text{Laser}}/V_{\text{Scan}}$ was often modified in combination to ascertain the effect of parameters with defined hatch spacing and layer thickness. A process map for continuous laser melting of different types of particles was determined, as shown in Fig. 4. The graphic showed that high scan speeds combined with high laser powers resulted in a smooth surface and less balling due to rapid solidification of the melt pool.

Calignano et al. [11] studied the influence of different parameters like scan speed, laser power and hatching distance on the surface roughness of direct metal laser sintering processed aluminum specimens. Taguchi's L18 orthogonal array was utilized which applies S/N ratio to determine the variations. Surface morphology was examined with the help of FESEM (Field emission scanning electron

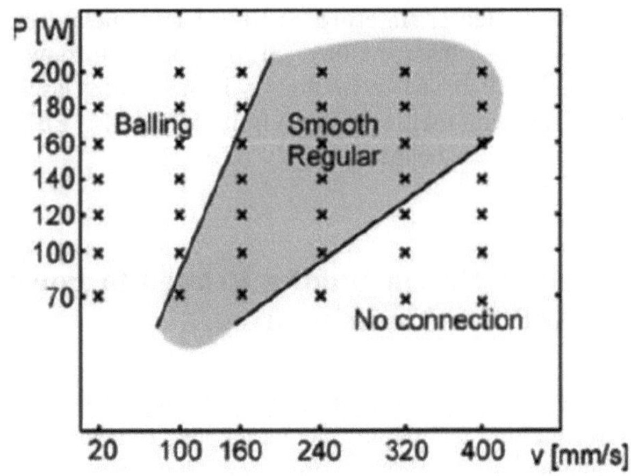

Fig. 4 Process map for continuous laser melting operation [10]

microscope). It was examined that the scan speed had greatest dependency on the surface roughness. The main effect plot for S/N ratios and R_a confirmed these results. As shown in Fig. 5, the most favorable surface roughness was attained when the scan speed was 900 mm/s, the laser power was 120 W, and the hatching distance was 0.10 mm.

Abd-Elghany and Bourell [12] studied the various properties of products manufactured by SLM using 304L stainless steel powders. Twenty four samples with different shapes were fabricated having layer thickness of 30, 50 and 70 mm and laser scanning speed as 70 and 90 mm/s so as to examine the effect of as-said parameters. Density was analyzed on the basis of weight and dimensions of the samples. Figure 6 show the average data of density for all samples at different layer thicknesses and scanning speeds. It was revealed that density decreased at large layer thickness.

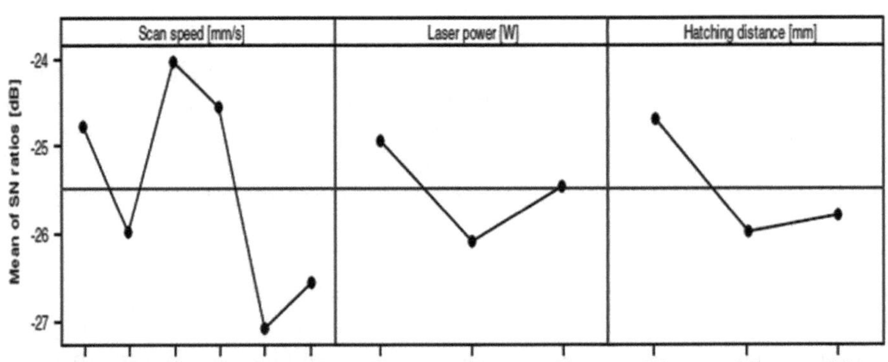

Fig. 5 Main effect plot for S/N ratios for R_a. [11]

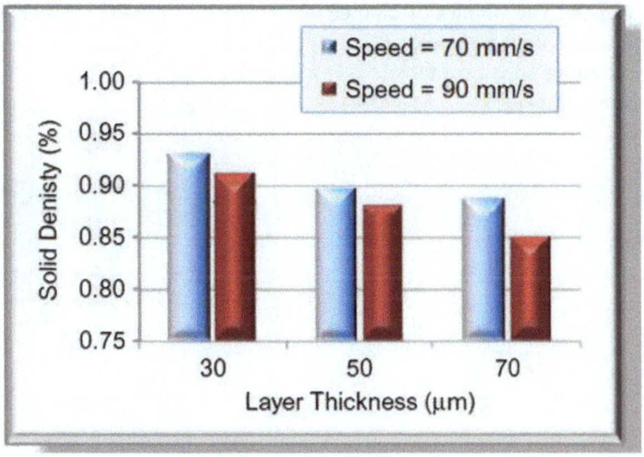

Fig. 6 The impact of the layer thickness and scanning speed on the solid density [12]

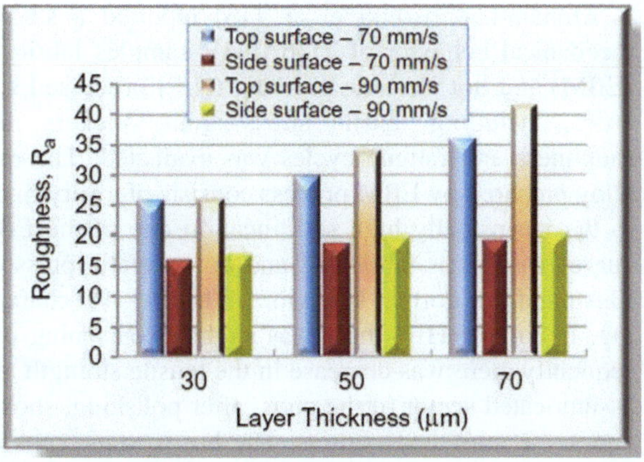

Fig. 7 The impact of the layer thickness and scanning speed on the surface roughness [12]

Figure 7 show the calculated values of surface roughness (R_a) at the upper surface and the side surface of each sample. The existence of large particles within the thick layers resulted in increase in surface roughness and decrease in the density due to the volume of particles present at the surface and the affinity to form voids when the particles are removed in finishing process.

In spite of the fact that there is reduction in processing time by half with a layer thickness of 70 μm, it was also reported that there is reduction in density and declination in the yield and ultimate tensile strength when compared with layer thickness 30 and 50 mm as can be seen from Fig. 8. The author reported density as 92% and hardness as 190 HV when thickness of layer was 30 μm and scan speed was 70 mm/s. Yield strength and ultimate tensile strengths were reported as 182 and 393 MPa, respectively, which were about 70% of the standard material properties. The XRD analysis revealed a consistent austenite phase and SEM micrographs showed that there was no precipitation of carbide near grain margins. It was because of the rapid melting and cooling cycles.

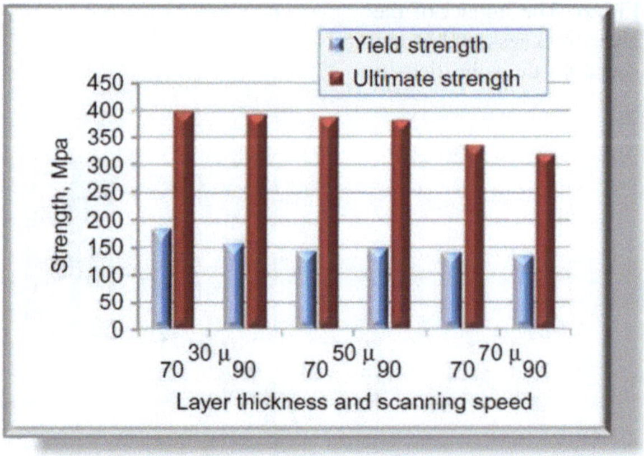

Fig. 8 The impact of change in layer thickness and scanning speed on tensile strength [12]

Mohammadhosseini et al. [13] reported a study on the microstructures and mechanical behavior of Ti–6Al–4V samples fabricated by electron beam melting (EBM) and hot isostatic pressing (HIP) processed samples. The mechanical properties, including tensile stress–strain, Vickers micro-hardness (HV), surface roughness and fatigue cycles was evaluated. The microstructure of a Ti–6Al–4V alloy prepared by EBM process consists of a very fine and acicular morphology due to the intrinsically high solidification rate of the EBM. The trend of stress–strain curves for the as-fabricated and hot isostatic pressed parts was almost same and illustrated the normal behavior of Ti–6Al–4V characterized by low strain hardening. In case of HIP, there was slightly coarsening of the microstructure, and consequently there was decrease in the tensile strength and increase in elongation. The as-fabricated sector of the parts, after polishing, showed some defects like porosity and some unmelted regions. The hot isostatic pressed parts consisted no defects. The micro-hardness of as-fabricated specimens was higher when compared with hot isostatic pressed parts. The increase in micro-hardness can be credited to microstructure coarsening. The fatigue strength was increased significantly because of the fact that densification increases the fatigue resistance.

Murr et al. [14] reported the microstructural architecture and mechanical behavior of as-fabricated and hot isostatic pressed Inconel 625 components produced by atomized powder using EBM. The authors investigated that at 538 °C temperature the tensile properties of EBAM cylinders didn't considerably change micro-hardness from as-fabricated cylinders tested at room temperature, but yield strength decreased from 0.33 to 0.30 GPa, and ultimate tensile strength dropped by 23%. The elongation resulted in increase from 44 to 53%. Similarly, at 538 °C the as-fabricated and hot isostatic pressed cylinders also showed no noteworthy alteration in micro-hardness, but both yield strength and ultimate tensile strength decreased slightly with little increase in ductility when compared with tests at room temperature. By comparison, wrought Ni superalloy when tested at 538 °C revealed yield strength of 280 MPa and ultimate tensile strength of 830 MPa.

Koike et al. [15] investigated the mechanical properties, the grindability and corrosion resistance of electron beam melting (EBM) fabricated Ti–6Al–4V ELI (extra low interstitial) specimens. The yield strength and tensile strength of the EBM samples were reported to be 735 and 775 MPa, respectively. The elongation of as-fabricated samples was found to be 2.4%. The above said values were similar to cast Ti–6Al–4V ELI. The cast samples showed slightly higher tensile strength (860 vs. 735 MPa) and elongation (4.4 vs. 2.3%) when compared with as-fabricated. These alterations were attributed to rippled, rough sample surfaces and high oxygen content in the EBM samples (0.34 vs. 0.22%). The authors reported that high micro-hardness of the electron beam melting samples was perhaps because of the finer α/β lamellar microstructures. The grindability of the electron beam melting Ti–6Al–4V ELI samples was higher than that of the wrought or cast specimens. The author found lower ductility of the electron beam melting samples responsible for better grindability.

Kircher et al. [16] performed investigations on the chemical, microstructural, and mechanical properties of several test specimens produced by EBM of a Co–Cr–Mo alloy before and after heat treatment. They found that heat treatments result in re-dissolving a portion of the carbides and increasing the grain size of the material. As-built specimens have tensile properties of: yield stress 717 MPa and UTS 1110 MPa in the horizontal direction, and yield stress 786 MPa and UTS 869 MPa in the vertical direction.

Gaytan et al. [17] investigated the microstructures and mechanical behavior of parts by EBM of Co–Cr–Mo powder. They found that different types of EBM Co–Cr–Mo structures had different mechanical properties. The corresponding hardness for the mesh struts increased from 5.6 to 6.8 GPa contrary to solid monolithic components which increased from 4.4 to 5.9 GPa. While a small increase in carbide content for the reticulated mesh structures accounts in large part for this hardness increase, a more rapid cooling rate especially for the reticulated mesh was also a contributing factor. The tensile test was conducted on samples prepared from cylindrical models manufactured by electron beam melting process. The authors found average ultimate tensile strength, yield stress, and elongation to be 1.45 GPa, 0.51 GPa, and 3.6%.

Rafi et al. [18] compared selective laser melting (SLM) and electron beam melting (EBM), on the basis of microstructure and mechanical property estimation of Ti–6Al–4V parts produced by these two processes. Bars in accordance with ASTM standards were manufactured using Ti6Al4V for tensile and fatigue testing. Optical and scanning electron microscopy were used to study the microstructure of the samples. The microstructure of SLM-fabricated parts consisted of an α' martensitic phase, whereas the EBM-fabricated parts contained mainly α and a minor amount of β phase. SLM-fabricated samples showed high tensile strength than EBM-fabricated samples. But EBM-fabricated samples have high ductility and inferior fatigue limit. The author also reported that the high tensile strength of SLM samples is endorsed to the martensitic α' microstructure and the higher ductility and inferior fatigue limit in EBM-produced samples is recognized to the lamellar α phase.

Zhao et al. [19] compared the microstructure and mechanical properties of selective laser melting and electron beam melting Ti–6Al–4V samples. The author also studied the effect of specimen size and direction on the defects, microstructure, tensile and fatigue properties. The selective laser melting specimens consisted of α' phase and electron beam melting consisted of $\alpha + \beta$ phases. The selective laser melting specimens showed high porosity as compared to electron beam melting specimens. Both the specimens showed higher strength and ductility in vertical direction as compared to horizontal direction. When compared with EBM-fabricated specimens, SLM-fabricated specimens showed higher tensile strength and lower ductility. Both the specimens had lower fatigue strength as compared to cast and annealed alloys because of the pores in specimens. However, HIP considerably increased the fatigue limits of SLM and EBM specimens owing to the reason of closing of pores.

4 Concluding Remarks

1. The microstructure of selective laser melting fabricated specimens consisted of α' martensitic phase, whereas the electron beam melting fabricated specimens consisted mainly α phase and a minor quantity of β phase.
2. Selective laser melting fabricated specimens showed high tensile strength than electron beam melting fabricated specimens. Higher tensile strength of selective laser melting specimens can be attributed to the martensitic α' microstructure.
3. Electron beam melting fabricated samples have higher ductility and inferior fatigue limit. Higher ductility and inferior fatigue limit in electron based melting produced specimens can possibly due to the lamellar α phase.

References

1. B.G. Seong, S.Y. Hwang, K.Y. Kimb, High-temperature corrosion of recuperators used in steel Mills. Surf. Coat. Technol. **126**(2–3), 256–265 (2000)
2. R.A. Mahesh, R. Jayaganthan, S. Prakash, Evaluation of hot corrosion behaviour of HVOF sprayed NiCrAl coating on superalloys at 900 °C. Mater. Chem. Phys. **111**(2–3), 524–533 (2008)
3. H.S. Sidhu, B.S. Singh, S. Prakash, The role of HVOF coatings in improving hot corrosion resistance of ASTM-SA210 GrA1 steel in the presence of Na_2SO_4–V_2O_5 salt deposits. Surf. Coat. Technol. **200**(18–19), 5386–5394 (2006)
4. T.S. Sidhu, R.D. Agarwal, S. Prakash, Studies of the metallurgical and mechanical properties of high velocity oxy-fuel sprayed stellite-6 coatings on Ni- and Fe-based superalloys. Surf. Coat. Technol. **201**(1–2), 273–281 (2006)
5. ASTM, ASTM F2792–12a Standard Terminology for Additive Manufacturing Technologies
6. Y. Song, Y. Yan, R. Zhang, D. Xu, F. Wang, Manufacturing of the die of an automobile deck part based on rapid prototyping and rapid tooling technology. J. Mater. Process. Technol. **120** (1–3), 237–242 (2002)

7. J. Giannatsis, V. Dedoussis, Additive fabrication technologies applied to medicine and health care: a review. Int. J. Adv. Manuf. Technol. **40**(1–2), 116–127 (2009)
8. E. Sachlos, J.T. Czernuszka, Making tissue engineering scaffolds work. Review: the application of solid freeform fabrication technology to the production of tissue engineering scaffolds. Eur. Cells Mater. **5**, 29–39 (2003)
9. J. Sun, Y. Yang, D. Wang, Parametric optimization of selective laser melting for forming Ti6Al4V samples by Taguchi method. Opt. Laser Technol. **49**, 118–124 (2013)
10. J.P. Kruth, L. Froyen, J. Van Vaerenbergh, P. Mercelis, M. Rombouts, B. Lauwers, Selective laser melting of iron-based powder. J. Mater. Process. Technol. **149**(1–3), 616–622 (2004)
11. F. Calignano, D. Manfredi, E.P. Ambrosio, L. Iuliano, P. Fino, Influence of process parameters on surface roughness of aluminum parts produced by DMLS. The International Journal of Advanced Manufacturing Technology **67**(9), 2743–2751 (2013)
12. K. Abd-Elghany, D.L. Bourell, Property evaluation of 304L stainless steel fabricated by selective laser melting. Rapid Prototyp. J. **18**(5), 420–428 (2012)
13. A. Mohammadhosseini, D. Fraser, S.H. Masood, M. Jahedi, Microstructure and mechanical properties of Ti–6Al–4 V manufactured by electron beam melting process. Mater. Res. Innov. **17**(2), 106–112 (2013)
14. L.E. Murr, E. Martinez, S.M. Gaytan, D.A. Ramirez, B.I. MacHado, P.W. Shindo, J.L. Martinez, F. Medina, J. Wooten, D. Ciscel, U. Ackelid, R.B. Wicker, Microstructural architecture, microstructures, and mechanical properties for a nickel-base superalloy fabricated by electron beam melting. Metall. Mater. Trans. A Phys Metall. Mater. Sci. **42** (11), 3491–3508 (2011)
15. M. Koike, K. Martinez, L. Guo, G. Chahine, R. Kovacevic, T. Okabe, Evaluation of titanium alloy fabricated using electron beam melting system for dental applications. J. Mater. Process. Technol. **211**, 1400–1408 (2011)
16. R.S. Kircher, A.M. Christensen, K.W. Wurth, Electron beam melted (EBM) Co-Cr-Mo alloy for orthopaedic implant applications, in *Solid Freeform Fabrication Symposium* (2009), pp. 428–436
17. S.M. Gaytan, L.E. Murr, E. Martinez, J.L. Martinez, B.I. Machado, D.A. Ramirez, F. Medina, S. Collins, R.B. Wicker, Comparison of microstructures and mechanical properties for solid and mesh cobalt-base alloy prototypes fabricated by electron beam melting. Metall. Mater. Trans. A **41**, 3216–3227 (2010)
18. H.K. Rafi, N.V. Karthik, H. Gong, T.L. Starr, B.E. Stucker, Microstructures and mechanical properties of Ti6Al4V parts fabricated by selective laser melting and electron beam melting. J. Mater. Eng. Perform. **22**(12), 3872–3883 (2013)
19. X. Zhao, S. Li, M. Zhang, Y. Liu, T.B. Sercombe, S. Wang, Y. Hao, R. Yang, L.E. Murr, Comparison of the microstructures and mechanical properties of Ti–6Al–4 V fabricated by selective laser melting and electron beam melting. Mater. Des. **95**, 21–31 (2016)
20. J. Sedlák, M. Ptáčková, J. Nejedlý, M. Madaj, J. Dvořáček, J. Zouhar, O. Charvát, M. Píška, L. Rozkošný, Material analysis of titanium alloy produced by direct metal laser sintering. Int. J. Metalcast. **7**(2), 43–50 (2013)

Distortion in Metal Additive Manufactured Parts

Hemnath Anandan Kumar and Senthilkumaran Kumaraguru

Abstract Metal based additive manufacturing (AM) techniques are continuously adopted by aerospace, automobile, defense, and healthcare industries. The primary concern for the components used in these applications is high precision. Despite being able to produce complex shapes, they lack in precision due to the distortions in shape and size of the part during and after fabrication. Distortion is the deviation of the part from its actual shape or dimension. Eventually, shape deviation has an unfavorable effect on the part functional performance which will hinder their use for critical technological applications. Even though, the temperature gradients during the process build affect these errors; they need to be studied in detail to fix the causation of these errors. In this paper, we review and classify the causation for shape and size distortion that occurs in metal additive processes. The approach to classification is multi-faceted and are based on the geometry of the fabricated part, the material used, process-related parameters, part orientation and physical phenomenon that occurs during the process. This work would help understand the root cause of distortion in major commercially successful metal AM processes and eliminate the need for costly trials.

Keywords Additive manufacturing · Powder bed fusion
Directed energy deposition · Defects · Distortions

1 Introduction

Additive Manufacturing (AM) exists for three decades and was initially used to make models and prototypes. Now, AM technology has developed into a rapid manufacturing. American Society for Testing and Materials defined AM as 'A process of joining materials to make objects from 3D model data, usually layer upon layer, as

---

H. Anandan Kumar · S. Kumaraguru (✉)
Department of Mechanical Engineering, Indian Institute of Information Technology, Design and Manufacturing Kancheepuram, Chennai 600127, India
e-mail: skumaran@iiitdm.ac.in

L. J. Kumar et al. (eds.), *3D Printing and Additive Manufacturing Technologies*,
https://doi.org/10.1007/978-981-13-0305-0_24

opposed to subtractive manufacturing methodologies' [1]. The field of AM has been developed in the past two decades due to the research and development of various technologies. The major advantages of AM process come from the ability to fabricate complex and customized design. AM is capable of processing different kind of materials like polymer, ceramic and metals. Metals can be processed by different technologies like directed energy deposition (DED), Powder Bed Fusion (PBF), Binder jetting and sheet lamination [2].

PBF is a technique in which powders are pre-laid down in a platform and laser power source is used to sinter or melt the metal powders. Whereas DED is different from PBF. PBF has many commercial names like Direct Metal Laser Sintering (DMLS), Selective Laser Melting (SLM), and Electron Beam Melting (EBM). In DED process, a simultaneous delivery of metallic powder and focusing of the energy beam (laser beam, arc or electron beam) occurs to melt the powder particles and form a molten pool into which the feedstock is deposited in a layered manner. DED has both powder-based-, and wire-based technology. Powder-based DED can be commercially called as Laser Engineered Net Shaping (LENS), Direct Metal Deposition (DMD), Laser deposition whereas wire-based DED can be called as Shape Metal Deposition (SMD), Hybrid Layer Manufacturing (HLM) and Electron Beam Direct Melting (EBDM) [3–8]. Application of DMD technology is used in the repairing of worn components like vanes, drive shafts [3], bearing seals [6], turbine blades [8], etc. Binder jetting process is entirely different from PBF and DED where no heat source is used. This process occurs by deposition of binder on metal powder followed by curing of the binder to powder particles together and sintering them. Then it is infiltrated with a second metal of low melting point. Sheet lamination process occurs by stacking up sheets of metal in layer wise manner and joined metallurgically by ultrasonic consolidation. Ultrasonic Additive Manufacturing (UAM) is one of the commercial Sheet lamination process [9, 10].

Though there are several metal AM techniques, PBF and DED are the predominantly used technologies in industries. Currently, researchers are actively involved in exploring the capabilities of metal additive manufacturing technology. This is because metal AM has revolutionized the industry by its ability to build complex and intricate features which were a difficult milestone to be achieved by the conventional machining process. Metal-based additive manufacturing techniques are continuously adopted by aerospace, automobile, defense and healthcare industries [11]. The primary concern for the components used in these applications is high precision. Despite being able to produce complex shapes, they lack in precision due to the distortions in shape and size of the part during and after fabrication. Distortion is the deviation of the part from its actual shape or dimension. Eventually, shape deviations have an unfavorable effect on the part functional performance which will hinder their use for the critical technological applications [12, 13]. This paper reviews about the defects caused in metal AM processes by different factors. In Sect. 2, defects and its causes will be discussed in detail under four different classifications. They are based on the geometry of the part, support structures, and orientation; process parameters like laser power, scan speed, scan pattern; the material used and physical phenomenon.

2 Defects in Metal Additive Manufacturing Process

PBF and DED processes suffer from many defects like porosity, ball formation, superelevated edges, distortion, residual stress, delamination, etc. prevailing during the fabrication of metal components using AM even though they are capable of producing complex shapes and structures. In this section, factors causing distortions in the parts will be discussed in detail. Some of the common defects that are present in metal deposition are depicted in Fig. 1.

2.1 Defects Due to Geometry of the Part, Orientation and Supports

In this section, defects caused by variation in geometry parameters and support structure are presented. Kleszczynski et al., has observed distortion in parts by forming super elevated edges and poor surface connection. They found that if these edges project out from the powder bed they will cause damage to the powder coater and will result in improper powder distribution and cause defects [14, 15]. Overhanging features or protruding features are some of the difficult features to achieve in metal AM parts as they are the most concentrated area of stress. This, in turn, will result in warpage, curling, and distortion of the entire shape of the part. Hence, support structures are very important features in AM technology when there is an overhang feature present in the part [16]. Some authors have pointed out that if rigid supports are used then a large amount of residual stress will be generated [17–19]. So support structures should be in such a way that it is easily removable from the substrate but should also be strong enough to withstand the stresses. Jhabvala et al., proposed a new method for producing support structure, which is not dense and can withstand the stresses and heat produced during the build [20]. However, Calignano suggests that overhanging features should be avoided while designing the parts [21]. Mohanty et al., studied the effect of scan

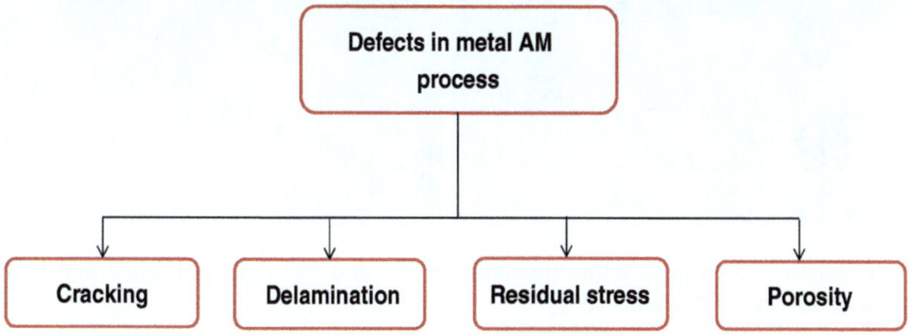

Fig. 1 Defects in metal AM process

pattern on the significance of overhanging features, with and without support as two separate cases and conducted a sensitivity analysis [22]. Zhang et al., investigated the optimal conditions for producing a curved overhanging feature using 316L SS powder and suggested that the quality of overhangs can be increased by decreasing energy level and increasing the angle of obliquity [23].

Zheng et al., also proposed that the support structures should be built in an optimized manner to reduce the distortion as improper positioning of support structures will lead to thermal effects which will negatively affect the part dimensions by thermal phenomenon [24]. In addition to the previous findings, Thomas et al., investigated distortions that occur due to different part orientation, chamfers, and radius present in parts. From the experiments, it was identified that minimum angle to be oriented for a simple flat angled part should be 45° [25, 26]. Parts successfully built with suitable orientation angle and failed parts due to improper orientation angle are shown in Fig. 2. Papadakis et al., did a parametric study by fabricating a cantilever beam and discovered that there is a complex relationship existing between the part thickness and input power which results in distortion based on the selection of these process/part parameters [27].

Grasso et al., discussed four different examples of defective parts generated during an SLM process on AISI 316L parts which were shown in Fig. 3. In Fig. 3a defect is caused by improper heat conduction in overhang features, in Fig. 3b it is caused by the improper powder deposition and spreading due to the presence of worn recoating blade. In the next one Fig. 3c, d defect occurred due to the improper heat conduction to the underlying powder layer which is the connection between the bottom layers of the part and the supports. They also proposed an approach in order to identify the areas of overheating and eliminating partially to produce defect free parts [28, 29].

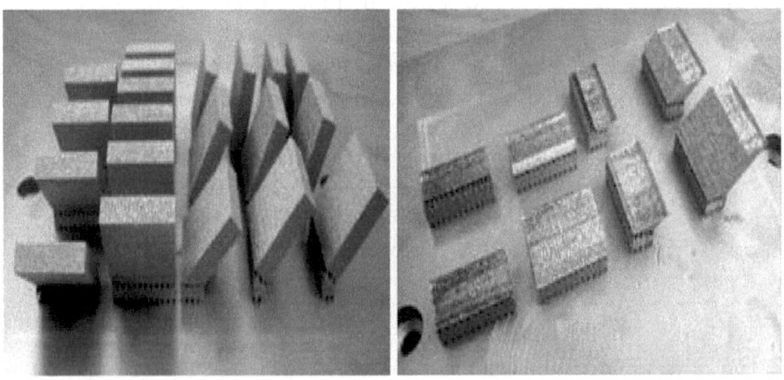

Fig. 2 Successfully built parts with orientation of 45° and 90° (left) and failed part builds with orientation of 25° and 40° (right) [25]

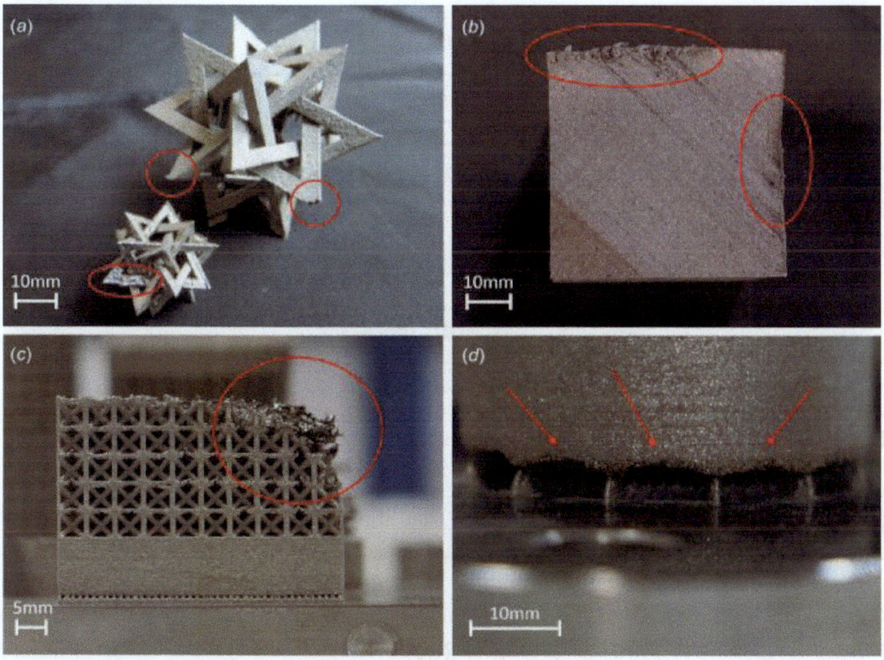

Fig. 3 Geometrical defects in metal AM parts due to **a** sharp corners, **b** poor contour, **c** lattice structures and **d** support interface with the part [28]

2.2 *Defects Due to Improper Process Control*

From the literature, it is vivid that process parameters or process variables are the main factors that cause defects much in a metal deposition process. Many researchers have found out the predominant process parameters like layer thickness, scan pattern, laser power takes part in defect occurrence [31–33]. The defects that occur mainly due to the process parameters include crack formation, balling phenomenon, and delamination. Carter et al., performed parametric studies to find out the effect of process parameters like scan speed, laser power and scan spacing on the crack formation. They related the defect formation with the energy density and scan spacing parameters [31]. Mcnutt conducted experiments on nickel-based superalloy CM247LC produced by laser deposition method to study crack formation. Experiments were conducted using different tool path patterns and their effect on the crack formation and found out that the most effective one in reducing the cracking phenomenon in the transverse axis was cross hatch pattern, while the lowest crack density in the longitudinal axis was resulted from long raster pattern [32]. Similarly, Zaeh et al., investigated the effect of scanning strategy in the fabrication of components using electron beam sintering. They used different scan

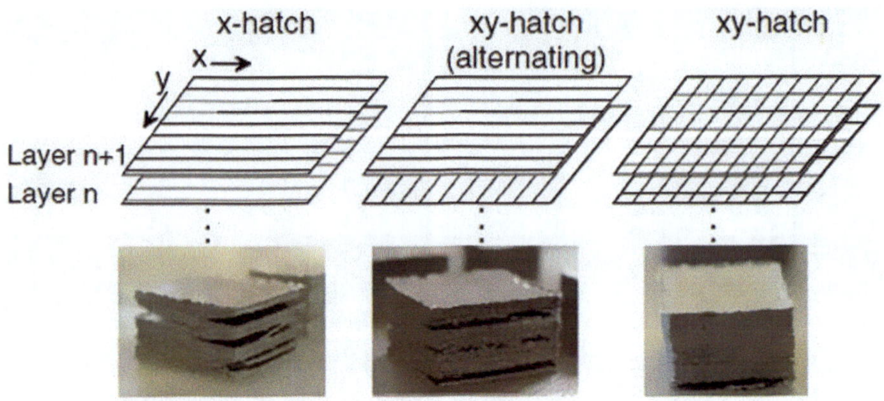

Fig. 4 Effect of scanning pattern on delamination [34]

pattern as shown in Fig. 4 and found that the parts resulted in poor cohesion between layers due to residual stress built up and scan strategy [34, 35].

Kobryn et al., studied the influence of transverse speed and laser power on porosity in laser deposited Ti–6Al–4V. As a result they found that different tool path strategies and powder quality highly influence the formation of porosity defects. In addition, the results reveal that if scan speed and laser power increases, lack of fusion and porosity will decrease [36]. In the same way, Nickel et al. studied the influence of the deposition patterns in laser metal deposited components and listed out the defects that were produced. They found that the very low deflections were observed while fabrication on beam substrate using raster scan pattern whereas while fabrication on plate substrate using spiral pattern low and uniform deflections was produced [37]. Meltball formation is another important phenomenal defect that occurs in SLM and EBM due to poor wettability. Ball formation in SLM can be of two types namely ellipsoidal and spherical balls where ellipsoidal balls are detrimental to part quality in larger dimensions and latter type in order of microns. When lower scan speed and higher laser power are used, the balling defect can be eliminated. It may be mitigated to a certain level by remelting the surface. A typical balling defect for different scan pattern is shown in Fig. 5 [30, 38–40]. This meltball formation defect was also observed by Zah et al., while conducting experiments using EBM. They listed out the important process parameters such as scan speed, beam power, layer thickness and scan pattern. [33]. To prevent the formation of meltball, Mumtaz et al., performed experiments using Inconel 625 fabricated by laser melting by varying the energy distribution. They found that this technique is useful in reducing the balling phenomenon and improve surface roughness [41, 42]. As well as Gu et al., from his experiments on stainless steel powder recommended that by increasing the energy density with suitable laser power and scan speed, balling phenomenon can be alleviated [43].

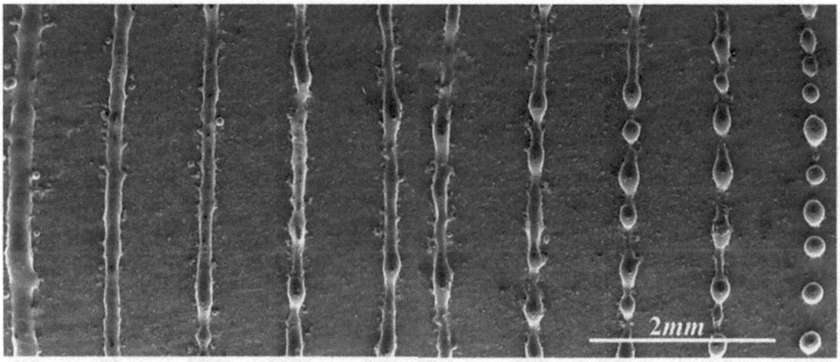

Fig. 5 Balling defects of single scan tracks under different scan speeds [38]

2.3 *Defects Based on the Materials Used*

This section reviews the defects that occur based on material used for part fabrication. Metal AM can process a wide range of materials like ferrous and non-ferrous metals and alloys.

Carter et al., and Mcnutt performed experiments using Nickel-based superalloy CM247LC to observe the cracking phenomenon using selective lase sintering and laser metal deposition process respectively. Carter et al., observed three different types of crack like solidification cracking, grain boundary cracking and void formation which occurred due to the process parameters like high energy density, low energy density, and very low energy density respectively [31]. While Mcnutt proposed that scan speed and laser power has a high influence on the formation of cracks whereas laser spot diameter and deposit dilution has less impact. It was found out that the particle size also influences the formation of cracks [32]. In general, cracking defects occur when the deposited material undergoes solidification. In some cases, porosity may also lead to the formation of macroscopical cracks [44].

Harrison et al., took a step to reduce the formation of micro-crack in nickel superalloys processed by SLM using an alloying approach. Hastelloy X was used for experimentation as it has the high susceptibility for crack formation. A different alloy combination was chosen resulting in a 65% reduction in cracking and an increase in performance of parts elevated temperature thus making a new approach to produce crack-resistant nickel super alloys [45].

Kempen et al., carried out an experimental study on high-density M2 High Speed Steel (HSS) parts by selective laser melting method. They found that preheating of the substrate will result in producing crack free parts [46]. From some research findings, it was found that the nickel-based super alloys and stainless steel are found to be the weakest materials which are able to generate cracks with high vulnerability. This primarily occurs because of their high thermal coefficient of expansion and low thermal conductivity [31, 47–49]. The solution to solve this type of cracking defect is preheating the substrate [49, 50].

In this section, material-based porosity defects will be discussed. Porosity may be like voids present within the part or on the surface. Mostly these pores will be present within the part and possess different pore geometry. Porosity formed within the parts can be classified into two types namely gas induced porosity which occurs when air or gas gets trapped inside the powder particles. While production and process induced porosity occurs when the applied energy is not adequate for thorough melting or when spatter ejection occurs [51]. The two types of porosity formed in EBM are shown in Fig. 6.

Porosity is highly undesirable in the case of high-stress applications because parts should be fully dense to prevent part failure. Edwards et al., and Gong et al., performed experiments using Ti–6Al–4V powder and analysed the defects. Edwards et al., evaluated the fatigue properties of Ti–6Al–4V parts fabricated by EBM. It was found that the fatigue properties of EBM specimens were very low than that of wrought material due to the presence of pores and also concluded that the pores can further lead to the formation of cracks [52].

Whereas, Gong et al., investigated the defect morphology of Ti–6Al–4V parts fabricated by both EBM and SLM. They found that the shape of the pore is spherical when using higher energy density and irregular in shape while using lower energy density [53]. Aboulkhair et al., made an effort to reduce the porosity in AlSi10Mg samples fabricated by SLM to enhance its mechanical properties. They were successfully able to produce parts with a good density of 99.8% by varying suitable process parameters and by using various scan strategies [13, 54]. Non-destructive techniques were employed by some researchers to determine the porosity level present within the parts [55–57].

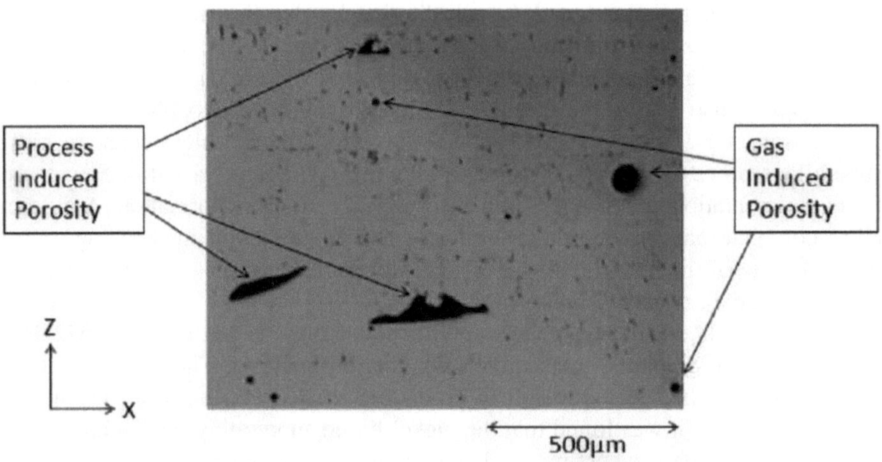

Fig. 6 Process induced porosity and gas induced porosity in Inconel 718 part processed by EBM [51]

2.4 Defects Due to Physical Phenomenon

Metal AM techniques are characterized by high-temperature melting followed by cooling during solidification which will lead to the accumulation of residual stresses. Defects are predominantly caused either directly or indirectly by residual stresses that prevail during the part fabrication. This section reviews the defects caused by residual stresses and other physical phenomenon and the measures carried out to mitigate the residual stress effect.

The formation of residual stresses in laser deposition was analysed by Pilloz et al. They proposed that formation of residual stresses is due to non-homogeneous expansion and contraction of material during the deposition thermal cycle and depend on the plastic deformation, temperature changes, and phase transformations during that cycle, in particular, during cooling down to room temperature [58]. The residual stresses in thin walls and bulk prismatic samples of square cross-section made of 316L and Inconel 718, produced by laser powder deposition, were determined by Rangaswamy et al. They concluded that throughout the entire part, residual stresses are prevailing in the build direction (axial) and shows compressive stress at the centre of the part and tensile stress at the edges [59].

Kruth et al., experimented on the parts fabricated from iron-based powder [30] and found that changing the scanning patterns resulted in a good amount of reduction in the residual stresses and deformations. Especially, bi-directional scanning and island scanning strategy, can highly reduce distortions produced by residual stresses [60, 61]. Mazumder et al., studied stress generation in parts made by H13 tool steel and identified that the tool path location is a reason that makes the residual stresses to cause distortion [62]. Denlinger studied the effect of interlayer dwell time on the residual stress and distortion in directed energy deposited nickel and titanium alloys. It was found that increase in the dwell time between layers results in the increase in residual stresses accumulation [63]. In [64, 65], the authors have mentioned that when there is a high-temperature gradient, formation of residual stresses takes place which will lead to crack formation or interface or interlayer debonding. In addition to previous research findings, Wang et al., studied formation of cracks in several materials fabricated by laser deposition and concluded that the common cracking mechanism is cold cracking which happens owing to the existence of high internal stresses [66]. Cracks formed due to residual stresses can be of two types in which macroscopic cracks are highly influencing factor for defect formation compared to the microscopic cracks. Microscopic cracks can be eliminated by HIP or stress-relieving process whereas macroscopic cracks cannot be eliminated completely by post processing heat treatment [4, 48]. Li et al., studied the densification behaviour of 316L stainless steel powder under different process settings. They proposed that the higher laser power, small hatch spacing, lower scan speed will result in a very fine and smooth molten surface without any cracks [65].

Some studies have been carried out to reduce residual stress in AM since it contributes much to the defect formation. Preheating is one method which substantially eliminates residual stresses [67, 68]. Some researchers suggest that the post processing heat treatment of fabricated parts can reduce the residual stress induced distortions [69–71]. However, post processing will be time-consuming and lead to increase in cost despite of residual stress elimination. The common post processing for SLM is hot isostatic pressing and heat treatment process [72–75]. Research on support structures have been done and also found to be a suitable method to eliminate the residual stress induced defects [76, 77]. Some in-process measurements exist which measures the residual stress. To measure the residual stresses, Ocelik et al., used a digital image correlation method [78], Plati et al., used Linear Variable Differential Transformer [79] and laser displacement sensor was used by Denlinger et al. Post process measurements were also done by Denlinger et al., using the CMM and hole drilling method [63]. While, Yadroitsava et al., employed X-ray diffraction technique to measure the residual stresses [80].

3 Distortion Characterisation

In the previous sections, we reviewed the various defects that occur in metal AM process. In this section, distortion prediction using simulation software is presented. As we have understood the reasons for distortion of metal AM parts which were presented in the previous section, we need to characterize the effect of different geometrical features on distortion by analysing the benchmark parts as shown in Fig. 7. As process parameters play a vital role in defect formation, analysis can be

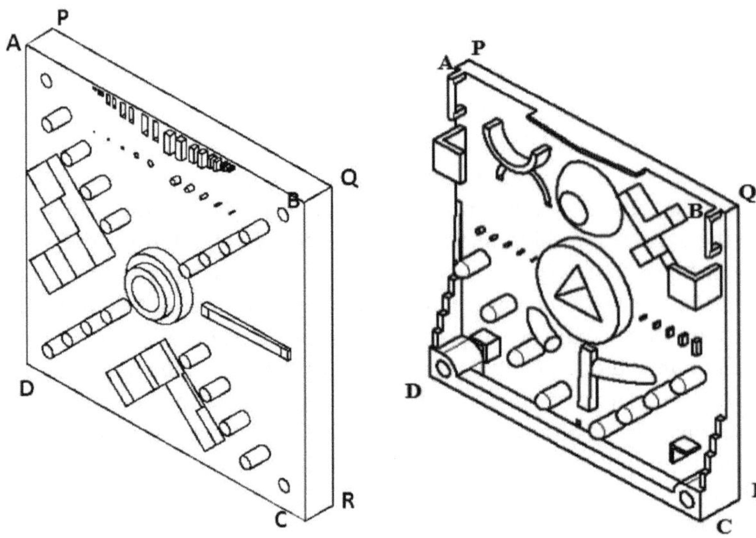

Fig. 7 **a** NIST test artefact and **b** IIITDM part

performed on different benchmark parts by varying different process parameters. It can also be analysed by fabricating the part in different possible orientations which may differ from the results obtained by varying process parameters. As distortion is material dependent, analysis can be performed with different materials available in the simulation software. Such kind of analysis will reduce the material wastage, time consumption and costs incurred in fabrication. The further detailed research can be done by analysing complex geometry parts and their respective distortions caused by various phenomenon.

4 Conclusion

This paper makes an attempt to review studies on defects present in metal additive manufacturing technologies under different classification based on the process parameter effect, the geometry of the part, support structures used, orientation of the part, physical phenomenon during the process and the material used in the process. Among the classification, we found that process parameters have a greater influence in defect formation. Hence, suitable process control methodologies should be adopted which can used in AM process as a preventive measure and will serve as an integral component to produce defect-free parts. As fabrication of AM parts cost higher, recently many simulation tools have been developed to simulate metal AM processes which can predict distortions and residual stresses. In simulation software tools, despite of their advantage like cost reduction and time saving, it may have certain limitations like reliability of predicted results. To overcome such limitation, fidelity modelling approaches should be developed. Reliable distortion prediction models and algorithms can be done which will help in developing analysing the complex parts used in various applications like defence, aerospace, automobiles, etc. Though this review does not cover all the research findings on entire DED and PBF processes, it is sufficient to acquire a basic knowledge about the defects in metal AM processes.

References

1. ASTM-International, ASTM Standard F2792-12a: Standard Terminology for Additive Manufacturing Technologies (2012), p. 2792
2. I. Gibson, D. Rosen, B. Stucker, *Additive Manufacturing Technologies: Rapid Prototyping to Direct Digital Manufacturing*, 1st edn. (Springer, New York, NY, USA, 2010)
3. M. Piazza, S. Alexander, *Additive Manufacturing : A Summary of the Literature* (Urban Publications, 2015)
4. D.D. Gu, W. Meiners, K. Wissenbach, R. Poprawe, Laser additive manufacturing of metallic components: materials, processes and mechanisms. Int. Mater. Rev. **57**(3), 133–164 (2012)
5. W.E. Frazier, Metal additive manufacturing: a review. J. Mater. Eng. Perform. **23**(6), 1917–1928 (2014)

6. H. Bikas, P. Stavropoulos, G. Chryssolouris, Additive manufacturing methods and modeling approaches: a critical review. Int. J. Adv. Manuf. Technol. **83**(1–4), 389–405 (2016)
7. E. Herderick, Additive manufacturing of metals: a review, in *Material Science and Technology Conference and Exhibition 2011, MS T'11*, vol. 2, no. 176252 (2011), pp. 1413–1425
8. C.Y. Yap et al., Review of selective laser melting: materials and applications. Appl. Phys. Rev. **2**(4) (2015)
9. R. Vilar, Laser powder deposition. Compr. Mater. Process. **10**, 163–216 (2014)
10. G. Tapia, A. Elwany, A review on process monitoring and control in metal-based additive manufacturing. J. Manuf. Sci. Eng. **136**(6), 60801 (2014)
11. W. Gao et al., The status, challenges, and future of additive manufacturing in engineering. Comput. Des. **69**, 65–89 (2015)
12. D. Gu et al., Densification behavior, microstructure evolution, and wear performance of selective laser melting processed commercially pure titanium. Acta Mater. **60**(9), 3849–3860 (2012)
13. N.T. Aboulkhair, N.M. Everitt, I. Ashcroft, C. Tuck, Reducing porosity in AlSi10Mg parts processed by selective laser melting. Add. Manuf. **1**, 77–86 (2014)
14. S. Kleszczynski, J. zur Jocobsmuhlen, J.T. Sehrt, Error detection in laser beam melting systems by high resolution imaging, in *Twenty Third Annual International Solid Freeform Fabrication Symposium*, no. May 2016 (2012)
15. J. zur Jacobsmuhlen, S. Kleszczynski, D. Schneider, G. Witt, High resolution imaging for inspection of laser beam melting systems, in *2013 IEEE International Instrumentation and Measurement Technology Conference* (2013), pp. 707–712
16. A.E. Patterson, S.L. Messimer, P.A. Farrington, Overhanging features and the SLM/DMLS residual stresses problem: review and future research need. Technologies **5**(2), 15 (2017)
17. A. Hussein, L. Hao, C. Yan, R. Everson, P. Young, Advanced lattice support structures for metal additive manufacturing. J. Mater. Process. Technol. **213**(7), 1019–1026 (2013)
18. J.-P. Kruth, J. Deckers, E. Yasa, R. Wauthle, Assessing and comparing influencing factors of residual stresses in selective laser melting using a novel analysis method. Proc. Inst. Mech. Eng. Part B J. Eng. Manuf. **226**(6), 980–991 (2012)
19. P. Vora, K. Mumtaz, I. Todd, N. Hopkinson, AlSi12 in-situ alloy formation and residual stress reduction using anchorless selective laser melting. Add. Manuf. **7**, 12–19 (2015)
20. J. Jhabvala, E. Boillat, C. André, R. Glardon, An innovative method to build support structures with a pulsed laser in the selective laser melting process. Int. J. Adv. Manuf. Technol. **59**(1–4), 137–142 (2012)
21. F. Calignano, Design optimization of supports for overhanging structures in aluminum and titanium alloys by selective laser melting. Mater. Des. **64**, 203–213 (2014)
22. S. Mohanty, J.H. Hattel, *Improving Accuracy of Overhanging Structures for Selective Laser Melting Through Reliability Characterization of Single Track Formation on Thick Powder Beds*, vol. 9738, p. 97381B (2016)
23. D. Wang, Y. Yang, Z. Yi, X. Su, Research on the fabricating quality optimization of the overhanging surface in SLM process. Int. J. Adv. Manuf. Technol. **65**(9–12), 1471–1484 (2013)
24. K. Zeng, Optimization of support structures for selective laser melting, Ph.D. dissertation, Department of Industrial Engineering, University of Louisville, Kentucky, US, 2015
25. B.R.D. Thomas, An investigation into the geometric constraints of selective laser melting for the development of design rules, in *9th National Conference on Rapid Design, Prototyping & Manufacture* (2008), pp. 11–20
26. Thomas, The Development of Design Rules for Selective Laser Melting, Ph.D. dissertation, University of Wales Institute, Cardiff, 2009
27. L. Papadakis, A. Loizou, J. Risse, J. Schrage, Numerical computation of component shape distortion manufactured by Selective Laser Melting. Procedia CIRP **18**, 90–95 (2014)
28. T. Craeghs, S. Clijsters, E. Yasa, F. Bechmann, S. Berumen, J.P. Kruth, Determination of geometrical factors in Layerwise Laser Melting using optical process monitoring. Opt. Lasers Eng. **49**(12), 1440–1446 (2011)

29. M. Grasso, V. Laguzza, Q. Semeraro, B.M. Colosimo, In-process monitoring of selective laser melting: spatial detection of defects via image data analysis. J. Manuf. Sci. Eng. **139**(5), 51001 (2016)
30. J.P. Kruth, L. Froyen, J. Van Vaerenbergh, P. Mercelis, M. Rombouts, B. Lauwers, Selective laser melting of iron-based powder. J. Mater. Process. Technol. **149**(1–3), 616–622 (2004)
31. L.N. Carter, M.M. Attallah, R.C. Reed, Laser powder bed fabrication of nickel-base superalloys: influence of parameters; characterisation, quantification and mitigation of cracking, in *Superalloys 2012*, no. October 2012 (2012), pp. 577–586
32. P.A. Mcnutt, M. Attallah, EngD Thesis An Investigation of Cracking in Laser Metal Deposited Nickel Superalloy CM247LC, no. February (2015)
33. M.F. Zäh, S. Lutzmann, Modelling and simulation of electron beam melting. Prod. Eng. **4**(1), 15–23 (2010)
34. M.F. Zaeh, M. Kahnert, The effect of scanning strategies on electron beam sintering. Prod. Eng. **3**(3), 217–224 (2009)
35. M. Kahnert, S. Lutzmann, M.F. Zaeh, Layer formations in electron beam sintering, in *Proceedings of the 18th Solid Freeform Fabrication Symposium* (2007), pp. 88–99
36. P.A. Kobryn, E.H. Moore, S.L. Semiatin, Effect of laser power and traverse speed on microstructure, porosity, and build height in laser-deposited Ti-6Al-4V. Scr. Mater. **43**(4), 299–305 (2000)
37. A.H. Nickel, D.M. Barnett, F.B. Prinz, Thermal stresses and deposition patterns in layered manufacturing. Mater. Sci. Eng. A **317**(1–2), 59–64 (2001)
38. R. Li, J. Liu, Y. Shi, L. Wang, W. Jiang, Balling behavior of stainless steel and nickel powder during selective laser melting process. Int. J. Adv. Manuf. Technol. **59**(9–12), 1025–1035 (2012)
39. N.K. Tolochko et al., Balling processes during selective laser treatment of powders. Rapid Prototyp. J. **10**(2), 78–87 (2004)
40. X. Zhou, X. Liu, D. Zhang, Z. Shen, W. Liu, Balling phenomena in selective laser melted tungsten. J. Mater. Process. Technol. **222**, 33–42 (2015)
41. K.A. Mumtaz, N. Hopkinson, Selective Laser Melting of thin wall parts using pulse shaping. J. Mater. Process. Technol. **210**(2), 279–287 (2010)
42. K. Mumtaz, N. Hopkinson, Top surface and side roughness of Inconel 625 parts processed using selective laser melting. Rapid Prototyp. J. **15**(2), 96–103 (2009)
43. D. Gu, Y. Shen, Balling phenomena in direct laser sintering of stainless steel powder: metallurgical mechanisms and control methods. Mater. Des. **30**(8), 2903–2910 (2009)
44. W.J. Sames, F.A. List, S. Pannala, R.R. Dehoff, S.S. Babu, The metallurgy and processing science of metal additive manufacturing. Int. Mater. Rev. **61**(5), 315–360 (2016)
45. N.J. Harrison, I. Todd, K. Mumtaz, Reduction of micro-cracking in nickel superalloys processed by Selective Laser Melting: a fundamental alloy design approach. Acta Mater. **94**, 59–68 (2015)
46. K. Kempen, L. Thijs, B. Vrancken, J. Van Humbeeck, J.-P. Kruth, Producing crack-free, high density M2 Hss parts by selective laser melting: pre-heating the baseplate, in *Proceedings of the 24th International Solid Freeform Fabrication Symposium* (2013), pp. 131–139
47. W. Huang, J. Chen, X. Lin, T. Wang, H. Yang, The hot cracking mechanism of 316L stainless steel cladding in rapid laser forming process. Rare Met. Mater. Eng. **32**(2), 183–186 (2003)
48. X. Zhao, X. Lin, J. Chen, L. Xue, W. Huang, The effect of hot isostatic pressing on crack healing, microstructure, mechanical properties of Rene88DT superalloy prepared by laser solid forming. Mater. Sci. Eng. A **504**, 129–134 (2009)
49. L.N. Carter, K. Essa, M.M. Attallah, Optimisation of selective laser melting for a high temperature Ni-superalloy. Rapid Prototyp. J. **21**(4), 423–432 (2015)
50. K. Kempen, B. Vrancken, S. Buls, L. Thijs, J. Van Humbeeck, J.-P. Kruth, Selective laser melting of crack-free high density M2 high speed steel parts by baseplate preheating. J. Manuf. Sci. Eng. **136**(6), 61026 (2014)

51. W.J. Sames, F. Medina, W.H. Peter, S.S. Babu, R.R. Dehoff, Effect of process control and powder quality on Inconel 718 produced using electron beam melting, in *8th International Symposium on Superalloy 718 and Derivatives* (2014), pp. 409–423
52. P. Edwards, A. O'Conner, M. Ramulu, Electron beam additive manufacturing of titanium components: properties and performance. J. Manuf. Sci. Eng. **135**(6), 61016 (2013)
53. H. Gong, K. Rafi, N.V. Karthik, T. Starr, B. Stucker, Defect morphology in Ti-6Al-4V parts fabricated by Selective Laser Melting and Electron Beam Melting, in *24th International SFF Symposium—An Additive Manufacturing Conference, SFF 2013* (2013), pp. 440–453
54. I. Maskery et al., Quantification and characterisation of porosity in selectively laser melted Al–Si10–Mg using X-ray computed tomography. Mater. Charact. **111**, 193–204 (2016)
55. S. Tammas-Williams, H. Zhao, F. Léonard, F. Derguti, I. Todd, P.B. Prangnell, XCT analysis of the influence of melt strategies on defect population in Ti-6Al-4V components manufactured by Selective Electron Beam Melting. Mater. Charact. **102**, 47–61 (2015)
56. R.J. Smith, M. Hirsch, R. Patel, W. Li, A.T. Clare, S.D. Sharples, Spatially resolved acoustic spectroscopy for selective laser melting. J. Mater. Process. Technol. **236**, 93–102 (2016)
57. J.A. Slotwinski, E.J. Garboczi, K.M. Hebenstreit, Porosity measurements and analysis for metal additive manufacturing process control. J. Res. Natl. Inst. Stand. Technol. **119**, 494 (2014)
58. J.M. Pelletier, A.B. Vannes, M. Pilloz, Residual stresses induced by laser coatings: phenomenological analysis and predictions. J. Mater. Sci. **27**, 1240–1244 (1992)
59. P. Rangaswamy et al., Residual stresses in LENS® components using neutron diffraction and contour method. Mater. Sci. Eng. A **399**(1–2), 72–83 (2005)
60. N.W. Klingbeil, J.L. Beuth, R.K. Chin, C.H. Amon, Residual stress-induced warping in direct metal solid freeform fabrication. Int. J. Mech. Sci. **44**(1), 57–77 (2002)
61. J.P. Kruth, M. Badrossamay, E. Yasa, J. Deckers, L. Thijs, J. Van Humbeeck, Part and material properties in selective laser melting of metals, in *16th International Symposium on Electromachining* (*ISEM 2010*), no. October 2016 (2010), pp. 3–14
62. J. Mazumder, J. Choi, K. Nagarathnam, J. Koch, D. Hetzner, The direct metal deposition of H13 tool steel for 3-D components. JOM **49**(5), 55–60 (1997)
63. E.R. Denlinger, *Thermo-Mechanical Model Development and Experimental Validation for Metallic Parts in Additive Manufacturing by Chair of Committee* (The Pennsylvania State University, 2015)
64. Q.S. Wei, S. Zhang, R.Z. Gui, Cracking behavior and formation mechanism of TC4 alloy formed by selective laser melting. J. Mech. Eng. **49**(23), 21–27 (2013)
65. R. Li, Y. Shi, Z. Wang, L. Wang, J. Liu, W. Jiang, Densification behavior of gas and water atomized 316L stainless steel powder during selective laser melting. Appl. Surf. Sci. **256**(13), 4350–4356 (2010)
66. L. Wang, S.D. Felicelli, P. Pratt, Residual stresses in LENS-deposited AISI 410 stainless steel plates. Mater. Sci. Eng. **496**, 234–241 (2008)
67. A. Vasinonta, J.L. Beuth, M. Griffith, Process maps for predicting residual stress and melt pool size in the laser-based fabrication of thin-walled structures. J. Manuf. Sci. Eng. **129**(1), 101 (2007)
68. D. Buchbinder, W. Meiners, N. Pirch, K. Wissenbach, J. Schrage, Investigation on reducing distortion by preheating during manufacture of aluminum components using selective laser melting. J. Laser Appl. **26**(1), 12004 (2014)
69. P. Mercelis, J. Kruth, Residual stresses in selective laser sintering and selective laser melting. Rapid Prototyp. J. **12**(5), 254–265 (2006)
70. M. Thöne, S. Leuders, Influence of heat-treatment on selective laser melting products—e.g. Ti6Al4V, in *SFF*, Austin, Texas (2012), pp. 492–498
71. T. Sercombe, N. Jones, R. Day, A. Kop, Heat treatment of Ti-6Al-7Nb components produced by selective laser melting. Rapid Prototyp. J. **14**(5), 300–304 (2008)
72. W. Tillmann, C. Schaak, J. Nellesen, M. Schaper, M.E. Aydinöz, K.P. Hoyer, Hot isostatic pressing of IN718 components manufactured by selective laser melting. Add. Manuf. **13**, 93–102 (2017)

73. S.L. Campanelli, N. Contuzzi, A.D. Ludovico, F. Caiazzo, F. Cardaropoli, V. Sergi, Manufacturing and characterization of Ti6Al4V lattice components manufactured by selective laser melting. Materials (Basel) **7**(6), 4803–4822 (2014)
74. W.M. Tucho, P. Cuvillier, A. Sjolyst-Kverneland, V. Hansen, Microstructure and hardness studies of Inconel 718 manufactured by selective laser melting before and after solution heat treatment. Mater. Sci. Eng. A **689**(December 2016), 220–232 (2017)
75. B. Song, S. Dong, Q. Liu, H. Liao, C. Coddet, Vacuum heat treatment of iron parts produced by selective laser melting: microstructure, residual stress and tensile behavior. Mater. Des. **54**, 727–733 (2014)
76. M.F. Zaeh, G. Branner, Investigations on residual stresses and deformations in selective laser melting. Prod. Eng. **4**(1), 35–45 (2010)
77. L. van Belle, G. Vansteenkiste, J.C. Boyer, Investigation of residual stresses induced during the selective laser melting process. Key Eng. Mater. **554–557**, 1828–1834 (2013)
78. V. Ocelík, J. Bosgra, J.T.M. de Hosson, In-situ strain observation in high power laser cladding. Surf. Coatings Technol. **203**(20–21), 3189–3196 (2009)
79. A. Plati, J.C. Tan, I.O. Golosnoy, R. Persoons, K. Van Acker, T.W. Clyne, Residual stress generation during laser cladding of steel with a particulate metal matrix composite. Adv. Eng. Mater. **8**(7), 619–624 (2006)
80. I. Yadroitsava, I. Yadroitsev, Residual stress in metal specimens produced by direct metal laser sintering. J. Chem. Inf. Model. **53**(9), 1689–1699 (2013)

Laser Metal Deposition of Titanium Parts with Increased Productivity

Vishnuu Jothi Prakash, Mauritz Möller, Julian Weber and Claus Emmelmann

Abstract Laser Metal Deposition (LMD), an additive manufacturing technique, is described here as an alternate for conventional manufacturing process to build aerospace components. Traditional milling of thin-walled, ribbed-, lightweight, high-valued Titanium structures generate machining wastes as high as 95%. This paper presents an LMD system setup along with an adapted manufacturing process chain for fabrication of near-net shaped Ti–6Al–4V components. Demonstrator parts built using the system setup are then shown.

Keywords Additive manufacturing · Laser metal deposition · Titanium alloys Thin-walled aerospace structures · Increased productivity

1 Introduction

Owing to the fact that fuel accounts to 40–50% of the operational cost of typical airliners, aviation industry is steered towards manufacturing lighter and more fuel-efficient aircrafts [1]. Subsequently, numerous new materials like carbon fiber and alloys of titanium (Ti), nickel and aluminum are being adopted. Because of good strength-to-weight ratio, resistance to high-temperature and chemical corrosion, Ti

V. Jothi Prakash (✉) · J. Weber · C. Emmelmann
Institut für Laser- und Anlagensystemtechnik, Technische Universität Hamburg, Denickestraße 17, 21073 Hamburg, Germany
e-mail: vishnuu.jothi.prakash@tuhh.de

J. Weber
e-mail: julian.weber@tuhh.de

C. Emmelmann
e-mail: c.emmelmann@lzn-hamburg.de

M. Möller · C. Emmelmann
Laser-Zentrum Nord GmbH, Bionic Production GmbH, Am Schleusengraben 14, 21029 Hamburg, Germany
e-mail: mauritz.moeller@lzn-hamburg.de

L. J. Kumar et al. (eds.), *3D Printing and Additive Manufacturing Technologies*,
https://doi.org/10.1007/978-981-13-0305-0_25

alloys have been the preferred choice especially in aviation industry, which accounts for 40% of the total Ti production [1]. On the other hand, components are designed to be lighter and stiffer. However, realization of lightweight and thin-walled structures through traditional manufacturing technique is challenging since they are associated with excessive tool wear rate and generation of high machining waste (more than 90% in some cases) [2, 3]. Additive Manufacturing (AM) technique allows realization of innovative, light weight, load-adapted design incorporating topology optimization and bionic design through layered manufacturing process [4].

Laser Metal Deposition (LMD) is an additive manufacturing process that allows placement of desired material exactly where it is needed [5, 6]. Energy from the laser beam heats the substrate, while a fine stream of atomized metal powder is injected through nozzle(s) right into the melt pool to form the designed structure in a layer-by-layer manner. The process results in very high deposition rates in comparison to other AM techniques, like Laser Beam Melting (LBM). However, the process generates near-net shaped structures, whose geometrical deviation increases with increasing part complexity. The technique still lacks reproducible process strategies and defined manufacturing tolerances that are required for post-processing and also industrial application [7].

To overcome the above-mentioned issues, shape dependent process strategies have been developed, which are discussed in this paper. Additionally, a part-adaptive manufacturing process chain has been charted. Various Ti–6Al–4V demonstrator components built with optimized process parameters and strategies are then shown. Geometrical deviation plot of the built components are presented, followed by porosity analysis of a test specimen.

2 System Setup and Methodology

The system setup consists of a 6 kW multi-mode continuous wave disk laser with a wavelength of 1.03 μm. A three-nozzle laser deposition head from Trumpf GmbH, attached to six-axis industrial robot, delivers atomized metal powder fed through a rotational powder hoper unit. Because of high reactivity of Ti above 800 °C, an inert gas (argon) atmosphere with less than 50 ppm of residual oxygen is ensured inside the fabricated building chamber, as shown in Fig. 1. The used Ti–6Al–4V powder is spherical in shape with diameter ranging between 80 and 150 μm.

A process chain suitable for LMD process has been developed to carry out the experimental investigations in a scientific approach. Figure 2 depicts pictorially the four major LMD phases. Conceptualization of the part begins from the result of topology optimization and consideration of certain manufacturing limitations. A CAD model is then designed and the component is divided into several weld-able sections. In the planning phase, these sections are sliced depending upon the process strategy, part complexity and orientation of the base plate. The planned tool head path is implemented in robot program, which is followed by the final manufacturing phase.

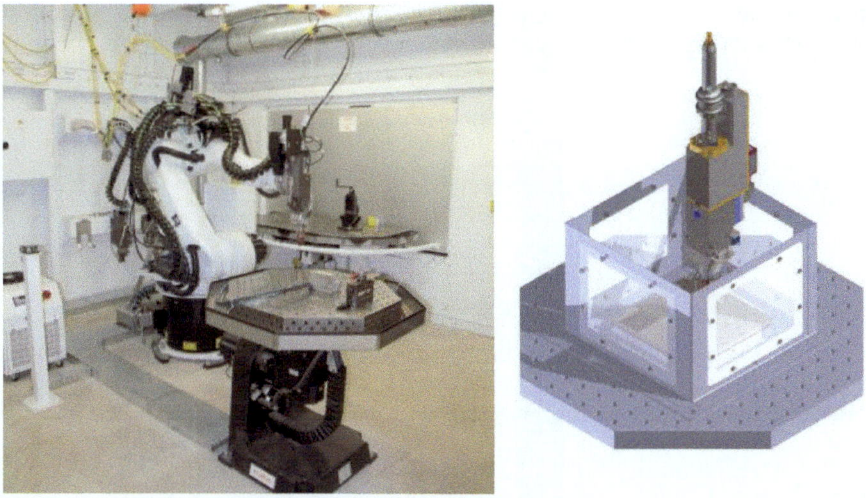

Fig. 1 Deposition head with robot (left), shielded process chamber (right)

For this study, three representative components were designed in the increasing order of shape complexity, which were later manufactured following the above proposed process chain. Table 1 shows the CAD model of the three demonstrator components along with the overall dimensions. Wall and orthogonal structures could be considered as 2D and 2.5D structures respectively, because, the layer geometry remains the same (or almost) over the building direction. On the other hand, freeform structures are characterized with completely varying layer geometry over the building direction. This imposes issues in terms of tool head programming, as there are no commercial software platforms for the afore-mentioned LMD system setup.[1]

The representative components considered are thin-wall structures with 4 mm wall thickness, which is achieved through a single pass of weld deposit.

Load-adaptive bionic designs developed from the results of topology optimization technique can be realized through Laser Beam Melting, irrespective of shape complexity [4]. However, the working principle of LMD imposes manufacturing limitations that restricts the design complexity. Design rules and rule-based process parameters are therefore required for successful realization of LMD components, specific to LMD system setup being used.

[1]Commercial software like Siemens NX is available for hybrid LMD process; however, it does not support a robotic LMD system.

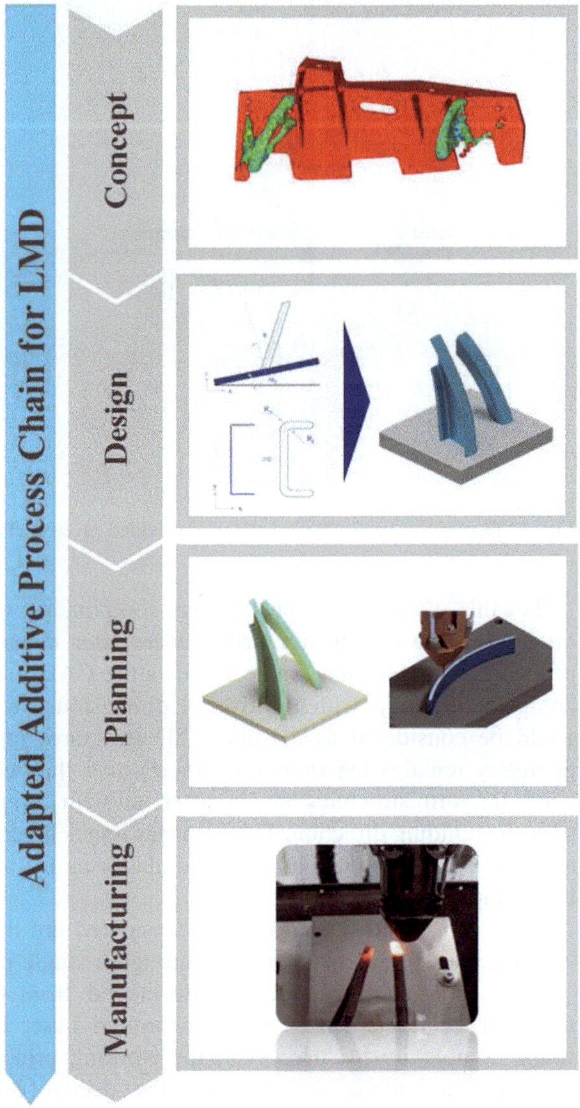

Fig. 2 Adapted manufacturing process chain for LMD

Table 1 LMD build part categorization

Component	1	2	3
Type	Wall structure	Orthogonal	Freeform
Shape complexity	Low	Medium	High
CAD model			
Dimensions (mm)	70 × 4.5 × 68	210 × 95 × 35	100 × 80 × 70

3 Design

The LMD process mechanics and the available system setup impose certain manufacturing restrictions. Components are built without support structures, which limits complexity of component's design. Also, the smallest feature possible is defined by the selected process parameters. With the above-mentioned system setup, minimum wall thickness of 4 mm is achievable. This leads to development of design rules, developed and verified primarily through experimental investigations. Even though there are many literatures regarding design rules in LBM, there isn't any literature concerning component design rules for LMD process, in the best knowledge of authors.

To begin with, wall structures and orthogonal structures (refer component 1 and 2 in Table 1), those are perpendicular to build platform needs no design considerations. Inclined wall structures, however, needs to be considered for maximum manufacturable inclination or adoption of different build strategies [7].

Freeform structures (component 3 in Table 1) also need design considerations and modifications. The representative component is designed with 30° inclination and divided into several sections, which depends upon distinguishable component features, its inclination and orientation with the base plate. Figure 3 shows the representative freeform component divided into several workpiece segments. Segment 1 and 2 will be fabricated through LMD process on a base plate. Segment 3 is achieved through wire-cut EDM process from base plate.

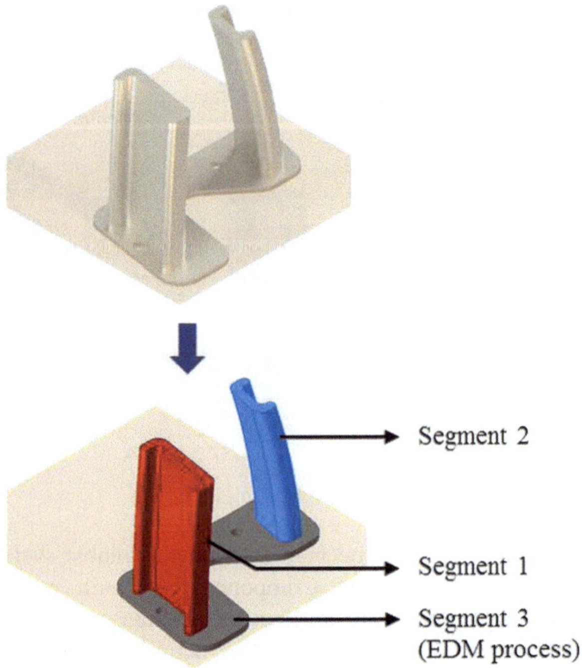

Fig. 3 Segmentation of freeform component

4 Part-Adaptive Process Planning

The first build trial of the demonstrator components were built with process parameters preliminarily developed by Möller et al. [7] and are shown in Fig. 4.

The preliminary process parameters developed in [7] have been further optimized by the authors using evolutionary algorithm technique to optimize process

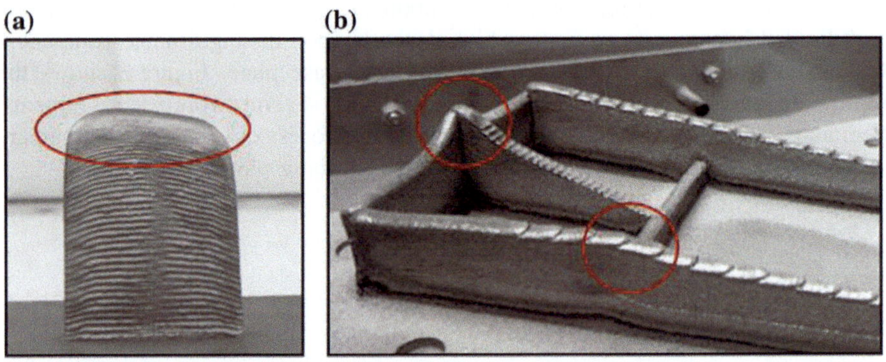

Fig. 4 First LMD build trials with circled areas of defects (red circles) in: **a** wall structure, **b** orthogonal structure

parameters like laser power, deposition velocity and powder flow rate for maximum geometrical accuracy and better optical appearance. Additionally, process strategies for wall structures have been developed that accounts for increasing geometrical deviation over build direction. The developed process parameters still fail to provide successful results when applied for complex structures like orthogonal and freeform. This leads to development of further part-adaptive process strategies for successful fabrication of complex workpiece geometries, which are discussed in the following section.

4.1 Manufacturing Tool Path Optimization

Orthogonal structures, like component 2, are characterized with intersecting sections and corners. Excessive heat input due to overlap at these hot spots results in accumulation of more material, thereby leading to excessive geometrical deviation, as shown in Fig. 4b. An empirical and experimental investigation to determine the appropriate offset distance (O_s) to minimize accumulated heat was carried out, refer Fig. 5. In case of thin-walled structures, the achievable geometrical accuracy is highly dependent on tool head programming. Additionally, the start and the end point of the deposition lines in every layer also influence the resulting build geometry.

4.2 Part-Building Strategies

With an increase in shape complexity, increases the process planning effort. Unlike Laser Beam Melting (LBM) process that builds component strictly in a layer-by-layer fashion, LMD lets one to develop several part-building strategies that are important in optimizing cost, time and quality of the build component. The process setup allows segmentation of component and deposition of individual segments at different rates. Figure 6 shows segmentation of the workpiece along with a cut-section view. Schematic representation of three different part-building strategies developed for the representative freeform component are shown as well and they are:

(1) *Layer-by-layer build strategy*: One layer of each segment deposited alternatively. Deposition pattern is similar to LBM process.
(2) *Set-wise build strategy*: Set of layers (20 layers in this case) of one particular segment to be deposited before alternating to other segment, where a set of 20 layers will be then deposited.
(3) *Segment-wise build strategy*: All layers of one segment (60 layers in this case) to be deposited before alternating to other segment.

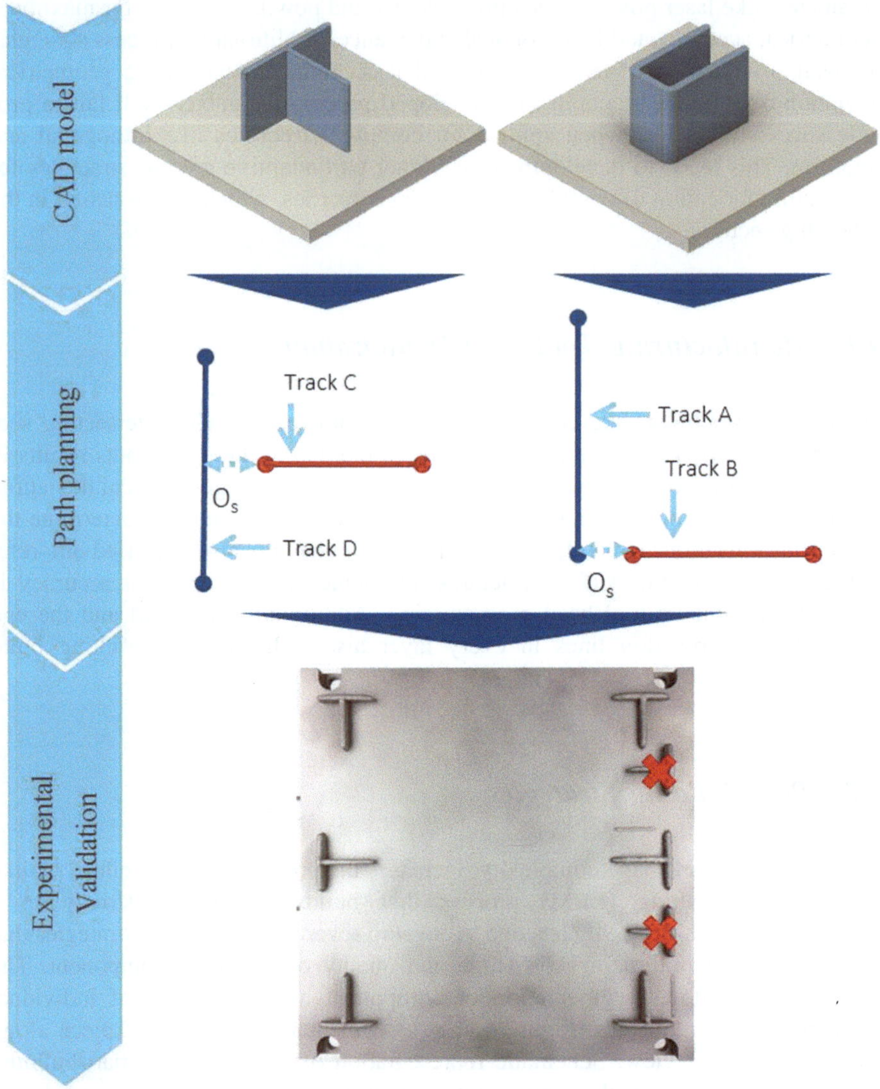

Fig. 5 Experimental validation of optimum manufacturing tool path

The representative components were built once again with the developed part-adaptive process parameters and photographs of the same are shown in Figs. 7 and 8. The deposition rate depends upon the selected process parameters and build strategy. The components shown here are manufactured at a rate of 150 cm^3/h. Higher deposition rates could be achieved with a different nozzle and a different set of process parameters.

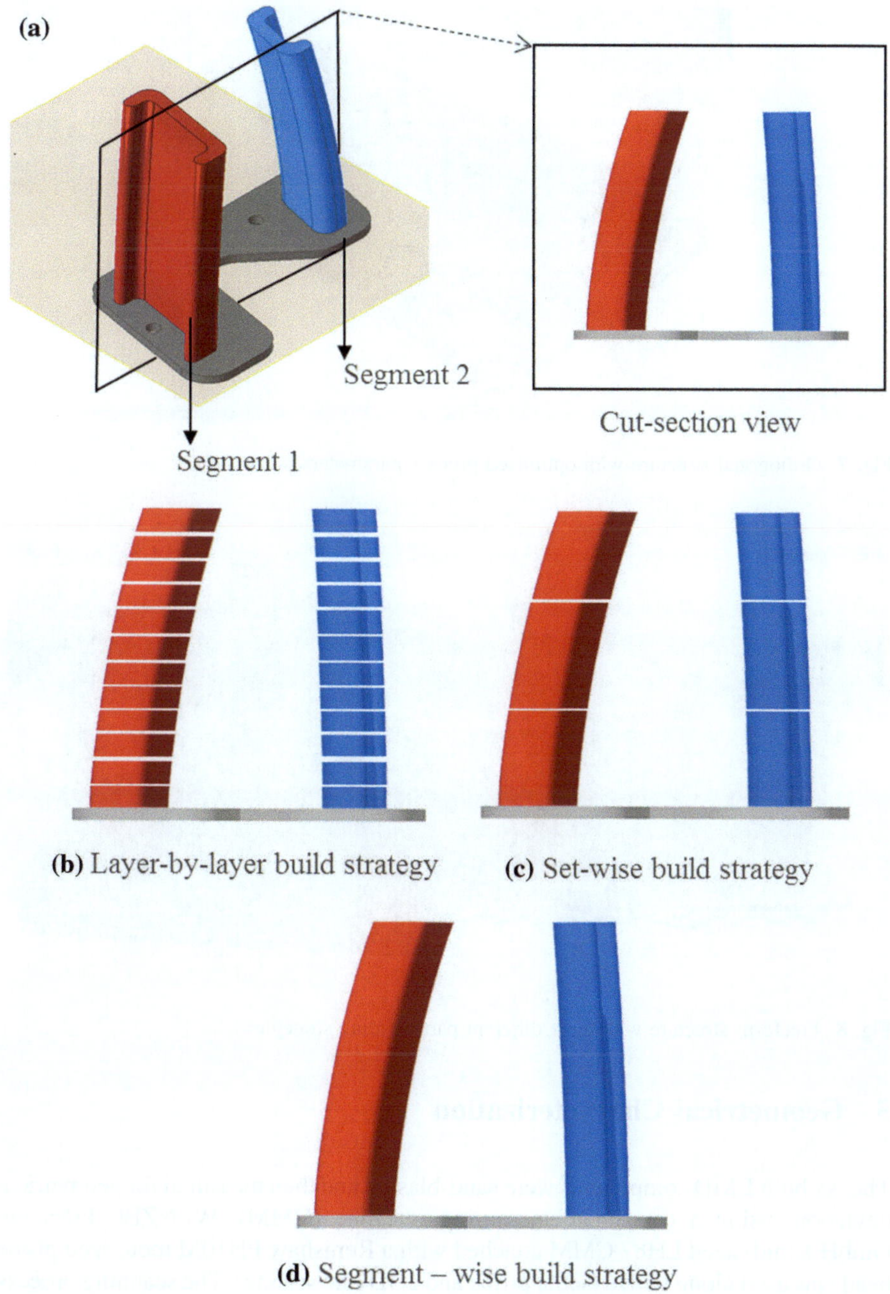

Fig. 6 **a** Component segmentation with cross-sectional view; **b–d** part-building strategies for freeform structures

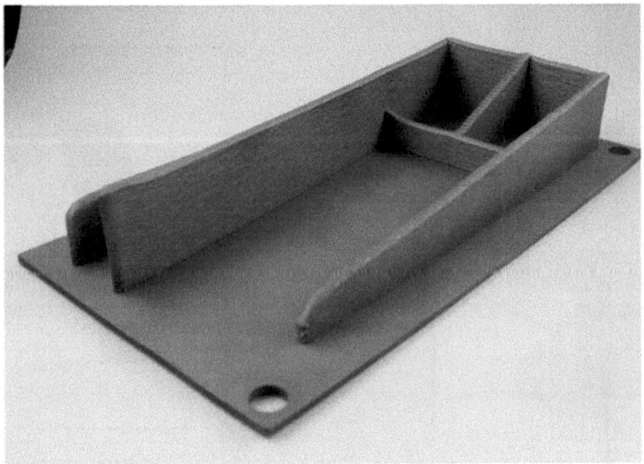

Fig. 7 Orthogonal structure with optimized process parameters

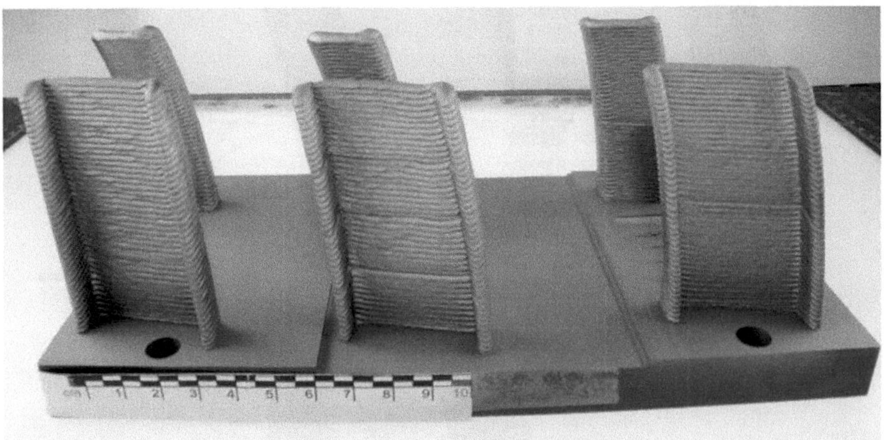

Fig. 8 Freeform structure with three different part-building strategies

5 Geometrical Characterization

The As-built LMD components were sand-blasted and then measured for geometrical deviation, using a coordinate measuring machine (CMM). WENZEL Präzision GmbH's mid-sized LH87 CMM attached with a Renishaw PH10M motorized probe head was used along with a tactile probe and a 3D line scanner. The scanning process generates a point cloud represented by x, y, and z coordinates at a rate of 48,000 points per second with an accuracy of 20 μm. The width of the wall structure (Component 1) was measured using the tactile probe and Component 2 and 3 were scanned using the line scanner to generate a point cloud. The point cloud was then compared with respective 3D CAD model using PointMaster 5.0 software, resulting in a 3D geometrical deviation plot. Figure 9 shows the measuring equipment used along with tactile probe and laser scanner. Figure 10 shows the 3D geometrical deviation plot of the orthogonal structure and freeform structure.

The measured geometrical accuracy of three representative components is tabulated in Table 2 in terms of measured deviation in comparison with CAD model. For component 3, all three different build strategies were measured and are tabulated. An increasing trend of geometrical deviation has been observed with increasing shape complexity. It has also been observed that different part-building strategies influences the build quality.

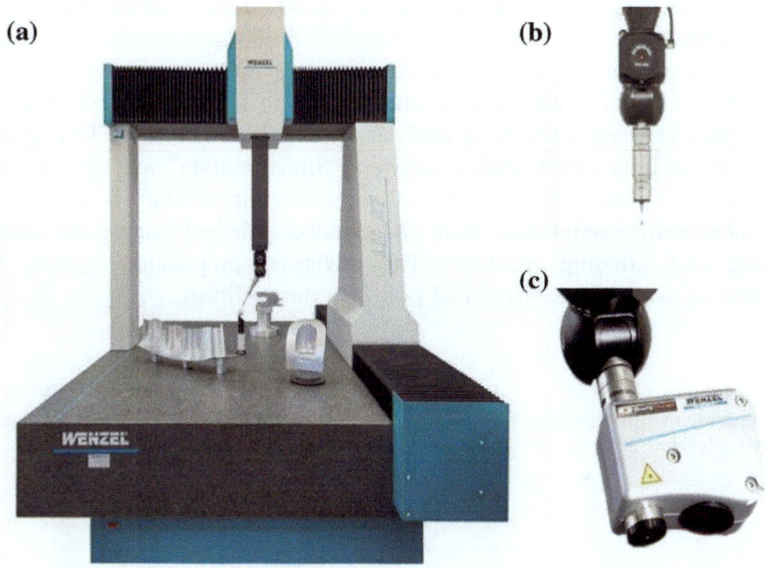

Fig. 9 **a** Coordinate measuring machine, **b** tactile probe, **c** laser scanner

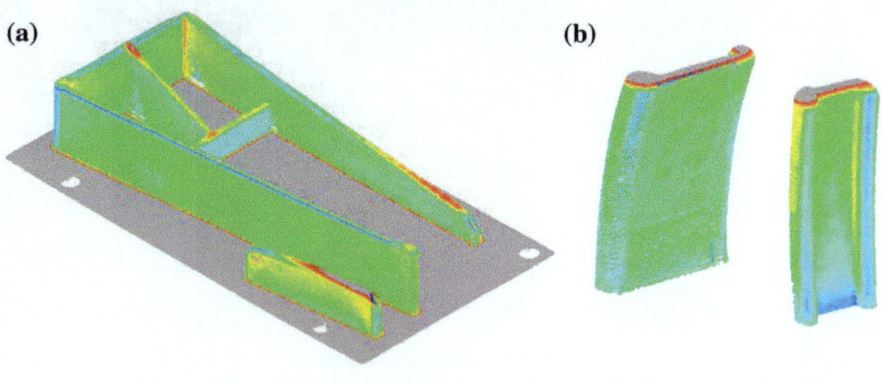

Fig. 10 Measured 3D geometrical deviation plot of: **a** orthogonal structure, **b** freeform structure

Table 2 Geometrical deviations

Geometrical deviation (mm)	Component 1	Component 2	Component 3		
			Strategy 1	Strategy 2	Strategy 3
90% C.I.	±0.1830	±0.5505	±0.6128	±0.9200	±0.6700
95% C.I.	±0.2560	±0.6714	±0.7474	±1.1240	±0.8184

6 Metallographic Characterization (Porosity)

Different types of porosity could be observed in LMD build components, like lack-of-fusion porosity, spherical porosity, gas entrapment and cracking [8]. However, cracking is considered as a build defect rather than porosity characteristic.

More component 1 type wall structures were built as samples for porosity measurement. These walls were sectioned with an abrasive cut-off wheel. Four sections were selected for porosity analysis. The optical microscope Olympus GX5, see Fig. 11, and the corresponding software "Streammotion" was used to examine the porosity. To prepare the sections for the metallographic analysis, all the samples were embedded in cold setting resin and grinded, polished using a semiautomatic polishing and grinding machine. The advanced preparation system Struers Tegramin–25 was used to grind and polish to mirror finish.

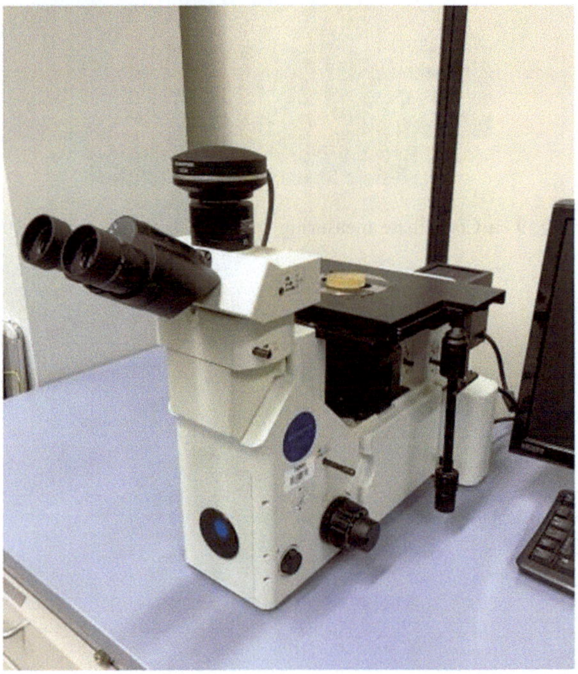

Fig. 11 Olympus GX5 optical microscope for porosity analysis

After capturing sharp images of each section, the Software Streammotion and its integrated measuring and counting function was used to analyze the images. The function processes a defined area of the image and indicates light pixels as solid material whereas dark pixels represent a cavity. Three identical areas are selected within every section, positioned at 5, 50 and 95% of building height. To examine the porosity of a 3D structure, typically the ratio of solid material to cavity is taken. Since the image of an optical microscope is two-dimensional, the ratio of the solid area to the hollow area has been used to calculate the dimensionless solid material percentage ϕ with the following equation.

$$\phi = 1 - \frac{A_h}{A_h + A_\upsilon}$$

As a result, a high solid material percentage means a low porosity and therefore a good result. Figure 12 pictures exemplary the process of measuring the porosity respectively the solid material percentage within section 1. A minimum solid material percentage of 99.87% was identified in the upper area of the deposited wall while the middle area was represented by the highest solid material percentage of 99.95%. The arithmetic mean calculates to 99.92% of solid material percentage.

The examination of the remaining sections shows comparable results. The best result was reached within section 2 and a solid material percentage of 99.96% while the lowest solid material percentage was measured in section 4, reaching 99.83%. All results are summarized in Table 3. Additionally, the data is shown graphically in Fig. 13, where the solid material percentages are plotted by the z-position of the deposited wall.

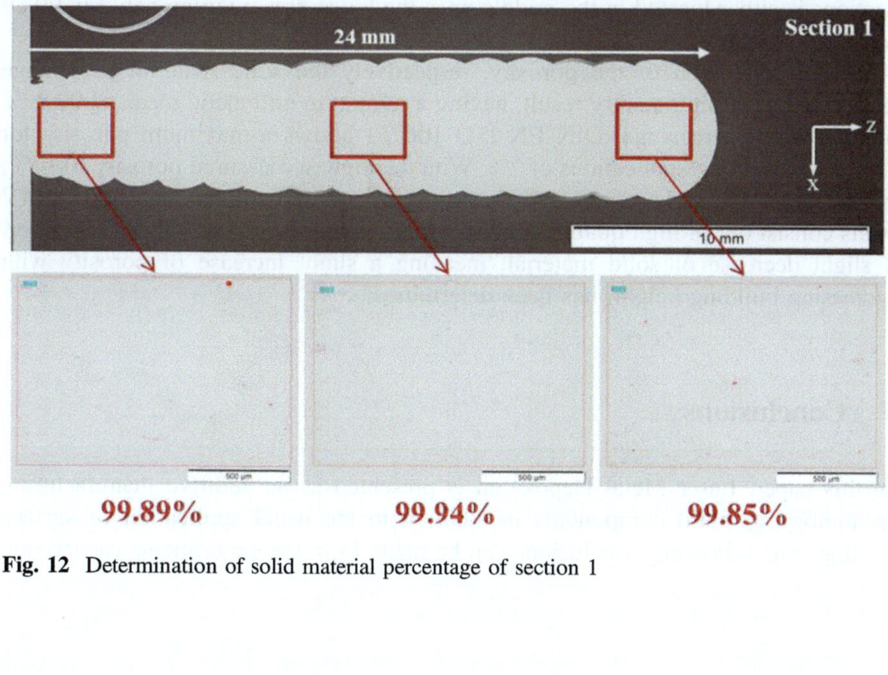

Fig. 12 Determination of solid material percentage of section 1

Table 3 Summary of solid material percentages

Section	5%—b.h. (%)	50%—b.h. (%)	95%—b.h. (%)	Mean
1	99.89	99.94	99.85	99.90
2	99.96	99.90	99.87	99.92
3	99.89	99.89	99.86	99.90
4	99.92	99.94	99.83	99.89

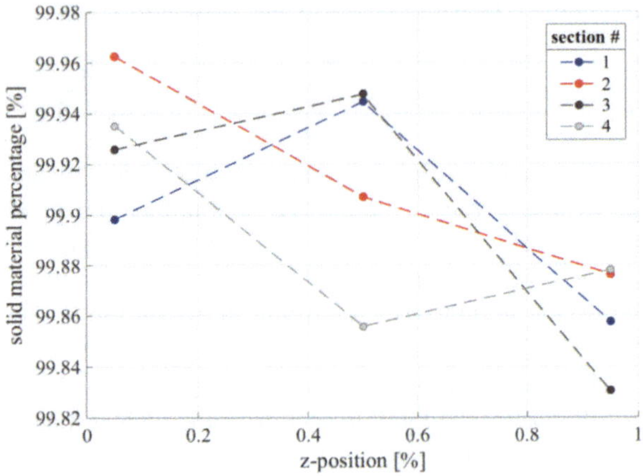

Fig. 13 Data plot for solid material percentage by *z*-position

The data plot clarifies, that the lowest solid material percentage is mostly located in the upper area of the wall, ranging between 99.83 and 99.87%. Only section four has its minimum value located at 50% building height. The maximum of the sections is either located in the middle or in the lower area, ranging between 99.89 and 99.96% solid material percentage.

The investigation of the porosity, respectively the solid material percentage overall shows a high quality result, having a minimum arithmetic mean of 99.89% solid material percentage. DIN EN ISO 106751 allows a maximum porosity for multilayer welding applications of 2%. With the highest measured porosity (0.17%) in section 4, it can be concluded that present laser metal deposited Ti–6Al–4V walls consist out of high quality solid material without significant pores. However, a slight decrease of solid material, meaning a slight increase of porosity with increasing building height, has been determined.

7 Conclusions

In this paper, Laser Metal Deposition is presented as an additive manufacturing technology to build components in addition to the usual application of surface coating. The following conclusions can be made from the experiments carried out.

- Representative components ranging from simple wall structures to freeform structures have been fabricated and presented in this paper
- Several part-adaptive process parameters and build strategies have been developed for higher geometrical accuracy
- Measured geometrical deviation shows an increasing trend of deviation value with increasing shape complexity
- Porosity measurements of sample wall structures shows that fabrication of fully dense components are possible with LMD technique.

References

1. K. Chi, A look at global metallic fastener development in two airline manufacturing giants, Boing and Airbus. Ind. Focus, 178–182
2. I. Alexander et al., Machining of thin-walled parts produced by additive manufacturing technologies. Procedia CIRP **41**, 1023–1026 (2016)
3. E. Abele, B. Fröhlich, High speed milling of titanium alloys. Adv. Prod. Eng. Manag. **3**, 131–140 (2008)
4. C. Emmelmann, P. Sander, J. Kranz, E. Wycisk, Laser additive manufacturing and bionics: redefining lightweight design. Phys. Procedia **12**, 364–368 (2011)
5. C.Y. Kong, R.J. Scudamore, J. Allen, High-rate laser metal deposition of Inconel 718 component using low heat-input approach. Phys. Procedia **5**, 379–386 (2010)
6. L. Tang, R.G. Landers, Layer-to-layer height control for laser metal deposition process. J. Manuf. Sci. Eng. **133**(2), 21009 (2011)
7. M. Möller, N. Baramsky, A. Ewald, C. Emmelmann, J. Schlattmann, Evolutionary-based design and control of geometry aims for AMD-manufacturing of Ti-6Al-4V parts. Phys. Procedia **83**, 733–742 (2016)
8. P.A. Kobryn, E.H. Moore, S.L. Semiatin, The effect of laser power and traverse speed on microstructure, porosity, and build height in laser-deposited Ti-6Al-4V. Scripta Mater. **43**(4), 299–305 (2000)

MIX
Papier aus verantwortungsvollen Quellen
Paper from responsible sources
FSC® C105338

If you have any concerns about our products,
you can contact us on
ProductSafety@springernature.com

In case Publisher is established outside the EU,
the EU authorized representative is:
Springer Nature Customer Service Center GmbH
Europaplatz 3, 69115 Heidelberg, Germany

Printed by Libri Plureos GmbH
in Hamburg, Germany